WITHDRAWN
University of
Illinois Library
at Urbana-Champaign

L

Applied Probability
Control
Economics
Information and Communication
Modeling and Identification
Numerical Techniques
Optimization

Applications of Mathematics

2

Editorial Board

A. V. Balakrishnan
Managing Editor

W. Hildenbrand

Advisory Board

K. Krickeberg
G. I. Marchuk
R. Radner

G. I. Marchuk

Methods of Numerical Mathematics

translated by Jiri Ružička

Springer-Verlag

New York   Heidelberg   Berlin

1975

G. I. Marchuk
Director
Computer Center
Novosibirsk 630090
USSR

Jiri Ružička
University of California
Systems Science Department
Los Angeles, California 90024

Editorial Board

A. V. Balakrishnan
University of California
Systems Science Department
Los Angeles, California 90024

W. Hildenbrand
Institut für Gesellschafts- und
Wirtschaftswissenschaften der
Universität Bonn
D-5300 Bonn
Adenauerallee 24–26
German Federal Republic

AMS Subject Classification
65-01

Library of Congress Cataloging in Publication Data

Marchuk, Guriĭ Ivanovich.
　　Methods of numerical mathematics.

　　(Applications of mathematics; v. 2)
　　Translation of Metody vychislitel'noi matematiki.
　　Bibliography: p. 295
　　Includes index.
　　1. Numerical analysis.　I. Title.
QA297.M3413　　　　519.4　　　75-15566

The original Russian edition METODY VYCHISLITEL'NOI MATEMATIKI was published in 1973 by NAUKA, Novosibirsk.

All rights reserved.

No part of this book may be translated or reproduced in any form without written permission from Springer-Verlag.

© 1975 by Springer-Verlag New York Inc.

Printed in the United States of America.

ISBN 0-387-90156-6　Springer-Verlag　New York Heidelberg Berlin

ISBN 3-540-90156-6　Springer-Verlag　Berlin Heidelberg New York

Preface to the English Edition

The English translation of the present book fully corresponds to the Russian original version. The minor changes that have been made consist only of some remarks on the text and the correction of misprints that have been discovered. The author hopes that the English edition will help foreign readers obtain first hand knowledge of the methods in computational mathematics being developed in the Soviet Union and become aware of some scientific trends resulting from the necessity of solving complicated problems by reducing them to elementary ones.

The author expresses his gratitude to Professor I. Babuška who made a number of remarks on the book that have been considered in its English version. He would also like to acknowledge his deep gratitude to Dr. J. Ružička, the translator of the book, and to Professor A. V. Balakrishnan, head of the System Science Department at UCLA, who called attention to the need for an English language edition. The author appreciates very much the cooperation of Springer-Verlag who, with the present book, started a new series of monographs on mathematical systems and economics.

Preface

The present volume is an adaptation of a series of lectures on numerical mathematics which the author has been giving to students of mathematics at the Novosibirsk State University during the span of several years. In dealing with problems of applied and numerical mathematics the author sought to focus his attention on those complicated problems of mathematical physics which, in the course of their solution, can be reduced to simpler and theoretically better developed problems allowing effective algorithmic realization on modern computers.

It is usually these kinds of problems that a young practicing scientist runs into after finishing his university studies. Therefore this book is primarily intended for the benefit of those encountering truly complicated problems of mathematical physics for the first time, who may seek help regarding rational approaches to their solution.

In writing this book the author has also tried to take into account the needs of scientists and engineers who already have a solid background in practical problems but who lack a systematic knowledge in areas of numerical mathematics and its more general theoretical framework.

Consequently, the author has selected a form of exposition which in his opinion helps to attract the attention of a wide range of researchers to problems of numerical mathematics. This style has required certain concessions in the exposition, thus allowing concentration only on basic ideas and approaches. As for the details (sometimes important) and the possible generalizations (such as minimal smoothness requirements, constraints on the input data, etc.), they are obvious to the specialist and present useful exercises for a beginner.

Chapter 8 is an expanded version of the paper given by the author at the International Congress of Mathematicians in Nice (1970). This chapter gives some idea both of the material considered in the previous chapters, and of various methods and problems of numerical mathematics that are of fundamental importance but have not found their way into this volume.

In the process of preparation for publication this book has undergone considerable changes in response to advice and comments obtained by the author from his colleagues and associates. Those whose help is gratefully acknowledged include M. M. Lavrentiev, V. I. Lebedev, I. Marek, M. K. Fage, and N. N. Yanenko. They have made a number of constructive comments regarding the exposition of individual chapters, especially the first and fifth. The changes in the second chapter, which are due to Yu. A. Kuznetsov, are so profound that the nature of his contribution in this part is essentially that of coauthorship. The author has also enjoyed valuable advice and comments from V. T. Vasil'ev, V. P. Il'in, A. N. Konovalov,

V. P. Kochergin, V. V. Penenko, V. V. Smelov, U. M. Sultangazin, and others. G. S. Rivin did considerable work in editing the manuscript. To all these, as well as M. S. Yudin who took part in preparing the book for publication, the author expresses his deep gratitude.

Contents

Chapter 1
Fundamentals of the Theory of Difference Schemes

1.1. Basic Equations and their Adjoints *1*

    1.1.1. Norm Estimates of Certain Matrices *5*
    1.1.2. Computing the Spectral Bounds of a Positive Matrix *6*
    1.1.3. Eigenvalues and Eigenfunctions of the Laplace Operator *9*
    1.1.4. Eigenvalues and Eigenvectors of the Finite-Difference Analogue of the Laplace Operator *11*

1.2. Approximation *15*

1.3. Countable Stability *22*

1.4. The Convergence Theorem *30*

Chapter 2
Methods of Constructing Difference Schemes for Differential Equations

2.1. Method of Constructing Difference Equations for Problems with Discontinuous Coefficients on the Basis of an Integral Identity *35*

2.2. Variational Methods in Mathematical Physics *42*

    2.2.1. The Ritz Method *43*
    2.2.2. The Galerkin Method *45*
    2.2.3. The Least-Squares Method *46*

2.3. Difference Schemes for Equations with Discontinuous Coefficients Based on Variational Principles *47*

    2.3.1. Simple Difference Equations for a Diffusion Based on the Ritz Method *48*
    2.3.2. Constructions of Simple Difference Schemes Based on the Galerkin (Finite Elements) Method *51*

2.4. General Approach to Variational-Difference Schemes for One-Dimensional Equations and Construction of Subspaces *53*

2.5. Variational-Difference Schemes for Two-Dimensional Equations of Elliptic Type *57*

    2.5.1. The Ritz Method *57*
    2.5.2. The Galerkin Method *64*

2.6. Variational Methods for Multidimensional Problems *67*

    2.6.1. Methods of Choosing the Subspaces *67*
    2.6.2. Coordinate-by-Coordinate Methods for Variational-Difference Schemes *69*

2.7. Interpolation of Solutions of Difference Equations by Means of Splines *71*

    2.7.1. Interpolation of Functions of One Variable *72*
    2.7.2. Piece-Wise Interpolation with Smoothing *76*
    2.7.3. Interpolation of Functions of Two Variables *84*

Chapter 3

Methods for Solving Stationary Problems of Mathematical Physics

3.1. Some Iterative Methods and their Optimization *88*

    3.1.1. The Simple Iterative Method *90*
    3.1.2. The Displacement Method *92*
    3.1.3. The Chebyshev Acceleration Method *92*
    3.1.4. The Over-Relaxation Method *97*
    3.1.5. A Comparison of the Asymptotic Rate of Convergence for Various Iterative Methods *103*

3.2. Gradient Iterative Methods *103*

    3.2.1. The Residual Method *104*
    3.2.2. The Two-Step Residual Method *106*
    3.2.3. The Method of Conjugate Gradients *108*

3.3. The Splitting-Up Method *113*

3.4. The Splitting-Up Method with Variational Optimization *123*

3.5. Equations with Singular Operators *126*

3.6. Iterative Methods for Inaccurate Input Data *130*

3.7. The Fast Fourier Transform *132*

3.8. Factorization of Difference Equations *139*

Chapter 4

Methods for Solving Non-Stationary Problems

4.1. Second-Order-Approximation Difference Schemes with Time-Varying Operators *142*

4.2. Nonhomogeneous Equations of Evolution Type *145*

4.3. Splitting-Up Methods for Nonstationary Problems *146*

    4.3.1. The Stabilization Method *147*
    4.3.2. The Predictor-Corrector Method *151*
    4.3.3. The Component-by-Component Splitting-Up Method *154*
    4.3.4. Some General Remarks *159*

4.4. Multicomponent Splitting *160*

    4.4.1. The Stabilization Method *160*
    4.4.2. The Predictor-Corrector Method *162*
    4.4.3. The Component-by-Component Splitting-Up Method Based on the Elementary Schemes *164*
    4.4.4. Splitting-Up of Quasilinear Problems *169*

4.5. General Approach to Component-by-Component Splitting *170*

4.6. Methods of Solving Equations of Hyperbolic Type *174*

    4.6.1. The Stabilization Method *174*
    4.6.2. Reduction of the Wave Equation to an Evolution Problem *178*

Chapter 5

Numerical Methods for Some Inverse Problems

5.1. Basic Definitions and Examples   *185*

5.2. Fourier Series Method for Inverse Evolution Problems   *189*

5.3. Inverse Evolution Problems with Time-Varying Operators   *193*

5.4. Methods of Perturbation Theory for Inverse Problems   *199*
   - 5.4.1. Some Problems of the Linear Theory of Measurements   *199*
   - 5.4.2. Conjugate Functions and the Notion of Value   *200*
   - 5.4.3. Perturbation Theory for Linear Functionals   *203*
   - 5.4.4. Numerical Methods for Inverse Problems and Design of Experiment   *205*

Chapter 6

The Simplest Problems of Mathematical Physics

6.1. The Poisson Equation   *211*
   - 6.1.1. The Dirichlet Problem for the One-Dimensional Poisson Equation   *211*
   - 6.1.2. The One-Dimensional Neumann Problem   *213*
   - 6.1.3. The Two-Dimensional Poisson Equation   *215*
   - 6.1.4. A Problem of Boundary Conditions   *222*

6.2. The Heat Equation   *224*
   - 6.2.1. The One-Dimensional Problem of Heat Conduction   *224*
   - 6.2.2. The Two-Dimensional Problem of Heat Conduction   *229*

6.3. The Wave Equation   *230*

6.4. The Equation of "Motion"
   - 6.4.1. The Simplest Equations of Motion   *234*
   - 6.4.2. The Two-Dimensional Equation of Motion with Variable Coefficients   *241*
   - 6.4.3. The Multidimensional Equation of Motion   *246*

6.5. On Increasing the Order of Approximation of Difference Schemes   *251*

Chapter 7

Numerical Methods in the Theory of Radiative Transfer

7.1. Problem Statement   *259*

7.2. The Transport Equation in Various Geometries   *261*

7.3. Numerical Solution of the Transport Equation in the Parallel-Plane Geometry   *263*

7.4. The Stationary Transport Problem   *272*

7.5. Nonisotropic Particle Scattering   *276*

Chapter 8
A Review of the Methods of Numerical Mathematics

8.1. The Theory of Approximation, Stability, and Convergence of Difference Schemes *279*

8.2. Numerical Methods for Problems of Mathematical Physics *281*

    8.2.1. Constructions of Difference Schemes *282*
    8.2.2. Variational Methods *282*
    8.2.3. Multidimensional Stationary Problems *283*
    8.2.4. Multidimensional Nonstationary Problems *284*

8.3. Conditionally Well-Posed Problems *287*

8.4. Numerical Methods in Linear Algebra *288*

    8.4.1. Direct Methods of Linear Algebra *288*
    8.4.2. Iterative Methods *289*
    8.4.3. Round-Off Error Analysis *291*
    8.4.4. Complexes of Standard Programs *291*

8.5. Optimization Problems in Numerical Mathematics *291*

8.6. Some Trends in Numerical Mathematics *293*

References *295*

Index *315*

Chapter 1

Fundamentals of the Theory of Difference Schemes

This chapter briefly surveys those fundamentals of the theory of difference schemes that are used extensively in the following chapters. We restrict our theoretical considerations to the simplest and most easily interpreted cases, since our main purpose is to achieve familiarity with certain modern concepts in the construction of numerical algorithms in mathematical physics. For more refined and more complex theoretical developments we refer the reader to the specialized bibliography given at the end of the book.

1.1. Basic Equations and their Adjoints

Let us consider a region D in the n-dimensional Euclidean space E_n. Observe at the very outset that the regions usually encountered in applied mathematics are of such a structure that they will possess a measure: area in the two-dimensional case, volume in three dimensions and so on. Nevertheless the theory of Lebesgue measure is vital in the subsequent definitions and, thus, the reader is assumed to be familiar with measure theory and the Lebesgue integral. (Smirnov [2], Sobolev [1], Vladimirov [2], Natanson [1], and others.)

Next let us define the Hilbert space $L_2(D)$ of all real measurable square integrable functions:

$$\int_D f^2(x)\,dx < \infty,$$

with the inner product

$$(f, g) = \int_D f(x)g(x)\,dx. \tag{1.1}$$

As usual, the norm of the function $f \in L_2(D)$ is defined by

$$\|f\|^2 = (f, f). \tag{1.2}$$

Let us now choose a subspace (a linear manifold) Φ of the Hilbert space $L_2(D)$ by imposing certain additional conditions which every element $\phi \in \Phi$ must satisfy. For example, we may require some specified smoothness conditions, conditions on the limit behavior at the boundary D, etc. These conditions, however, must be sufficient to guarantee that an operator A, if given, maps the subspace Φ into $L_2(D)$.

A linear operator A, defined on the linear manifold Φ, is called positive semidefinite if

$$(A\phi, \phi) \geq 0 \tag{1.3}$$

for all $\phi \in \Phi$, with the equality sign possibly holding for a nonzero element ϕ. It is customary to write $A \geq 0$ in this case. If the equality sign above can not hold for nonzero elements, that is

$$(A\phi, \phi) > 0, \qquad \phi \neq 0, \tag{1.4}$$

then we say that the operator A is positive and write $A > 0$. Finally in the case of the stronger inequality

$$(A\phi, \phi) \geq \gamma(\phi, \phi), \qquad \phi \in \Phi, \tag{1.5}$$

where $\gamma > 0$ is a positive constant independent of ϕ, the operator A is called positive definite.

Note that if A is a positive symmetric matrix, then it is positive definite (Faddeev, Faddeeva [8]).

The subspace Φ will be called the *domain* of the operator A and denoted by $\Phi(A)$.

Consider next the adjoint operator A^* defined by the Lagrange identity

$$(Ag, h) = (g, A^*h), \tag{1.6}$$

where $g \in \Phi(A)$, $h \in \Phi(A^*)$.

The subspaces $\Phi(A)$ and $\Phi(A^*)$ of the Hilbert space $L_2(D)$ do not coincide in general, despite the fact that their elements are defined on the same region D in E_n. In what follows we will assume that the adjoint operator exists and is *closed* in the following sense:

Consider a sequence $\phi_n^* \to \psi$ and let $A^*\phi_n^* \to \chi$. Then $\psi \in \Phi(A^*)$ and the limit relation $A^*\psi = \chi$ holds. The operator A is called *selfadjoint* if $A = A^*$ and $\Phi(A) = \Phi(A^*)$.

Let us note one important consequence regarding the properties of adjoint operators. Namely, if $\Phi(A) \equiv \Phi(A^*)$, then $A > 0$ implies $A^* > 0$.

A considerable role in analyzing algorithms is played by the Fourier expansions with respect to the eigenfunctions of operators and their adjoints.

Consider the following two spectral problems for $A \geq 0$:

$$Au = \lambda(A)u, \qquad A^*u^* = \lambda(A^*)u^*. \tag{1.7}$$

Assume that each of the homogeneous equations (1.7) generates a complete set of eigenfunctions, $\{u_n\}$ and $\{u_n^*\}$, which are normalized as follows:

$$(u_n, u_m^*) = \begin{cases} 1, n = m, \\ 0, n \neq m, \end{cases} \tag{1.8}$$

and the corresponding eigenvalues $\lambda_n(A)$ belong to the interval

$$\alpha(A) \leq \lambda_n(A) \leq \beta(A).$$

Basic Equations and their Adjoints

This complete set of eigenfunctions will be called a *biorthogonal basis*. Thus, under the assumption of completeness, arbitrary functions $f \in \Phi$ and $f^* \in \Phi^*$ can be represented in the form of a Fourier series

$$f = \sum_n f_n u_n, \qquad f^* = \sum_n f_n^* u_n^* \qquad (1.9)$$

where

$$f_n = (f, u_n^*), \qquad f_n^* = (f^*, u_n). \qquad (1.10)$$

(In what follows, we use Φ, Φ^* instead of $\Phi(A)$, $\Phi(A^*)$ for the sake of simplicity).

Of great importance in the analysis of numerical algorithms are the norm estimates of operators. A *norm of an operator* A is defined as follows:

$$\|A\|^2 = \sup_{\substack{\phi \in \Phi \\ \phi \neq 0}} \frac{(A\phi, A\phi)}{(\phi, \phi)} \qquad (1.11)$$

(in order to simplify notation the qualification $\phi \neq 0$ will not be explicitly mentioned again). Since

$$(A\phi, A\phi) = (\phi, A^*A\phi),$$

the square of the norm of A can also be expressed in the following way:

$$\|A\|^2 = \sup_{\phi \in \Phi} \frac{(\phi, A^*A\phi)}{(\phi, \phi)}. \qquad (1.12)$$

The operator A^*A is symmetric and positive semidefinite. Consider the spectral problem

$$A^*A\Omega = \lambda(A^*A)\Omega \qquad (1.13)$$

This problem defines a family of eigenfunctions $\{\Omega_n\}$ and eigenvalues $\lambda_n(A^*A) \geq 0$. We will assume that $\{\Omega_n\}$ is a complete set. Then the function ϕ has the following Fourier expansion:

$$\phi = \sum_n \phi_n \Omega_n, \qquad (1.14)$$

where

$$\phi_n = (\phi, \Omega_n). \qquad (1.15)$$

Using the orthonormality of the functions Ω_n, the substitution of Series (1.14) into (1.12) yields

$$\|A\|^2 = \sup_{\{\phi_n\} \in Q} \frac{\sum_n \lambda_n(A^*A)\phi_n^2}{\sum_n \phi_n^2}, \qquad (1.16)$$

where Q is the space of Fourier coefficients. It is easy to see that

$$\frac{1}{\|A^{-1}\|^2} = \lambda_{\min}(A^*A) = \alpha(A^*A),$$
$$\|A\|^2 = \lambda_{\max}(A^*A) = \beta(A^*A), \quad (1.17)$$

where $\lambda_{\min}$ is the smallest and $\lambda_{\max}$ is the greatest eigenvalue respectively in the set $\{\lambda_n(A^*A)\}$ for (spectral) Problem (1.13). The quantity $\beta(A^*A) = \lambda_{\max}(A^*A)$ is usually called the *spectral radius* of the operator A^*A. In general, the spectral radius is defined as $\beta(A) = \sup\{|\lambda(A)|\}$. Note that for $\lambda(A) > 0$ the spectral radius $\beta(A) = \sup\{\lambda(A)\}$.

In the case of a selfadjoint operator A consider the spectral problem

$$Au = \lambda u. \quad (1.18)$$

We have

$$\|A\| = \beta(A). \quad (1.19)$$

It is not difficult to see that for a selfadjoint operator

$$\beta(A^2) = [\beta(A)]^2. \quad (1.20)$$

Consider a fixed closed positive operator C on the Hilbert space $L_2(D)$. We will call it the *energy operator*. Thus

$$(C\phi, \phi) > 0 \quad (1.21)$$

for all $\phi \in \Phi$, the domain of C dense in $L_2(D)$. In other words, for any element $f \in L_2(D)$ there is an element $g \in \Phi$ such that $\|f - g\| \leq \varepsilon$, where ε is an arbitrarily small, positive constant. Denote by $\Phi^* = \Phi(C^*)$ the domain of definition of the adjoint operator C^*. Assume Φ^* coincides with Φ. Then $C^*\phi$ exists for all $\phi \in \Phi$ and $(C^*\phi, \phi) = (\phi, C\phi) = (C\phi, \phi)$. Consequently $(C\phi, \phi) = (\frac{1}{2}[C + C^*]\phi, \phi)$, where $\frac{1}{2}[C + C^*]$ is now a symmetric, positive operator. This allows one to introduce a new inner product in Φ, namely

$$(f, g)_C = (Cf, g),$$

and the norm

$$\|\phi\|_C^2 = (C\phi, \phi) = (\bar{C}\phi, \phi),$$

where $\bar{C} = \frac{1}{2}[C + C^*]$. This norm will be called the *energy norm*. One can obtain the following significant estimate:

$$\|\phi\|_C^2 = \|\phi\|_{\bar{C}}^2 \leq \|\bar{C}\| \|\phi\|^2 = \beta(\bar{C})\|\phi\|^2, \quad (1.22)$$

where $\beta(\bar{C})$ is the largest eigenvalue of the operator $\bar{C}$.

In conclusion let us note that in dealing with problems of mathematical physics and their adjoints it is often convenient to use functions from the *Sobolev space* $W_2^l(D)$. This space is a Hilbert space of $L_2(D)$ functions whose generalized derivatives up to and including lth order are square integrable in

Basic Equations and their Adjoints

D. The inner product in such a space is defined by the formula (see Sobolev [1], Vladimirov [2])

$$(u, v)_{W_2^l} = \sum_{k=0}^{l} \sum_{(k)} \int_D \frac{\partial^k u}{\partial x^k} \frac{\partial^k v}{\partial x^k} \, dD. \qquad (1.23)$$

Here we have used the following notation for the partial derivatives:

$$\frac{\partial^k \phi}{\partial x^k} = \frac{\partial^{\alpha_1 + \cdots + \alpha_n}}{\partial x_1^{\alpha_1} \cdots \partial x_n^{\alpha_n}} \phi, \quad \alpha_1 + \cdots + \alpha_n = k.$$

The norm in the space $W_2^l(D)$ is defined by the relation

$$\|\phi\|_{W_2^l}^2 = (\phi, \phi)_{W_2^l}. \qquad (1.24)$$

1.1.1. Norm Estimates of Certain Matrices

Let us consider a positive semidefinite matrix $A \geq 0$ on the Euclidean space. Then for any value of the parameter $\sigma \geq 0$ we have the following relation:

$$\|(E + \sigma A)^{-1}\| \leq 1. \qquad (1.25)$$

For the proof of this important proposition we exploit the formula

$$\|(E + \sigma A)^{-1}\| = \max_\phi \frac{((E + \sigma A)^{-1}\phi, (E + \sigma A)^{-1}\phi)}{(\phi, \phi)}. \qquad (1.26)$$

Let us introduce new elements

$$\psi = (E + \sigma A)^{-1}\phi.$$

Then

$$\|(E + \sigma A)^{-1}\| = \max_\psi \frac{(\psi, \psi)}{((E + \sigma A)\psi, (E + \sigma A)\psi)}$$

$$= \frac{1}{\min_\psi \left[1 + 2\sigma \frac{(A\psi, \psi)}{(\psi, \psi)} + \sigma^2 \frac{(A\psi, A\psi)}{(\psi, \psi)} \right]}.$$

Since $A \geq 0$ on the elements ϕ, ψ, the last relation implies (1.25). If $A > 0$, then for $\sigma > 0$ we have immediately

$$\|(E + \sigma A)^{-1}\| < 1. \qquad (1.27)$$

Kellogg's lemma [15]. For any matrix $A \geq 0$ and for any $\sigma \geq 0$ one has

$$\|(E - \sigma A)(E + \sigma A)^{-1}\| \leq 1. \qquad (1.28)$$

For the proof let us define T by

$$T = (E - \sigma A)(E + \sigma A)^{-1},$$

and consider the expression for $\|T\|^2$:

$$\|T\|^2 = \max_{\phi} \frac{((E-\sigma A)(E+\sigma A)^{-1}\phi, (E-\sigma A)(E+\sigma A)^{-1}\phi)}{(\phi,\phi)}$$

$$= \max_{\psi} \frac{((E-\sigma A)\psi, (E-\sigma A)\psi)}{((E+\sigma A)\psi, (E+\sigma A)\psi)}$$

$$= \max_{\psi} \frac{(\psi,\psi) - 2\sigma(A\psi,\psi) + \sigma^2(A\psi, A\psi)}{(\psi,\psi) + 2\sigma(A\psi,\psi) + \sigma^2(A\psi, A\psi)} \le 1.$$

Here the crucial role has been played by the positive semidefinitness of the matrix A. The lemma is proved.

In the case when the matrix A is positive and $\sigma > 0$, the expression (1.28) is replaced by

$$\|(E - \sigma A)(E + \sigma A)^{-1}\| < 1. \tag{1.29}$$

1.1.2. Computing the Spectral Bounds of a Positive Matrix

Consider the problem of finding the largest and smallest eigenvalues of a matrix $A > 0$ with a positive spectrum. The approach below is due to Lyusternik [4].

Assume that the spectral problem

$$Au = \lambda u \tag{1.30}$$

defines a complete set of eigenfunctions $u_k \in \Phi$, and a set of eigenvalues $\lambda_k(A)$. (A fairly complete treatment of spectral problems can be found in the papers by Marek [8].) Consider the iterative process

$$\phi^{(n+1)} = (1/c_n)A\phi^{(n)},$$
$$\phi^{(0)} = g,$$

where g is an arbitrary nonzero vector, and c_n is a normalizing factor which can be conveniently chosen in the form

$$c_n = \|\phi^{(n)}\| = \sqrt{\sum_p |\phi_p^{(n)}|^2}.$$

Here $\phi_p^{(n)}$ is the pth component of the vector $\phi^{(n)}$. Thus

$$\phi^{(n+1)} = A \frac{\phi^{(n)}}{\|\phi^{(n)}\|}. \tag{1.31}$$

Let $0 < \alpha(A) = \lambda_1 \le \cdots \le \lambda_{m-1} < \lambda_m = \beta(A)$. Clearly, the following relation holds:

$$\beta(A) = \lim_{n\to\infty} \|\phi^{(n)}\|. \tag{1.32}$$

Basic Equations and their Adjoints

Indeed, because of the assumption of completeness of the system $\{u_n\}$ we have the representation

$$\phi^{(0)} = \sum_k g_k u_k,$$

where $g_k = (g, u_k^*)$, and $\{u_k^*\}$ are the eigenvectors of the matrix A^*. Using the recursive relation

$$\phi^{(n+1)} = A \frac{\phi^{(n)}}{\|\phi^{(n)}\|} = \cdots = \frac{A^n g}{\|A^{n-1} g\|}$$

we obtain

$$\lim_{n \to \infty} \|\phi^{(n+1)}\| = \lim_{n \to \infty} \frac{\|A^n g\|}{\|A^{n-1} g\|}.$$

Since

$$A^n g = \sum_k [\lambda_k(A)]^n g_k u_k,$$

then for large enough n

$$A^n g = \beta^n(A) g_m u_m \left\{ 1 + 0\left[\left(\frac{\lambda_{m-1}}{\lambda_m}\right)^n\right] \right\},$$

where $\beta(A) = \lambda_m$ is the largest eigenvalue of the matrix A.

If we choose the vector of initial approximation g as an arbitrary linear combination of the eigenvectors, which correspond to the eigenvalues distinct from $\beta(A)$, the actual computer-implemented process of sequential approximations still allows one to obtain $\lambda(A)$, but at the expense of having all the basis components in the expansion of ϕ_n (because of the round-off errors).

From the last relation we have

$$\frac{\|A^n g\|}{\|A^{n-1} g\|} = \beta(A) + 0\left[\left(\frac{\lambda_{m-1}}{\lambda_m}\right)^n\right],$$

and therefore

$$\beta(A) = \lim_{n \to \infty} \frac{\|A^n g\|}{\|A^{n-1} g\|}. \tag{1.33}$$

Next let us compute the smallest eigenvalue of the matrix A. Consider a new matrix

$$B = \beta(A)E - A \tag{1.34}$$

and the spectral problem

$$Bu = \lambda(B)u. \tag{1.35}$$

Clearly, $B \geq 0$. Considering Relation (1.34), we see that A and B share the common basis $\{u_k\}$. Similarly as before, consider the iterative process

$$\psi^{(n+1)} = B \frac{\psi^{(n)}}{\|\psi^{(n)}\|}. \tag{1.36}$$

As a result we obtain

$$\beta(B) = \lim_{n \to \infty} \|\psi^{(n)}\|. \tag{1.37}$$

Note that from Relation (1.34) and from the generality of the basis for the matrices A and B it follows that

$$\beta(B) = \beta(A) - \alpha(A).$$

Hence

$$\alpha(A) = \beta(A) - \beta(B). \tag{1.38}$$

It is to be remarked, however, that in the case of ill conditioned matrices A the smallest eigenvalue $\alpha(A)$ is obtained as a difference of large numbers $\beta(A)$ and $\beta(B)$. For this reason the actual numerical algorithm can make errors not only in the magnitude of $\alpha(A)$, but even in the sign. In order to avoid such errors, the computation of $\alpha(A)$ will be slightly changed. With this in mind, consider the iterative process

$$\psi^{n+1} = B\psi^n, \quad \psi^0 = h.$$

Note that the system of eigenvectors u_k of the matrix A, ordered through the natural ordering of the eigenvalues, is transformed into the ordered system of eigenvectors v_k of the matrix B in such a way, that $v_k = u_{m-k+1} (k = 1, 2, \ldots, m)$. Consider the expressions

$$A\psi^n = \sum_{k=1}^{m} \lambda_k^n(B) h_k A v_k, \quad \psi^n = \sum_{k=1}^{m} \lambda_k^n(B) h_k v_k,$$

where $h_k = (h, v_k^*)$. For $n \gg 1$ we have approximately

$$A\psi^n = \beta^n(B) h_m A v_m, \quad \psi^n = \beta^n(B) h_m v_m.$$

Since $A v_m = A u_1 = \alpha(A) u_1 = \alpha(A) v_m$, the following algorithm is obtained:

$$\psi^{(n+1)} = B \frac{\psi^{(n)}}{\|\psi^{(n)}\|}, \tag{1.39}$$

$$\alpha(A) = \lim_{n \to \infty} \frac{\|A\psi^{(n)}\|}{\|\psi^{(n)}\|}. \tag{1.40}$$

The last formula no longer includes differences of large numbers and, as a rule, can be effectively used to find the spectral bounds with aid of a computer.

It should be pointed out, however, that Iterations (1.31), (1.36), and (1.39) converge slowly. In order to accelerate the convergence one can use various

Basic Equations and their Adjoints 9

methods, the most feasible of which are the Chebyshev accelerations and the shift of origin method (Faddeev, Faddeeva [8], Gavurin [9], and Wilkinson [8]).

Let us note that in the case of symmetric matrices the computation of $\alpha(A)$ and $\beta(A)$ is handled more conveniently by using the energy norm.

Optimization of numerical processes and various theoretical estimates of algorithms often require knowledge of the norm of the operator A and of its inverse. The following relations hold for arbitrary operators:

$$\|A\|^2 = \sup_{\phi \in \Phi} \frac{(A\phi, A\phi)}{(\phi, \phi)} = \sup_{\phi \in \Phi} \frac{(A^*A\phi, \phi)}{(\phi, \phi)} = \beta(A^*A)$$

and

$$\|A^{-1}\|^2 = \sup_{\phi \in \Phi} \frac{(A^{-1}\phi, A^{-1}\phi)}{(\phi, \phi)} = \sup_{\phi \in \Phi} \frac{((AA^*)^{-1}\phi, \phi)}{(\phi, \phi)} = [\alpha(A^*A)]^{-1}.$$

Hence

$$\|A\| = \sqrt{\beta(A^*A)}, \tag{1.41}$$

$$\|A^{-1}\| = (\sqrt{\alpha(A^*A)})^{-1}. \tag{1.42}$$

The numbers $\alpha(A^*A)$ and $\beta(A^*A)$ are obtained by means of the successive approximation method described above.

In conclusion, the following remark seems necessary: The algorithms for computing the spectral bounds of positive matrices, which are described above, facilitate the possibility of optimizing the iterative processes for solving problems of mathematical physics. They are based on well-developed methods which will be dealt with in Chapter 3. Such processes become constructive and allow for effective procedures in solving various problems of mathematical physics.

1.1.3. Eigenvalues and Eigenfunctions of the Laplace Operator

Let

$$A = -\Delta, \tag{1.43}$$

where $\Delta = \partial^2/\partial x^2 + \partial^2/\partial y^2$ is the Laplace operator. The operator A is defined on the set Φ of elements ϕ satisfying the following requirements. First,

$$\phi = 0 \quad \text{on} \quad \partial D, \tag{1.44}$$

where ∂D stands for the boundary of the region D. For simplicity we take $D = \{(x, y), 0 < x < 1, 0 < y < 1\}$.

Second, the functions $\phi(x)$ together with their first and second derivatives are continuous on the closed region $D + \partial D$.

Third, the set Φ of elements ϕ is a subspace of the Hilbert space $L^2(D)$ with the inner product

$$(a, b) = \int_D ab\,dD, \tag{1.45}$$

where $a \in L_2(D)$, $b \in L_2(D)$, and with the norm

$$\|\phi\| = \sqrt{(\phi, \phi)}. \tag{1.46}$$

We now show that under these conditions the operator A is symmetric. To this end consider a function $\phi^* \in L_2(D)$ and the functional

$$(A\phi, \phi^*) = -\int_D \phi^* \Delta\phi\,dD. \tag{1.47}$$

Note that these conditions guarantee boundedness of the functional $(A\phi, \phi^*)$ for any $\phi^* \in L_2(D)$. Assume next that the function ϕ^* is smooth enough, so that one can use the second formula of Green. Then

$$(A\phi, \phi^*) = -\int_{\delta D} \left(\phi^* \frac{\partial \phi}{\partial n} - \phi \frac{\partial \phi^*}{\partial n}\right) ds - \int_D \phi \Delta\phi^*\,dD, \tag{1.48}$$

where n is the external normal of D. If the function ϕ^* satisfies the boundary condition

$$\phi^* = 0 \quad \text{on} \quad \partial D, \tag{1.49}$$

then, using Conditions (1.44) and (1.49),

$$(A\phi, \phi^*) = -\int_D \phi \Delta\phi^*\,dD = (\phi, A\phi^*). \tag{1.50}$$

This means $A = A^*$, that is, A is symmetric. An analysis of the above shows that the function ϕ^* must be assumed to have continuous derivatives. Finally we conclude that the operator A is selfadjoint on Φ. (The above development presupposes quite strong constraints on the problem to be solved. It can be shown that the Green formula, and hence all of our conclusions, hold true for any function ϕ from the Sobolev space $\overset{0}{W}{}_2^2$.)

Next let us investigate the positivity of A. For this purpose consider the functional

$$(A\phi, \phi) = -\int_D \phi \Delta\phi\,dD. \tag{1.51}$$

With the help of the first formula of Green we get

$$(A\phi, \phi) = -\int_{\partial D} \phi \frac{\partial \phi}{\partial n} ds + \int_D \left[\left(\frac{\partial \phi}{\partial x}\right)^2 + \left(\frac{\partial \phi}{\partial y}\right)^2\right] dD. \tag{1.52}$$

Basic Equations and their Adjoints 11

Since ϕ satisfies (1.44), it follows that for any $\phi \in \Phi$ not identically zero

$$(A\phi, \phi) = \int_D \left[\left(\frac{\partial \phi}{\partial x}\right)^2 + \left(\frac{\partial \phi}{\partial y}\right)^2 \right] dD > 0. \tag{1.53}$$

Let us use this example to illustrate the eigenvalue problem. It is well known (see Courant [2], Sobolev [1]) that an orthonormalized eigenfunction system of the problem

$$Au = \lambda u \quad \text{in} \quad D, \tag{1.54}$$

$$u = 0 \quad \text{on} \quad \partial D, \tag{1.55}$$

is complete and has the form

$$u_{mp} = 2 \sin m\pi x \sin p\pi x, \tag{1.56}$$

where $m = 1, 2, \ldots$ and $p = 1, 2, \ldots$. The eigenvalues of the operator A are of the form

$$\lambda_{mp}(A) = (m^2 + p^2)\pi^2 > 0. \tag{1.57}$$

Hence

$$2\pi^2 \leq \lambda_{mp}(A) \leq \infty.$$

Thus

$$\alpha(A) = 2\pi^2, \quad \beta(A) = \infty \tag{1.58}$$

and $(A\phi, \phi) \geq 2\pi^2(\phi, \phi)$. Consequently the operator A is positive definite. Since the system of eigenfunctions is complete, any function from Φ can be represented in terms of Fourier series

$$\phi(x, y) = \sum_m \sum_p \phi_{mp} u_{mp}(x, y) = \sum_i \phi_i u_i(x, y), \tag{1.59}$$

and, in addition, since the system $\{u_{mp}\}$ is orthonormalized,

$$\phi_i = (\phi, u_i), \tag{1.60}$$

where i is a new ordering index for the series.

1.1.4. Eigenvalues and Eigenvectors of the Finite-Difference Analogue of the Laplace Operator

Let us denote by $\phi_{k,l}$ the values of the function ϕ at the points (x_k, y_l) such that these points cover uniformly with the step h the region D, i.e. $x_{k+1} = x_k + h$, $y_{l+1} = y_l + h$. In addition it is assumed that the boundary of D consists of parallels to the coordinate lines. The set of points $\{(x_k, y_l)\}$ is called a *net*, the individual points are called *net points* and h is called the *mesh size*. Further let us denote by D_h the domain of definition of the *net functions* (as it is customary to call the functions defined on the net points), and let ∂D_h denote the boundary points of D_h. The symbol Φ_h will represent the set of net functions ϕ^h. To every function $\phi \in \Phi$ there corresponds a net function which we denote $(\phi)_h$ and which is defined by the following rule: the value of $(\phi)_h$ at the net point (x_k, y_l) is equal to $\phi(x_k, y_l)$. This correspondence is a linear operator from the subspace Φ into Φ_h, the set of net functions on D_h; this operator is called a *projection of the function ϕ on the net*.

Next, let A be a linear operator defined on the functions $\phi \in \Phi$. Then the function $\psi = A\phi$ can also be projected on the net by taking $(\psi)_h = (A\phi)_h$. The correspondence $(\phi)_h \to (A\phi)_h$ is again a linear operator, defined on the net functions $(\phi)_h$. We will call the obtained operator a *projection of A on the net* and denote it $(A)_h$. Projections of this kind lead to the finite difference analogs of the equations. Methods of their construction—as well as approximation problems, countable stability, and convergence of solutions of the approximating problems to the exact ones—will be considered in what follows.

Let ϕ^h be a net function with the components $\phi_{k,l}$ and let Δ^h be the finite difference analogue of the Laplace operator on the uniform net $\Delta x = \Delta y = h$, defined as follows:

$$(\Delta^h \phi^h)_{k,l} = \frac{\phi^h_{k+1,l} + \phi^h_{k-1,l} + \phi^h_{k,l+1} + \phi^h_{k,l-1} - 4\phi^h_{k,l}}{h^2}. \quad (1.61)$$

Let us assume that the net function $\phi^h \in \Phi_h$ vanishes on the boundary of the net region, i.e.

$$(\phi^h)_{k,l} = 0 \quad \text{on} \quad \partial D_h. \quad (1.62)$$

Next, let us introduce the difference operators indexed by k and l, as follows:

$$(\Delta_k \phi^h)_{k,l} = \frac{1}{h}(\phi^h_{k+1,l} - \phi^h_{k,l}),$$

$$(\nabla_k \phi^h)_{k,l} = \frac{1}{h}(\phi^h_{k,l} - \phi^h_{k-1,l}),$$

$$(\Delta_l \phi^h)_{k,l} = \frac{1}{h}(\phi^h_{k,l+1} - \phi^h_{k,l}),$$

$$(\nabla_l \phi^h)_{k,l} = \frac{1}{h}(\phi^h_{k,l} - \phi^h_{k,l-1}).$$

Consider the new difference operators A^h, A_k, and A_l which are defined by the following relations:

$$\begin{aligned} A^h &= A_k + A_l, \\ A_k &= -\Delta_k \nabla_k, \\ A_l &= -\Delta_l \nabla_l. \end{aligned} \quad (1.63)$$

Then we have

$$-\Delta^h = A_k + A_l = A^h.$$

Basic Equations and their Adjoints

The set of net points for which $k \in \{0, n\}$ or $l \in \{0, n\}$ is the boundary ∂D_h. Let us recall that at these net points the function ϕ^h vanishes [see Conditions (1.62)].

Let us next consider the inner product

$$(a, b) = h^2 \sum_{k=1}^{n-1} \sum_{l=1}^{n-1} a_{k,l} b_{k,l}$$

and let

$$\|\phi\| = \sqrt{(\phi, \phi)}.$$

Consider the functional

$$(A^h\phi, \phi^*) = -h^2 \sum_{k=1}^{n-1} \sum_{l=1}^{n-1} [(\Delta_k \nabla_k \phi)_{k,l} + (\Delta_l \nabla_l \phi)_{k,l}] \phi^*_{k,l};$$

(here and below we omit the index h in the net functions ϕ and ϕ^* for simplicity). The following analogs of the first and second Green formulas hold true (Ladyzenskaya [2], Samarskyi [3]):

$$-\sum_{k=1}^{n-1} (\Delta_k \nabla_k \phi)_{k,l} \phi^*_{k,l} = \sum_{k=1}^{n} (\nabla_k \phi)_{k,l} (\nabla_k \phi^*)_{k,l},$$
$$-\sum_{k=1}^{n-1} (\Delta_k \nabla_k \phi)_{k,l} \phi^*_{k,l} = \sum_{k=1}^{n-1} (\Delta_k \nabla_k \phi^*)_{k,l} \phi_{k,l}. \tag{1.64}$$

The above identities are valid only for $\phi \in \Phi_h$ which satisfy the condition (1.62) and for $\phi^* \in \Phi_h^*$ which satisfy the relation

$$\phi^*_{k,l} = 0 \quad \text{on} \quad \partial D_h. \tag{1.65}$$

Similar identities hold also for sums over the index l. With the help of the second relation in Equations (1.64) we obtain

$$(A^h\phi, \phi^*) = (\phi, A^h\phi^*).$$

From here we conclude that A^h is selfadjoint, i.e.

$$A^h = (A^h)^* \quad \text{and} \quad \Phi(A) = \Phi(A^*).$$

Next consider the functional

$$(A^h\phi, \phi) = -h^2 \sum_{k=1}^{n-1} \sum_{l=1}^{n-1} [(\Delta_k \nabla_k \phi)_{k,l} + (\Delta_l \nabla_l \phi)_{k,l}] \phi_{k,l}.$$

The first identity in Equations (1.64) in k and l yields

$$(A^h\phi, \phi) = h^2 \sum_{k=1}^{n} \sum_{l=1}^{n} [((\nabla_k \phi)_{k,l})^2 + ((\nabla_l \phi)_{k,l})^2].$$

Hence

$$(A^h\phi, \phi) > 0,$$

provided ϕ is not the zero vector.

Finally consider the spectral problem

$$A^h u = \lambda u \quad \text{in} \quad D_h, \tag{1.66}$$
$$u = 0 \quad \text{on} \quad \partial D_h.$$

The components of the corresponding orthonormalized eigenvectors are of the form

$$u_{mp}^{kl} = 2 \sin m\pi kh \sin p\pi lh, \tag{1.67}$$
$$m = 1, 2, \ldots, n-1; \quad p = 1, 2, \ldots, n-1.$$

Recall that

$$(u_{m_1 p_1}, u_{m_2 p_2}) = h^2 \sum_{k=1}^{n-1} \sum_{l=1}^{n-1} u_{m_1 p_1}^{kl} u_{m_2 p_2}^{kl}.$$

The indices k, l in Equation (1.67) indicate the components, and m, p identify the eigenvectors which can be ordered by writing

$$u_{m,p} = u_i \quad (i = 1, 2, \ldots).$$

Since

$$-(\Delta_k \nabla_k u_{mp})_{k,l} = \left(\frac{4}{h^2} \sin^2 \frac{m\pi h}{2} \sin m\pi kh\right) \sin p\pi lh$$

and

$$-(\Delta_l \nabla_l u_{mp})_{k,l} = \sin m\pi kh \left(\frac{4}{h^2} \sin^2 \frac{p\pi h}{2} \sin p\pi lh\right),$$

the eigenvalues become

$$\lambda_{mp}(A^h) = \frac{4}{h^2}\left(\sin^2 \frac{m\pi h}{2} + \sin^2 \frac{p\pi h}{2}\right). \tag{1.68}$$

Let us note that m and p are between unity and $n-1$. Consequently $h = 1/n \leq mh \leq (n-1)h = 1-h$ and $h \leq ph \leq 1-h$; therefore

$$\frac{8}{h^2} \sin^2 \frac{\pi h}{2} \leq \lambda_i(A^h) \leq \frac{8}{h^2} \cos^2 \frac{\pi h}{2}.$$

Here $\lambda_i(A^h)$ are the ordered eigenvalues $\lambda_{mp}(A^h)$. Since, as a rule, $(\pi h/2) \ll 1$, one can estimate

$$\sin^2 \frac{\pi h}{2} = \frac{\pi^2 h^2}{4} - O(h^4), \quad \cos^2 \frac{\pi h}{2} = 1 - O(h^2),$$

and

$$\alpha(A^h) \leq \lambda_i \leq \beta(A^h), \tag{1.69}$$

where

$$\alpha(A^h) = \frac{8}{h^2}\sin^2\frac{\pi h}{2}, \qquad \beta(A^h) = \frac{8}{h^2}\cos^2\frac{\pi h}{2}, \qquad (1.70)$$

hence

$$\alpha(A^h) = \frac{1}{\|(A^h)^{-1}\|} \approx 2\pi^2, \qquad \beta(A^h) = \|A^h\| \approx \frac{8}{h^2}.$$

Using the basis of eigenvectors from Equation (1.67), the vector ϕ can be expanded into the series

$$\phi = \sum_i \phi_i u_i, \qquad (1.71)$$

where

$$\phi_i = (\phi, u_i). \qquad (1.72)$$

1.2. Approximation

Consider a problem of mathematical physics in the operator form

$$\begin{aligned} A\phi &= f \quad \text{in} \quad D, \\ a\phi &= g \quad \text{on} \quad \partial D, \end{aligned} \qquad (2.1)$$

where A is a linear operator, $\phi \in \Phi$ and $f \in F$. Here Φ and F are Hilbert spaces, the elements of which are defined on $D + \partial D$ and D respectively, a is a linear operator which represents the boundary conditions, and $g \in G$, where G is a Hilbert space with elements defined on ∂D.

Along with Equation (2.1) let us also consider the following equations in a finite-dimensional Euclidean space:

$$\begin{aligned} A^h \phi^h &= f^h \quad \text{in} \quad D_h, \\ a^h \phi^h &= g^h \quad \text{on} \quad \partial D_h, \end{aligned} \qquad (2.2)$$

where A^h is a linear operator depending on the mesh size h; $\phi^h \in \Phi_h, f^h \in F_h$, and Φ_h, F_h are Euclidean spaces. D_h is the set of interior netpoints of the region D and ∂D_h is the set of net points which are used to approximate the boundary conditions; a^h is a linear operator, $g^h \in G_h$, G_h is a Euclidean space whose vectors are defined on ∂D_h.

Let us introduce the norms in the *net spaces* F_h, G_h, Φ_h. Also, denote by $(\xi)_h$ the vector obtained by projecting the function ξ on the corresponding net region. We say that Equation (2.2) is an *n-order approximation* of Equation (2.1) on the solution ϕ if

$$\begin{aligned} \|(A\phi)_h - A^h(\phi)_h\|_{F_h} &\leq M_1 h^n, \\ \|(f)_h - f^h\|_{F_h} &\leq M_2 h^n, \\ \|(a\phi)_h - a^h(\phi)_h\|_{G_h} &\leq M_3 h^n, \\ \|(g)_h - g^h\|_{G_h} &\leq M_4 h^n, \end{aligned}$$

where M_i are constants independent of the constant h.

Recall once more that A^h is a net operator which approximates the operator A. It is defined on the net functions ϕ^h. In particular, it is also defined on $(\phi)_h$, i.e. on the projection of the solution of Equation (2.1) on the net region. This means that both operations $A^h\phi^h$ and $A^h(\phi)_h$ make sense. On the other hand, since $A\phi$, $\phi \in \Phi$, is defined in the region D, we can project it on the net and obtain the net function $(A\phi)_h$. The difference $(A\phi)_h - A^h(\phi)_h$ is the one appearing in the first formula of (2.3). The index F_h indicates that the norm is taken in the space F_h. The second formula is constructed the same way. The last two formulas in (2.3) have similar meaning, but this time on D_h. A more detailed exposition of these matters can be found in the books by Godunov, Ryabenkii [3], Richtmyer [3], Kantorovich, Akilov [1], and Samarskii [3].

If the solution of Equation (2.1) is sufficiently smooth, the approximation order can be conveniently found with the help of a natural norm on the space of continuous and differentiable functions. For this purpose one can usually use the Taylor expansions for the solution and other functions entering the problem statement.

In what follows we will assume that Problem (2.1) has already been reduced to (2.2), and moreover that the boundary condition from (2.2) has been used to eliminate the solution at the boundary points of the region $D_h + \partial D_h$. As a result we obtain the equivalent problem

$$A^h \phi^h = f^h, \tag{2.4}$$

where the domain of definition of the solution ϕ^h is now D_h. The behavior of the solution at the boundary is determined by Equation (2.2) and by the solution of (2.4).

In some cases it is convenient to use form (2.4) of the approximation problem; otherwise we use (2.2).

Thus, as a result of the indicated reduction, and with the required approximation taken into account, the continuous problem (2.1) has been transformed into (2.4), a problem in linear algebra.

Example. Let us consider the following problem:

$$\begin{aligned} -\Delta \phi &= f \quad \text{in} \quad D, \\ \phi &= 0 \quad \text{on} \quad \partial D. \end{aligned} \tag{2.5}$$

We assume the domain of definition $\bar{D}$ to be the square $\{0 < x < 1, 0 < y < 1\}$, and f to be a smooth function. Let $\bar{D}$ (the closure of D) be covered by a uniform net with the mesh size h. The net points of the region will be identified by the pair of indices (k, l), where the first index k ($0 \le k \le n$) corresponds to discretization of the x coordinate, and similarly l ($0 \le l \le n$) corresponds to the discretization of the y coordinate. Consider the following approximations:

$$\phi_{xx} \to \Delta_k \nabla_k (\phi)_h, \qquad \phi_{yy} \to \Delta_l \nabla_l (\phi)_h,$$

Approximation

where the difference operators $\Delta_k, \Delta_l, \nabla_k$, and ∇_l have been defined in Section 1.1.4. The problem (2.5) can be approximated by the following one:

$$-[\Delta_k \nabla_k \phi^h + \Delta_l \nabla_l \phi^h] = f^h \quad \text{in} \quad D_h,$$
$$\phi^h = 0 \quad \text{on} \quad \partial D_h, \tag{2.6}$$

where ∂D_h is the set of net points which belong to the boundary. Now (2.6) can be rewritten as follows:

$$-\Delta^h \phi^h = f^h \quad \text{in} \quad D_h,$$
$$\phi^h = 0 \quad \text{on} \quad \partial D_h, \tag{2.7}$$

where ϕ^h and f^h are vectors with the components $\phi^h_{k,l}$ and $f^h_{k,l}$, and

$$(\Delta^h \phi^h)_{k,l} = h^{-2}(\phi^h_{k+1,l} + \phi^h_{k-1,l} + \phi^h_{k,l+1} + \phi^h_{k,l-1} - 4\phi^h_{k,l}),$$

$$f^h_{k,l} = h^{-2} \int_{x_{k-1/2}}^{x_{k+1/2}} \int_{y_{l-1/2}}^{y_{l+1/2}} f \, dx dy,$$

$$x_{k \pm 1/2} = x_k \pm (h/2), \quad y_{l \pm 1/2} = y_l \pm (h/2).$$

Let us introduce a solution space Φ_h. Let the elements from Φ_h be defined in the domain $D_h + \partial D_h = \{(x_k, y_l); 0 \leq k \leq n, 0 \leq l \leq n\}$. The vector f^h belongs to F_h with the domain of definition $D_h = \{(x_k, y_l); 1 \leq k \leq n-1, 1 \leq l \leq n-1\}$. By expanding the solution in the Taylor series in the vicinity of (x_k, y_l) and assuming that the derivatives with respect to (x, y) up to and including the fourth order are bounded, we obtain

$$\phi(\bar{x}, \bar{y}) = \sum_{n=0}^{3} \frac{1}{n!} \left\{ \left[(\bar{x} - x_k) \frac{\partial}{\partial x} + (\bar{y} - y_l) \frac{\partial}{\partial y} \right]^n \phi \right\}_{k,l}$$
$$+ \frac{1}{4!} \left\{ \left[(\bar{x} - x_k) \frac{\partial}{\partial x} + (\bar{y} - y_l) \frac{\partial}{\partial y} \right]^4 \phi \right\}_{k+\Theta_1, l+\Theta_2},$$

where $x_{k+\Theta_1} \in (\bar{x}, x_k)$ and $y_{l+\Theta_2} \in (\bar{y}, y_l)$.

A similar expansion is obtained also for the function $f(x, y)$. Consider the expansions for ϕ and f in the region $\{x_{k-1} \leq x \leq x_{k+1}, y_{l-1} \leq y \leq y_{l+1}\}$ and substitute in (2.7). We have the following norm estimate:

$$\|(\Delta \phi)_h - \Delta^h(\phi)_h\|_{F_h} \leq M_1 h^2,$$
$$\|(f)_h - f^h\|_{F_h} \leq M_2 h^2, \tag{2.8}$$

where M_1 and M_2 are constants. Let us note that if $f^h_{k,l}$ is chosen to be equal to $f(x_k, y_l)$, the second relation in (2.8) holds with $M_2 = 0$ and therefore we obtain an accurate approximation of the right-hand side of Equation (2.5) in the given metric.

Some simple analysis shows that Problem (2.7) represents the second-order approximation of the initial Problem (2.5).

So far we have considered only the approximation problem with respect to the space variables. However a similar procedure can be used to approximate the evolution equation

$$\frac{\partial \phi}{\partial t} + A\phi = f \quad \text{in} \quad D,$$
$$a\phi = g \quad \text{on} \quad \partial D, \quad (2.9)$$
$$\phi = \phi^0 \quad \text{in} \quad D \quad \text{for} \quad t = 0.$$

(The term *evolution equation* always designates an equation of the above type, with A containing no partials with respect to time.) The approximation procedure for Problem (2.9) will be split into two stages. First let us approximate the problem with respect to the space variables in the region $D_h + \partial D_h$. The result is a difference–differential equation: "difference" in the space variables and "differential" in the time variable.

Consider a new evolution equation, namely

$$\frac{d\phi^h}{dt} + \Lambda\phi^h = f^h, \quad (2.10)$$

where Λ, f^h, ϕ^h are functions of time. From now on we will drop the index h in problem (2.10) as unsignificant, assuming we deal with a difference analog of the original problem of mathematical physics with respect to the space variables.

System (2.10) is clearly a system of ordinary differential equations for the components of the vector ϕ^h.

Thus, consider the following Cauchy problem:

$$\frac{d\phi}{dt} + \Lambda\phi = f,$$
$$\phi = g \quad \text{for} \quad t = 0. \quad (2.11)$$

Assume that the operator Λ does not depend on time. Consider the simplest approximation methods for Problem (2.11) with respect to time. The most convenient are the difference schemes with first- and second-order approximation in t.

Let us start with the simplest explicit first-order approximation scheme on the net D_τ:

$$\frac{\phi^{j+1} - \phi^j}{\tau} + \Lambda\phi^j = f^j, \quad \phi^0 = g, \quad (2.12)$$

where $\tau = t_{j+1} - t_j$, f^j is a projection of f. For simplicity we will take $f^j = f(t_j)$.

The simplest implicit scheme is of the form

$$\frac{\phi^{j+1} - \phi^j}{\tau} + \Lambda\phi^{j+1} = f^j, \quad \phi^0 = g, \quad (2.13)$$

Approximation

where we choose $f^j = f(t_{j+1})$. The approximations with respect to t in (2.12), (2.13) are first-order approximations, as can be easily seen from the Taylor series expansion (assuming, of course, the existence of bounded first- and second-order derivatives of the solutions with respect to time).

Solving for ϕ^{j+1} in (2.12) and (2.13) yields the recursive relation

$$\phi^{j+1} = T\phi^j + \tau S f^j, \tag{2.14}$$

where the *transition operator* T and the *source operator* S are defined as follows: for Scheme (2.12), $T = E - \tau \Lambda$, $S = E$; for Scheme (2.13) $T = (E + \tau \Lambda)^{-1}$, $S = T$.

The above kind of difference schemes for evolution equations will be called *two-layer* schemes.

Of great interest in applications is the second-order approximation scheme of Crank and Nicholson

$$\frac{\phi^{j+1} - \phi^j}{\tau} + \Lambda \frac{\phi^{j+1} + \phi^j}{2} = f^j, \quad \phi^0 = g, \tag{2.15}$$

where $f^j = f(t_{j+1/2})$. Equation (2.15) can also be written in the form of (2.14) by taking

$$T = \left(E + \frac{\tau}{2}\Lambda\right)^{-1}\left(E - \frac{\tau}{2}\Lambda\right),$$

$$S = \left(E + \frac{\tau}{2}\Lambda\right)^{-1}.$$

The difference Equations (2.12), (2.13), and (2.15) are in certain cases conveniently written as a system of two equations: one, which approximates just the equation itself in $D_{h\tau}$, and the other which approximates the boundary conditions on $\partial D_{h\tau}$. In this case the difference analog of Problem (2.9) becomes

$$\begin{aligned} L^{h\tau}\phi^{h\tau} &= f^{h\tau} \quad \text{in} \quad D_{h\tau}, \\ l^{h\tau}\phi^{h\tau} &= g^{h\tau} \quad \text{on} \quad \partial D_{h\tau}, \end{aligned} \tag{2.16}$$

where $D_{h\tau} = D_h \times D_\tau$, $\partial D_{h\tau} = \partial D_h \times D_\tau$. It is assumed that $L^{h\tau}$ approximates the operator

$$L = \frac{\partial}{\partial t} + \Lambda,$$

and $l^{h\tau}$ approximates l on the interval $0 \le t \le T$. Similarly, $f^{h\tau}$ and $g^{h\tau}$ approximate f and g in the corresponding (different in general) norms, that is

$$\begin{aligned} \|(L\phi)_{h\tau} - L^{h\tau}(\phi)_{h\tau}\|_{F_{h\tau}} &\le M_1 h^n + N_1 \tau^p, \\ \|(l\phi)_{h\tau} - l^{h\tau}(\phi)_{h\tau}\|_{G_{h\tau}} &\le M_2 h^n + N_2 \tau^p, \\ \|(f)_{h\tau} - f^{h\tau}\|_{F_{h\tau}} &\le M_3 h^n + N_3 \tau^p, \\ \|(g)_{h\tau} - g^{h\tau}\|_{G_{h\tau}} &\le M_4 h^n + N_4 \tau^p. \end{aligned} \tag{2.17}$$

The operator $(\ )_{h\tau}$ in these inequalities, as well as in Equation (2.3), projects on the corresponding net-space.

The canonical form (2.14) of the difference equations can also be written as

$$L\phi = f \qquad (2.18)$$

by introducing vector functions and new operators with domains in $D_h \times D_\tau$, where D_τ is the set $\{t_j\}$.

In this manner the evolution equation with its boundary conditions and initial data is reduced to a problem (2.18) in linear algebra. Note that the approximation schemes can be analyzed in terms of either the net D_h, or $D_h \times D_\tau$, depending on the choice. In particular (2.18) may represent a boundary problem of elliptic type, an integral equation, etc., while the approximation condition can again be written in the form (2.17) with the approximation index h alone (h being the maximum from the set $\{\Delta x_i\}$ of steps in the space variables).

Example. Consider the problem:

$$A\phi \equiv \frac{\partial \phi}{\partial t} - \Delta \phi = f \quad \text{in} \quad D \times D_t,$$

$$\phi = 0 \quad \text{on} \quad \partial D \times D_t, \qquad (2.19)$$

$$\phi = g \quad \text{in} \quad D \quad \text{for} \quad t = 0.$$

Solutions are assumed to be defined on $(D + \partial D) \times D_t$, where D is a square as before, and $D_t = \{0 \le t \le T\}$. Consider $D_h, \partial D_h$, and D_τ along with $D, \partial D, D_t$. Let D_τ be the set of points $\{t_j\}, t_{j+1} - t_j = \tau$. Then Problem (2.19) can be approximated as follows:

$$A^{h\tau}\phi^j = f^j \quad \text{in} \quad D_h \times D_\tau,$$

$$\phi^j = 0 \quad \text{on} \quad \partial D_h \times D_\tau, \qquad (2.20)$$

$$\phi^0 = g \quad \text{in} \quad D_h.$$

Consider the simplest explicit approximation

$$(A^{h\tau}\phi)^j_{k,l} \equiv \frac{\phi^{j+1}_{k,l} - \phi^j_{k,l}}{\tau} - \Delta^h \phi^j_{k,l}, \qquad (2.21)$$

$$f^j_{k,l} = \frac{1}{h^2} \int_{x_{k-1/2}}^{x_{k+1/2}} \int_{y_{l-1/2}}^{y_{l+1/2}} f(x, y, t_j)\, dx dy, \qquad (2.22)$$

$$g_{k,l} = \frac{1}{h^2} \int_{x_{k-1/2}}^{x_{k+1/2}} \int_{y_{l-1/2}}^{y_{l+1/2}} g(x, y)\, dx dy. \qquad (2.23)$$

The following scheme will be called *explicit*:

$$\phi^{j+1}_{k,l} = \phi^j_{k,l} + \tau \Delta^h \phi^j_{k,l} + \tau f^j_{k,l} \quad \text{in} \quad D_h \times D_\tau. \qquad (2.24)$$

Also,

$$\phi^j_{k,l} = 0 \quad \text{on} \quad \partial D_h \times D_\tau,$$
$$\phi^0_{k,l} = g_{k,l} \quad \text{in} \quad D_h. \qquad (2.25)$$

Approximation

The recursive relation (2.24) can be written as

$$\phi_{k,l}^{j+1} = T\phi_{k,l}^{j} + \tau f_{k,l}^{j}, \tag{2.26}$$

where $T = E + \tau\Delta^h = E - \tau(A_1 + A_2)$ is the transition operator, and the operators A_i ($A_1 = A_k$, $A_2 = A_l$) are defined in (1.63). Let us compute the norm of T. For that let us find the largest eigenvalue of T:

$$\begin{aligned} Tu &= \lambda(T)u \quad \text{in} \quad D_h \\ u &= 0 \quad \text{on} \quad \partial D_h. \end{aligned} \tag{2.27}$$

The following relation holds true:

$$\lambda_n(T) = 1 + \tau\lambda_n(\Delta^h).$$

Consequently, the norm of the operator T is expressed as

$$\|T\| = \max\left\{\left|1 - \frac{8\tau}{h^2}\cos^2\frac{\pi h}{2}\right|, \left|1 - \frac{8\tau}{h^2}\sin^2\frac{\pi h}{2}\right|\right\}, \tag{2.28}$$

and if $(\tau/h^2) < \frac{1}{4}$, then $\|T\| < 1$.

Along with the first-order explicit approximation with respect to τ it is possible to consider the *implicit* first-order approximation with respect to τ and the second-order approximation with respect to h. (2.21) is then replaced by

$$(A^{h\tau}\phi)_{k,l}^{j} \equiv \frac{\phi_{k,l}^{j+1} - \phi_{k,l}^{j}}{\tau} - \Delta^h\phi_{k,l}^{j+1}, \tag{2.29}$$

$f_{k,l}^{j}$ and $g_{k,l}$ are defined by (2.22), (2.23) respectively. Now Equations (2.20) can no longer be solved explicitly, and instead we have to solve the operator equation

$$(E - \tau\Delta^h)\phi_{k,l}^{j+1} = \phi_{k,l}^{j} + \tau f_{k,l}^{j} \quad \text{in} \quad D_h \times D_\tau, \tag{2.30}$$

with

$$\begin{aligned} \phi_{k,l}^{j} &= 0 \quad \text{on} \quad \partial D_h \times D_\tau, \\ \phi_{k,l}^{0} &= g_{k,l} \quad \text{in} \quad D_h. \end{aligned} \tag{2.31}$$

Let us write (2.30) in the form

$$\phi_{k,l}^{j+1} = T(\phi_{k,l}^{j} + \tau f_{k,l}^{j}), \tag{2.32}$$

where

$$T = (E - \tau\Delta^h)^{-1}.$$

The norm of T in this case becomes

$$\|T\| = \max\left\{\frac{1}{1 + \frac{8\tau}{h^2}\cos^2\frac{\pi h}{2}}, \frac{1}{1 + \frac{8\tau}{h^2}\sin^2\frac{\pi h}{2}}\right\}, \tag{2.33}$$

and hence $\|T\| < 1$ for any τ and h.

Finally let us consider the approximation scheme of Crank and Nicholson. In this case the operators and functions in (2.20) will be defined as follows:

$$(A^{h\tau}\phi)_{k,l}^j \equiv \frac{\phi_{k,l}^{j+1} - \phi_{k,l}^j}{\tau} - \Delta^h \frac{\phi_{k,l}^j + \phi_{k,l}^{j+1}}{2} \tag{2.34}$$

and

$$f_{k,l}^j = \frac{1}{h^2} \int_{x_{k-1/2}}^{x_{k+1/2}} \int_{y_{l-1/2}}^{y_{l+1/2}} f(x, y, t_{j+1/2})\, dx dy,$$

$$g_{k,l} = \frac{1}{h^2} \int_{x_{k-1/2}}^{x_{k+1/2}} \int_{y_{l-1/2}}^{y_{l+1/2}} g(x, y)\, dx dy. \tag{2.35}$$

As a result we arrive at the following problem:

$$\left(E - \frac{\tau}{2}\Delta^h\right)\phi_{k,l}^{j+1} = \left(E + \frac{\tau}{2}\Delta^h\right)\phi_{k,l}^j + \tau f_{k,l}^j \quad \text{in} \quad D_h \times D_\tau, \tag{2.36}$$

$$\phi_{k,l}^j = 0 \quad \text{on} \quad \partial D_h \times D_\tau,$$
$$\phi_{k,l}^0 = g_{k,l} \quad \text{in} \quad D_h. \tag{2.37}$$

Equation (2.36) can be formally solved with respect to the unknowns $\phi_{k,l}^{j+1}$ in the form

$$\phi_{k,l}^{j+1} = T\phi_{k,l}^j + \tau S f_{k,l}^j, \tag{2.38}$$

where

$$T = \left(E - \frac{\tau}{2}\Delta^h\right)^{-1}\left(E + \frac{\tau}{2}\Delta^h\right),$$

$$S = \left(E - \frac{\tau}{2}\Delta^h\right)^{-1}.$$

The norm of the transition operator is given by

$$\|T\| = \max\left\{\left|\frac{1 - \frac{8\tau}{h^2}\cos^2\frac{\pi h}{2}}{1 + \frac{8\tau}{h^2}\cos^2\frac{\pi h}{2}}\right|, \left|\frac{1 - \frac{8\tau}{h^2}\sin^2\frac{\pi h}{2}}{1 + \frac{8\tau}{h^2}\sin^2\frac{\pi h}{2}}\right|\right\}. \tag{2.39}$$

Since $\tau > 0$, $\|T\| < 1$.

1.3. Countable Stability

We will not strive for generality in defining the notion of countable stability. The reason is that our main objective is to study simple algorithmic methods and the properties of difference approximations of problems in mathematical physics. Various aspects of stability theory and important generalized results can be found in a number of sources (Ryabenkii and

Filippov [6], Lax [6], Richtmyer [3], Godunov and Rabenkii [7], Yanenko [3], Isaacson and Keller [3], Richtmyer and Morton [3], Samarskii [3], and others).

Basic definitions and methods in the theory of stability will be clarified at first with the explicit difference scheme (2.12):

$$\phi^{j+1} = (E - \tau\Lambda)\phi^j + \tau f^j, \quad \phi^0 = g, \tag{3.1}$$

the solution of which is sought for $0 \leq \tau j \leq T$.

Assume that the operator $\Lambda > 0$ induces a complete set of eigenfunctions $\{u_n\}$ along with the corresponding eigenvalues $\{\lambda_n > 0\}$, according to the spectral problem

$$\Lambda u = \lambda u.$$

We introduce the following Fourier series:

$$\phi^j = \sum_n \phi_n^j u_n, \quad f^j = \sum_n f_n^j u_n, \quad g = \sum_n g_n u_n, \tag{3.2}$$

where

$$\phi_n^j = (\phi^j, u_n^*), \quad f_n^j = (f^j, u_n^*), \quad g^n = (g, u_n^*),$$

and u_n^* are the eigenfunctions of the adjoint spectral problem. Using (3.2) in (3.1) and taking next the inner product of the result with the vectors u_n^*, we obtain the following expression for the Fourier coefficients:

$$\phi_n^{j+1} = (1 - \tau\lambda_n)\phi_n^j + \tau f_n^j. \tag{3.3}$$

Assuming that

$$\phi^0 = \sum_n g_n u_n,$$

we obtain the initial condition

$$\phi_n^0 = g_n. \tag{3.4}$$

Equations (3.3) and (3.4) can be solved by successive elimination of the unknowns. As a result, we have

$$\phi_n^j = r_n^j g_n + \tau \sum_{i=1}^{j} r_n^{j-i} f_n^{i-1}, \tag{3.5}$$

where

$$r_n = 1 - \tau\lambda_n. \tag{3.6}$$

From (3.5) it follows that for $\tau > 0$

$$|\phi_n^j| \leq |r_n|^j |g_n| + \tau \sum_{i=1}^{j} |r_n|^{j-i} |f_n^{i-1}|.$$

Hence, taking $|f_n| = \max_j |f_n^j|$ rather than $|f_n^{i-1}|$ under the summation symbol, we have

$$|\phi_n^j| \leq |r_n|^j |g_n| + \frac{1 - |r_n|^j}{1 - |r_n|} \tau |f_n|. \tag{3.7}$$

John Neumann has introduced the so-called spectral criterion of stability, the essence of which is as follows (Richtmyer [3]): If for every Fourier coefficient ϕ_n^j from (3.2) one has

$$|\phi_n^j| \leq C_{1n}|g_n| + C_{2n}|f_n| \quad (n = 1, 2, \ldots), \tag{3.8}$$

where C_{1n}, C_{2n} are constants with a uniform bound for $0 \leq j\tau \leq T$, then the difference scheme (3.1) is countably stable. Let us see what hypothesis regarding the parameters in the difference scheme (2.12) is enough to guarantee the validity of Relation (3.8). An analysis of (3.7) shows that the stability criterion (3.8) is satisfied if we require the following constraint on the parameter r_n:

$$|r_n| < 1 \quad (n = 1, 2, \ldots). \tag{3.9}$$

(Later we introduce a weaker assumption on the norm of the transition operator.)

Assume that the spectrum of the operator Λ is contained in the interval

$$0 < \alpha(\Lambda) \leq \lambda_n(\Lambda) \leq \beta(\Lambda).$$

According to (3.6), Relation (3.9) will then hold true provided

$$\tau < 2/\beta(\Lambda). \tag{3.10}$$

Thus (3.10) becomes a constructive condition for stability of the difference scheme (3.1). Let us note that Condition (3.10) is only sufficient; the scheme remains stable, for instance, when

$$\tau = 2/\beta(\Lambda).$$

Relation (3.7) in this latter case becomes

$$|\phi_n^j| \leq |g_n| + j\tau|f_n| \tag{3.11}$$

as can be easily seen. But $j\tau \leq T$, where T is fixed. This means that for a small τ a large number of steps j is required: $j \to \infty$ as $\tau \to 0$, the upper end point of the time interval T being fixed. Again, we arrive at schemes stable in the sense of Neumann.

Consider now some other difference schemes which are based on the implicit difference approximations. The implicit first-order approximation scheme (2.13) leads to an expression similar to (3.7):

$$|\phi_n^j| \leq |r_n|^j |g_n| + \frac{1 - |r_n|^j}{1 - |r_n|} \tau |r_n| |f_n|, \tag{3.12}$$

where

$$r_n = \frac{1}{1 + \tau\lambda_n(\Lambda)}.$$

If $\lambda_n(\Lambda) > 0$, this difference scheme is clearly stable for any $\tau > 0$, since

$$|r_n| < 1 \quad (n = 1, 2, \ldots).$$

Stability of this kind will be called the *absolute stability*.

In the case of the Crank–Nicholson Scheme (2.15) one obtains the following estimates for the Fourier coefficients of the solution:

$$|\phi_n^j| \leq |r_n|^j |g_n| + \frac{1 - |r_n|^j}{1 - |r_n|} \tau \mu_n |f_n|, \tag{3.13}$$

where

$$r_n = \frac{1 - \frac{\tau}{2}\lambda_n(\Lambda)}{1 + \frac{\tau}{2}\lambda_n(\Lambda)}, \quad \mu_n = \frac{1}{1 + \frac{\tau}{2}\lambda_n(\Lambda)}.$$

Hence $|r_n| < 1$ for an arbitrary $\tau > 0$, provided that $\lambda_n(\Lambda) > 0$.

We make the following comments at this point. First of all, stability in the sense of Neumann is based on the spectral analysis of the operator defined by the problem at hand. This means that in this approach the algorithm necessarily involves computation of the largest eigenvalue or the estimate of its upper bound. Secondly, the spectral stability criterion establishes stability of the solution with respect to each of the harmonics from the Fourier series, while nothing at all is said about the stability of solutions in terms of energy norms. At the same time the norm of the solution ϕ^j happens to be its only characteristics, as is often the case. All of this has triggered an effort to give new definitions of stability which would be related to the norms of the operators. It is to be emphasized, however, that up to the present, stability analysis of the Neumann type continues to play a prominent role in applications.

Let us now turn to a more general definition of the notion of countable stability. To this end, consider the following problem:

$$\frac{\partial \phi}{\partial t} + A\phi = f \quad \text{in} \quad D \times D_t \tag{3.14}$$

$$\phi = g \quad \text{for} \quad t = 0;$$

it can be approximated by the difference problem as follows:

$$\phi^{j+1} = T\phi^j + \tau Sf^j \quad \text{on} \quad D_h \times D_\tau, \tag{3.15}$$
$$\phi^0 = g.$$

We will say that the difference scheme (3.15) is *stable*, if for any h (the parameter characterizing the difference approximation), and $j \leq T/\tau$, one has

$$\|\phi^j\|_{\Phi_h} \leq C_1 \|g\|_{G_h} + C_2 \|f\|_{F_h}, \tag{3.16}$$

where the constants C_1 and C_2 are uniformly bounded on $0 \leq t \leq T$ and are independent of τ, h, g, and f.

The definition of countable stability is closely related to the notion of well-posed problems with a continuous argument. (Godunov [2], Lavrent'ev [2], Yanenko [3]). One may say that countable stability (for problems with a discrete argument) implies continuous dependence of the solutions on the input data.

Indeed, let $f = f_*, g = g_*$ be the input data for Problem (3.15). Denote by ϕ_* the corresponding solution. Similarly, let ϕ_{**} correspond to the input data $f = f_* + \xi, g = g_* + \delta$. The difference $\varepsilon = \phi_* - \phi_{**}$ will satisfy

$$\varepsilon^{j+1} = T\varepsilon^j + \tau S\xi^j, \qquad \varepsilon^0 = \delta.$$

Along with this, the stability criterion assumes the form

$$\|\varepsilon^{j+1}\|_{\Phi_h} \leq C_1 \|\delta\|_{G_h} + C_2 \|\xi\|_{F_h}.$$

Hence it follows that a small variation in the input data f, g results in a small variation of the solution ϕ.

It is easy to see that the definition of stability in the form of (3.16) already relates the solution itself with the *a priori* knowledge concerning the input data for the problem. Although less specific, this definition is often more suitable for analyzing stability than the definition in the sense of Neumann. From this point of view, let us consider the stability of Scheme (2.12). First we rewrite the recursive Relation (3.1) as follows:

$$\phi^{j+1} = T\phi^j + \tau f^j, \qquad \phi^0 = g, \tag{3.17}$$

$$T = E - \tau \Lambda. \tag{3.18}$$

The formal solution of (3.17) has the form

$$\phi^{j+1} = T^j g + \tau \sum_{i=1}^{j} T^{j-i} f^{i-1}. \tag{3.19}$$

Hence we obtain the norm-estimate

$$\|\phi^j\| \leq \|T\|^j \|g\| + \tau \sum_{i=1}^{j} \|T\|^{j-i} \|f^{i-1}\|. \tag{3.20}$$

Let

$$\|f\| = \max_j \|f^j\|,$$

and use this number instead of $\|f^{i-1}\|$ in (3.20); then

$$\|\phi^j\| \leq \|T\|^j \|g\| + \frac{1 - \|T\|^j}{1 - \|T\|} \tau \|f\|. \tag{3.21}$$

If we assume that

$$\|T\| < 1, \tag{3.22}$$

then Scheme (2.12) will be stable in the sense of Definition (3.16). Of course, (3.22) is a sufficient condition of stability. Sharper conditions could be obtained by exploiting the norms of powers of the transition operator, $\|T^i\|$ ($i = 1, 2, \ldots$). In this generality the problem was investigated by Lax and Richtmyer [7]. Weakening of the condition, however, brings additional difficulties in the constructive procedure of establishing the stability criteria. As a rule, it is the sufficient Condition (3.22) that is used in practice.

Consider the case where the operator $\Lambda = \Lambda^* > 0$, and denote

$$J[\phi] = \frac{(T\phi, T\phi)}{(\phi, \phi)}. \tag{3.23}$$

Then

$$J[\phi] = 1 - 2\tau \frac{(\Lambda\phi, \phi)}{(\phi, \phi)} + \tau^2 \frac{(\Lambda\phi, \Lambda\phi)}{(\phi, \phi)}.$$

Let

$$\phi = \sum_n \phi_n u_n,$$

where $\{u_n\}$ is the basis of the operator Λ. Then

$$J[\phi] = 1 - 2\tau\bar{\lambda} + \tau^2 \overline{\lambda^2}, \tag{3.24}$$

where

$$\bar{\lambda} = \frac{\sum\limits_n \lambda_n(\Lambda)\phi_n^2}{\sum\limits_n \phi_n^2}, \quad \overline{\lambda^2} = \frac{\sum\limits_n [\lambda_n(\Lambda)]^2 \phi_n^2}{\sum \phi_n^2},$$

Let us find out which conditions must be satisfied by τ in order that $J[\phi] \leq 1$, that is

$$1 - 2\tau\bar{\lambda} + \tau^2 \overline{\lambda^2} \leq 1,$$

Hence

$$\tau \leq 2\frac{\bar{\lambda}}{\overline{\lambda^2}} = 2\frac{\sum \lambda_n \phi_n^2}{\sum \lambda_n^2 \phi_n^2},$$

and thus if $\beta(\Lambda) = \|\Lambda\| = \max_n \lambda_n(\Lambda) = \lambda_1(\Lambda)$, then

$$\tau \leq \frac{2}{\lambda_1(\Lambda)} \frac{\phi_1^2 + \sum\limits_{n \neq 1} \frac{\lambda_n(\Lambda)}{\lambda_1(\Lambda)} \phi_n^2}{\phi_1^2 + \sum\limits_{n \neq 1} \frac{[\lambda_n(\Lambda)]^2}{[\lambda_1(\Lambda)]^2} \phi_n^2}. \tag{3.25}$$

Since

$$\frac{\phi_1^2 + \sum\limits_{n \neq 1} \frac{\lambda_n(\Lambda)}{\lambda_1(\Lambda)} \phi_n^2}{\phi_1^2 + \sum\limits_{n \neq 1} \frac{[\lambda_n(\Lambda)]^2}{[\lambda_1(\Lambda)]^2} \phi_n^2} \geq 1,$$

we obtain the following sufficient condition for $J[\phi] \leq 1$:

$$\tau \leq \frac{2}{\beta(\Lambda)}.$$

In this case we have (in agreement with the definition (1.11) of the norm of an operator)
$$\|T\|^2 = \sup_{\phi}(J[\phi]) \leq 1,$$
and hence the computation is stable in the sense of definition (3.16). Let us note, that the two definitions of stability [i.e. the stability in the sense of Neumann and the one defined by (3.16)] coincide if the operator is self-adjoint. The relation between these two definitions is studied in the monographs by Godunov and Ryabenkii [3] and Richtmyer and Morton [3].

The stability of the implicit difference Equations (2.13) and (2.15) can be handled in a similar fashion. In these cases we have

$$\|\phi^j\| \leq \|T\|^j \|g\| + \frac{1 - \|T\|^j}{1 - \|T\|} \tau \|S\| \|f\|,$$

where $T = (E + \tau\Lambda)^{-1}$, $S = (E + \tau\Lambda)^{-1}$ for Scheme (2.13), and $T = (E + \tau\Lambda/2)^{-1}(E - \tau\Lambda/2)$, $S = (E + \tau\Lambda/2)^{-1}$ for Scheme (2.15).

It is not difficult to show that the difference schemes above are absolutely stable in the sense of Definition (3.16), provided $\Lambda = \Lambda^* > 0$.

Let us briefly discuss the limiting behavior. In dealing with difference analogs for evolution-type problems of mathematical physics we have to consider approximations with respect to time (the step size τ) and also with respect to the space variables (the grid size h). In other words, the transition operator $T = T(\tau, h)$ depends both on τ and h.

Construction of a stable algorithm for a given approximation method usually reduces to the problem of how to relate τ and h so as to achieve countable stability. The difference scheme becomes absolutely stable if it is stable for arbitrary choice of $\tau > 0$, $h > 0$. If, however, a certain dependence between τ and h is required in order to ensure the stability, the scheme will be termed *conditionally stable*.

Assume that τ and h are related according to the inequality

$$\tau \leqslant Ch^p, \qquad (3.26)$$

where the constants C, p are given and independent of τ and h. Let us note that such relations usually arise when considering the "shortest" perturbation. As a rule, they reflect the dependence between the minimal spatial and time scales of the events to be described by means of the difference scheme.

Of course, larger perturbations (say, of the order of several h's) will then be described more precisely.

Assume we need to increase the accuracy of the solution by formally refining the grid size h. Then we must simultaneously decrease the step size τ so that the above inequality is again satisfied. This means that we can even allow passing to the limit as $\tau \to 0$, $h \to 0$, provided (3.26) is not violated, that is

$$\frac{\tau}{h^p} = \text{const.}$$

Under such circumstances the norm of the transition operator T remains usually unchanged. Even if the scheme under consideration is absolutely stable, it is recommended that the limit as $\tau \to 0$, $h \to 0$ not be taken independently, but in such a way that the norm of the transition operator T stays constant. This ensures both that the process is stable and that the approximation is proper for the typical scales of the events considered.

The definitions of stability discussed above are not the only ones which are being used in the literature. For example, the scheme is stable if

$$\|T\| \leq 1 + O(\tau). \tag{3.27}$$

For small τ, such a definition allows for the exponential growth of round-off errors as time increases (Yanenko [3], Rozhdestvenskii and Yanenko [2]). There are yet other notions of stability (Strang [6], Godunov and Ryabenkii [3], Kreiss [6], Yanenko and Shokin [7], Samarskii [3], and others), which allows us to enlarge the class of difference schemes of interest in applications.

So far, our considerations regarding countable stability have used the assumption that the operator Λ does not depend on time. This is a natural assumption in many problems of mathematical physics. It also permits us to take into consideration a number of additional constructive approaches which are frequently used in numerical mathematics. Indeed, the stability problem is reduced to that of estimating the norm of the transition operator T. As shown in Section 1.1, the square of the norm of T coincides with the spectral radius of the positive selfadjoint operator T^*T, which can then be determined by the Kellogg iterative process, i.e.

$$\|T\|^2 = \lim_{k \to \infty} \frac{(T^*T\phi^{(k)}, \phi^{(k)})}{(\phi^{(k)}, \phi^{(k)})}.$$

Here the elements $\phi^{(k)}$ are defined by

$$\phi^{(k+1)} = T\phi^{(k)}. \tag{3.28}$$

In this manner the problem of finding the norm of T is reduced to the sequential procedure defined by the recursive Relation (3.28). This is also the route along which most of the constructive computer-oriented work has been done. If T is selfadjoint, then

$$\|T\| = \beta(T).$$

Let us now make some particular comments. For problems with periodic boundary conditions, the estimate of the spectral radius should clearly be made by means of Kellogg's method, the operators T being already constructed to account for the actual boundary conditions.

If the operator Λ changes with time, the problem of stability becomes considerably more difficult. This is because the norm of the operator T also changes with time and so does the spectral radius. Therefore T has to be determined at each step in general. The best way to handle this situation is to try for absolutely stable analogs. Such schemes will be considered in Chapter 4.

In conclusion let us note that if the approximation of the evolution equation is investigated in terms of the space $D_h \times D_\tau$, then it is also useful to define stability in these terms. To be specific, let the original evolution problem be approximated by (2.16):

$$L^{h\tau}\phi^{h\tau} = f^{h\tau} \quad \text{in} \quad D_h \times D_\tau,$$
$$l^{h\tau}\phi^{h\tau} = g^{h\tau} \quad \text{on} \quad \partial D_h \times D_\tau. \tag{3.29}$$

The stability criterion may then be taken in the following form:

$$\|\phi^{h\tau}\|_{\Phi_{h\tau}} \leq C_1\|f^{h\tau}\|_{F_{h\tau}} + C_2\|g^{h\tau}\|_{G_{h\tau}}, \tag{3.30}$$

where the constants C_1, C_2 are uniformly bounded for $0 \leq t \leq T$ and do not depend on h, τ, f, or g.

Such a quite general approach to stability analysis has been presented by Ryabenkii and Filippov [6].

Assume that the original problem is approximated by the difference equation with the boundary conditions already taken into account. Then a convenient form of the stability criterion is as follows:

$$\|\phi^{h\tau}\|_{\Phi_{h\tau}} \leq C\|f^{h\tau}\|_{F_{h\tau}}, \tag{3.31}$$

where C is bounded on the interval $0 \leq t \leq T$.

1.4. The Convergence Theorem

Let us discuss one of the important theorems in numerical mathematics (the final form of which is apparently due to Ryabenkii and Filippov [6]). We will follow Godunov and Ryabenkii [3] in explaining its content. Loosely speaking, the theorem says that the approximation and stability of the difference scheme are enough to imply the convergence of the solution of the approximating problem to the solution of the exact problem.

Consider the evolution problem

$$\frac{\partial \phi}{\partial t} + A\phi = f \quad \text{in} \quad D \times D_t \tag{4.1}$$

with the boundary conditions

$$a\phi = g \quad \text{on} \quad \partial D \times D_t, \tag{4.2}$$

and the initial data

$$\phi = \phi^0 \quad \text{for} \quad t = 0. \tag{4.3}$$

Here D is a region of a Euclidean space, and D_t is the time interval $[0 \leq t \leq T]$.

Let us write Problem (4.1)–(4.3) in the following form:

$$L\phi = f \quad \text{in} \quad D \times D_t,$$
$$l\phi = g \quad \text{on} \quad \partial D \times D_t, \quad \phi = \phi^0 \quad \text{for} \quad t = 0 \quad \text{in} \quad D, \tag{4.4}$$

where $L = \partial/\partial t + A$ and l are linear operators.

The Convergence Theorem

Let us introduce a net $(D_h + \partial D_h) \times D_\tau$ in the space $(D + \partial D) \times D_t$. Further, let us project the solution of (4.4) on the net space $D_h \times D_\tau$ and consider the approximate problem corresponding to (4.4) in the following form:

$$\begin{aligned} L^{h\tau}\phi^{h\tau} &= f^{h\tau} \quad \text{in} \quad D_h \times D_\tau, \\ l^{h\tau}\phi^{h\tau} &= g^{h\tau} \quad \text{on} \quad \partial D_h \times D_\tau. \end{aligned} \qquad (4.5)$$

Assume now that the approximation is as follows:

$$\begin{aligned} \|(L\phi)_{h\tau} - L^{h\tau}\phi_{h\tau}\|_{F_{h\tau}} &\leq M_1 h^k + N_1 \tau^p, \\ \|(l\phi)_{h\tau} - l^{h\tau}\phi_{h\tau}\|_{G_{h\tau}} &\leq M_2 h^k + N_2 \tau^p, \\ \|f_{h\tau} - f^{h\tau}\|_{F_{h\tau}} &\leq M_3 h^k + N_3 \tau^p, \\ \|g_{h\tau} - g^{h\tau}\|_{G_{h\tau}} &\leq M_4 h^k + N_4 \tau^p. \end{aligned} \qquad (4.6)$$

Note that the lower index $h\tau$ indicates projection on the net space. M_k, N_k in (4.6) are constant on any bounded interval $0 \leq t \leq T$. Suppose next that the difference problem

$$\|\phi^{h\tau}\|_{\Phi_{h\tau}} \leq C_1 \|f^{h\tau}\|_{F_{h\tau}} + C_2 \|g^{h\tau}\|_{G_{h\tau}}, \qquad (4.7)$$

is countably stable, where C_1, C_2 are independent of h, τ, f^h, and g^h for all $0 \leq t \leq T$. Then from the approximation assumptions (4.6), stability (4.7), and linearity of L, l, $L^{h\tau}$, $l^{h\tau}$ we have that

$$\|\phi_{h\tau} - \phi^{h\tau}\|_{\Phi_{h\tau}} \leq Mh^k + N\tau^p, \qquad (4.8)$$

i.e. convergence.

For proof of this theorem consider the identities

$$\begin{aligned} \|L^{h\tau}\phi^{h\tau} - L^{h\tau}\phi_{h\tau}\|_{F_{h\tau}} &= \|L^{h\tau}\phi^{h\tau} - (L\phi)_{h\tau} + (L\phi)_{h\tau} - L^{h\tau}\phi_{h\tau}\|_{F_{h\tau}}, \\ \|l^{h\tau}\phi^{h\tau} - l^{h\tau}\phi_{h\tau}\|_{G_{h\tau}} &= \|l^{h\tau}\phi^{h\tau} - (l\phi)_{h\tau} + (l\phi)_{h\tau} - l^{h\tau}\phi_{h\tau}\|_{G_{h\tau}}, \end{aligned}$$

which can be rewritten as follows [see (4.4) and (4.5)]:

$$\begin{aligned} \|L^{h\tau}\phi^{h\tau} - L^{h\tau}\phi_{h\tau}\|_{F_{h\tau}} &= \|f^{h\tau} - f_{h\tau} + (L\phi)_{h\tau} - L^{h\tau}\phi_{h\tau}\|_{F_{h\tau}}, \\ \|l^{h\tau}\phi^{h\tau} - l^{h\tau}\phi_{h\tau}\|_{G_{h\tau}} &= \|g^{h\tau} - g_{h\tau} + (l\phi)_{h\tau} - l^{h\tau}\phi_{h\tau}\|_{G_{h\tau}}. \end{aligned} \qquad (4.9)$$

Using well-known inequalities we have

$$\begin{aligned} \|L^{h\tau}\phi^{h\tau} - L^{h\tau}\phi_{h\tau}\|_{F_{h\tau}} &\leq \|f^{h\tau} - f_{h\tau}\|_{F_{h\tau}} + \|(L\phi)_{h\tau} - L^{h\tau}\phi_{h\tau}\|_{F_{h\tau}}, \\ \|l^{h\tau}\phi^{h\tau} - l^{h\tau}\phi_{h\tau}\|_{G_{h\tau}} &\leq \|g^{h\tau} - g_{h\tau}\|_{G_{h\tau}} + \|(l\phi)_{h\tau} - l^{h\tau}\phi_{h\tau}\|_{G_{h\tau}}. \end{aligned} \qquad (4.10)$$

From (4.6) we further obtain

$$\begin{aligned} \|L^{h\tau}(\phi^{h\tau} - \phi_{h\tau})\|_{F_{h\tau}} &\leq (M_1 + M_3)h^k + (N_1 + N_3)\tau^p, \\ \|l^{h\tau}(\phi^{h\tau} - \phi_{h\tau})\|_{G_{h\tau}} &\leq (M_2 + M_4)h^k + (N_2 + N_4)\tau^p. \end{aligned} \qquad (4.11)$$

Consider now the identities

$$L^{h\tau}\varepsilon^{h\tau} = \theta^{h\tau}, \qquad l^{h\tau}\varepsilon^{h\tau} = \eta^{h\tau}, \qquad (4.12)$$

where
$$\varepsilon^{h\tau} = \phi^{h\tau} - \phi_{h\tau}, \quad \theta^{h\tau} = L^{h\tau}(\phi^{h\tau} - \phi_{h\tau}), \quad \eta^{h\tau} = l^{h\tau}(\phi^{h\tau} - \phi_{h\tau}). \quad (4.13)$$

Since, according to the assumptions of the theorem, the difference scheme (4.12) is stable, we have

$$\|\varepsilon^{h\tau}\|_{\Phi_{h\tau}} \leq C_1 \|\theta^{h\tau}\|_{F_{h\tau}} + C_2 \|\eta^{h\tau}\|_{G_{h\tau}}, \quad (4.14)$$

and taking into account (4.13) and (4.11),

$$\|\phi^{h\tau} - \phi_{h\tau}\|_{\Phi_{h\tau}} \leq [C_1(M_1 + M_3) + C_2(M_2 + M_4)]h^k$$
$$+ [C_1(N_1 + N_3) + C_2(N_2 + N_4)]\tau^p.$$

Finally
$$\|\phi^{h\tau} - \phi_{h\tau}\|_{\Phi_{h\tau}} \leq Mh^k + N\tau^p. \quad (4.15)$$

The convergence theorem is proved.

The assumptions of the theorem include rather strong requirement that C_1 and C_2 are independent of h and τ. Of particular inconvenience is the condition that C_1 and C_2 are independent of h, since in certain cases C_1, C_2 may well go to infinity as $h \to 0$. Let

$$C_1^h = C_1/h^m, \quad C_2^h = C_2/h^m,$$

where $m \geq 0$. The convergence of the approximate solution to the exact one will then be estimated as follows:

$$\|\phi^{h\tau} - \phi_{h\tau}\|_{\Phi_{h\tau}} \leq Mh^{k-n} + N\tau^p h^{-m}.$$

If $k > m$ and $\tau^p h^{-m} \to 0$ with $\tau \to 0$, $h \to 0$, then convergence follows. Of course, the convergence theorem can be formulated even when C_1, C_2 depend both on h and τ (Strang [6, 7]).

Let us next turn to the case of stationary problems of mathematical physics. Consider

$$\begin{aligned} A\phi &= f \quad \text{in} \quad D, \\ a\phi &= g \quad \text{on} \quad \partial D. \end{aligned} \quad (4.16)$$

This problem is approximated by the following difference scheme:

$$\begin{aligned} A^h\phi^h &= f^h \quad \text{in} \quad D_h, \\ a^h\phi^h &= g^h \quad \text{on} \quad \partial D_h. \end{aligned} \quad (4.17)$$

Assume that the approximation satisfies

$$\begin{aligned} \|(A\phi)_h - A^h\phi_h\|_{F_h} &\leq M_1 h^k, \\ \|(a\phi)_h - a^h\phi_h\|_{G_h} &\leq M_2 h^k, \\ \|f_h - f^h\|_{F_h} &\leq M_3 h^k, \\ \|g_h - g^h\|_{G_h} &\leq M_4 h^k. \end{aligned}$$

The Convergence Theorem

In addition there is an *a priori* estimate of the solution of (4.17) as follows:

$$\|\phi^h\|_{\Phi_h} \leq C_1 \|f^h\|_{F_h} + C_2 \|g^h\|_{G_h}, \tag{4.19}$$

where the constants C_1, C_2 are independent of h. Again, convergence is assured:

$$\|\phi_h - \phi^h\| \leq Mh^k. \tag{4.20}$$

Thus in stationary problems of mathematical physics the role of stability is played by a near concept of well-posed problems, based on the *a priori* estimates. This fact makes for a deep intrinsic relation between the difference equations for stationary problems and evolution problems. As soon as the approximation and the *a priori* estimates (or stability in the case of evolution equations) are established, these problems become essentially equivalent and can be studied by the same methods.

In some cases it is more convenient to define the approximation globally, rather than taking separate definitions of approximation for operators and right-hand sides of the corresponding equations. The global definition is as follows: The difference scheme (4.5) is said to approximate the original Problem (4.4) with orders n in h and p in τ on the solution ϕ, if there exist constants M_i, N_i (independent of h and τ) such that

$$\|L^{h\tau}(\phi)_{h\tau} - f^{h\tau}\|_{F_{h\tau}} \leq M_1 h^n + N_1 \tau^p,$$
$$\|l^{h\tau}(\phi)_{h\tau} - g^{h\tau}\|_{G_{h\tau}} \leq M_2 h^n + N_2 \tau^p.$$

Using the definition of stability given earlier, one can easily prove the corresponding convergence theorem in this case.

Chapter 2

Methods of Constructing Difference Schemes for Differential Equations

The development of difference approximation schemes for various problems of mathematical physics has been approached by a number of well-known techniques. The most complete picture has been obtained for equations with sufficiently smooth coefficients and solutions, in which case it is possible to obtain high-accuracy approximation schemes. These particular schemes have been the subject of an ever increasing interest, because the speed with which new complicated problems are emerging in science and technology has had a definite bearing on the evolutionary pace of computational means. In many problems it seems therefore reasonable to seek the approximate solutions (with a given accuracy) not at the expense of a formal increase in the dimensionality of the subspaces involved (for instance by decreasing the mesh size) but rather by means of constructing more accurate approximations of the original problem using the *a priori* information about the smoothness of the solution (see also Section 6.5). This point of view has turned out to be quite fruitful in many cases and has led to satisfactory and quite universal methods based on the Ritz and Galerkin variational methods and the least-square method. It is to be noted however that the class of problems which possess smooth solutions is somewhat small, and therefore our main effort must be directed towards the approximation methods suitable for problems with discontinuous coefficients. These problems come up, for instance, when studying diffusion, heat conduction, and hydrodynamics.

Therefore, we will sacrifice the opportunity to describe a number of original and fairly general results concerning difference approximations with high accuracy and will rather pursue the idea of building a general framework for constructing the difference analogs of the equations which do not possess high smoothness properties. Naturally, all the approximations to be discussed later are automatically applicable to the problems with smooth solutions and parameters.

In order to become more familiar with the ways the scientific ideas in the area of difference approximation schemes evolve, we start with a detailed exposition of boundary problems for ordinary differential equations. After that we will turn to more or less general approaches for solving two-dimensional and multidimensional problems of mathematical physics. We hope that the references to the original sources will help the reader to get a deeper and broader understanding of the theory and the algorithms.

2.1. Method of Constructing Difference Equations for Problems with Discontinuous Coefficients on the Basis of an Integral Identity

As of now there are several methods for solving boundary problems for ordinary differential equations. We are going to look at some of those which have been discussed at length in the literature. We will illustrate these methods on very simple and conveniently chosen problems because the main purpose of the present monograph is to familiarize the reader with certain fundamentals of numerical mathematics. Let us note that the difference analogs of boundary problems for ordinary differential equations with smooth coefficients are understood with sufficient completeness by now. Our attention will be directed towards the equations with discontinuous coefficients. Problems of this kind are encountered in many significant applications. The first schemes were obtained by Tikhonov and Samarskii [4].

In this section we will derive the finite difference equations for a diffusion, based on an integral identity due to the author [17].

Consider the diffusion equation for one-dimensional regions. It has the form

$$-\frac{d}{dx} p \frac{d\phi}{dx} + q\phi = f, \qquad (1.1)$$

where $p = p(x) \geq p_0 > 0$ is the diffusion coefficient, $q = q(x) \geq 0$ is the absorption coefficient, and $f = f(x)$ is the source of the diffusing substance. Let us suppose that the functions are piece-wise continuous with discontinuities of the first kind.

We wish to find a solution to (1.1) (over the range [0, 1]) which has a differentiable "flow"

$$J = p \frac{d\phi}{dx}$$

and which satisfies the boundary conditions

$$\phi(0) = 0, \qquad \phi(1) = 0. \qquad (1.2)$$

Let us choose two systems of net points over the range [0, 1] of the variable x: the basic system $\{x_k\}_{k=0}^n$ and the auxiliary system $\{x_{k+1/2}\}_{k=0}^{n-1}$. The points from these two systems are mutually alternating in succession, i.e., $x_k < x_{k+1/2} < x_{k+1}$ (in this case $x_0 = 0$ and $x_n = 1$). In what follows we will assume that $x_{k+1/2} = (x_{k+1} + x_k)/2$.

Integrate (1.1) with respect to x from $x_{k-1/2}$ to $x_{k+1/2}$. As a result we obtain the equilibrium relation

$$-J_{k+1/2} + J_{k-1/2} + \int_{x_{k-1/2}}^{x_{k+1/2}} (q\phi - f) \, dx = 0, \qquad (1.3)$$

where

$$J_{k \pm 1/2} = J(x_{k \pm 1/2}).$$

In order to find $J_{k\pm 1/2}$ procede as follows: Integrate (1.1) from $x_{k-1/2}$ to x:

$$p\frac{d\phi}{dx} = J_{k-1/2} + \int_{x_{k-1/2}}^{x} (q\phi - f)\, d\xi. \tag{1.4}$$

Divide Expression (1.4) by p and then integrate in the limits (x_{k-1}, x_k). We obtain

$$\phi_k - \phi_{k-1} = J_{k-1/2}\int_{x_{k-1}}^{x_k} \frac{dx}{p} + \int_{x_{k-1}}^{x_k} \frac{dx}{p}\int_{x_{k-1/2}}^{x} (q\phi - f)\, d\xi. \tag{1.5}$$

Solving for $J_{k-1/2}$ in (1.5) yields

$$J_{k-1/2} = \frac{1}{\int_{x_{k-1}}^{x_k} \frac{dx}{p}}\left[\phi_k - \phi_{k-1} - \int_{x_{k-1}}^{x_k} \frac{dx}{p}\int_{x_{k-1/2}}^{x} (q\phi - f)\, d\xi\right]. \tag{1.6}$$

A similar expression is obtained for $J_{k\pm 1/2}$ by taking $k+1$ rather than k in (1.6). In this way we have managed to express the flows $J_{k+1/2}$ by means of known functions and the solution of the problem. The relation (1.6) is exact. A substitution of (1.6) and the corresponding $J_{k+1/2}$ into (1.3) results in

$$-\frac{\phi_{k+1}-\phi_k}{\int_{x_k}^{x_{k+1}} \frac{dx}{p}} + \frac{\phi_k - \phi_{k-1}}{\int_{x_{k-1}}^{x_k} \frac{dx}{p}} + \int_{x_{k-1/2}}^{x_{k+1/2}} (q\phi - f)\, dx$$

$$= -\frac{1}{\int_{x_k}^{x_{k+1}} \frac{dx}{p}}\int_{x_k}^{x_{k+1}} \frac{dx}{p}\int_{x_{k+1/2}}^{x} (q\phi - f)\, d\xi$$

$$+\frac{1}{\int_{x_{k-1}}^{x_k} \frac{dx}{p}}\int_{x_{k-1}}^{x_k} \frac{dx}{p}\int_{x_{k-1/2}}^{x} (q\phi - f)\, d\xi. \tag{1.7}$$

Equation (1.7) is our basic identity to be used for obtaining the finite-difference equations.

Define the operator A on the domain Φ of the solutions of (1.1) as follows:

$$(A\phi)_k = -\frac{1}{\Delta x_k}\left(\frac{\phi_{k+1}-\phi_k}{\int_{x_k}^{x_{k+1}} \frac{dx}{p}} - \frac{\phi_k-\phi_{k-1}}{\int_{x_{k-1}}^{x_k} \frac{dx}{p}} - \int_{x_{k-1/2}}^{x_{k+1/2}} q\phi\, dx\right.$$

$$-\frac{1}{\int_{x_k}^{x_{k+1}} \frac{dx}{p}}\int_{x_k}^{x_{k+1}} \frac{dx}{p}\int_{x_{k+1/2}}^{x} q\phi\, d\xi$$

$$\left.+\frac{1}{\int_{x_{k+1/2}}^{x_k} \frac{dx}{p}}\int_{x_{k-1}}^{x_k} \frac{dx}{p}\int_{x_{k-1/2}}^{x} q\phi\, d\xi\right). \tag{1.8}$$

Method of Constructing Difference Equations 37

Also, consider the vector f with the components

$$(f)_k = \frac{1}{\Delta x_k}\int_{x_{k+1/2}}^{x_{k+1/2}} f\,dx + \frac{1}{\Delta x_k}\left(\frac{1}{\int_{x_k}^{x_{k+1}}\frac{dx}{p}}\int_{x_k}^{x_{k+1}}\frac{dx}{p}\int_{x_{k+1/2}}^{x} f\,d\xi\right.$$

$$\left. - \frac{1}{\int_{x_{k-1}}^{x_k}\frac{dx}{p}}\int_{x_{k-1}}^{x_k}\frac{dx}{p}\int_{x_{k-1/2}}^{x} f\,d\xi\right),$$

where

$$\Delta x_k = x_{k+1/2} - x_{k-1/2} \qquad (k = 1, \ldots, n-1).$$

(Do not confuse the above $(f)_k$ with $f(x_k)$.)

For the sake of simplicity we will assume from now on that the solutions of (1.1) are chosen from the class Φ, each function of which has certain smoothness properties and satisfies the boundary condition $\phi = 0$.

Using a more compact notation, Equations (1.7) for $k = 1, \ldots, n-1$ can be written as

$$A\phi = f. \tag{1.9}$$

Consider further various approximations of Equation (1.9): To this end let us introduce the Euclidean norm

$$\|\phi^h\|_{F_h}^2 = \sum_{k=1}^{n-1} \phi_k^2 \,\Delta x_k, \tag{1.10}$$

where F_h is the space of net functions of the form $\phi^h = (\phi_1^h, \ldots, \phi_{n-1}^h)'$, defined at points $x_1, x_2, \ldots, x_{n-1}$. Consider the following approximating problem:

$$A^h\phi^h = f^h, \tag{1.11}$$

where

$$(A^h\phi^h)_k = -\frac{1}{\Delta x_k}\left(\frac{\phi_{k+1}-\phi_k}{\int_{x_k}^{x_{k+1}}\frac{dx}{p}} - \frac{\phi_k-\phi_{k-1}}{\int_{x_{k-1}}^{x_k}\frac{dx}{p}} - \phi_k\int_{x_{k-1/2}}^{x_{k+1/2}} q\,dx\right) \tag{1.12}$$

for $k = 1, \ldots, n-1$, and $\phi_0 = \phi_n = 0$. Using the triangle inequality we have

$$\|(A\phi)_h - A^h(\phi)_h\|_{F_h} \leq \|\xi^h\|_{F_h} + \|\eta^h\|_{F_h}, \quad \|(f)_h - f^h\|_{F_h} = \|\Theta^h\|_{F_h}, \tag{1.13}$$

where

$$(\xi^h)_k = \frac{1}{\Delta x_k}\left(\int_{x_{k-1/2}}^{x_{k+1/2}} q\phi\,dx - \phi_k \int_{x_{k-1/2}}^{x_{k+1/2}} q\,dx\right),$$

$$(\eta^h)_k = -\frac{1}{\Delta x_k}\left(\frac{1}{\int_{x_k}^{x_{k+1}}\frac{dx}{p}}\int_{x_k}^{x_{k+1}}\frac{dx}{p}\int_{x_{k+1/2}}^{x} q\phi\,d\xi\right.$$

$$\left. -\frac{1}{\int_{x_{k-1}}^{x_k}\frac{dx}{p}}\int_{x_{k-1}}^{x_k}\frac{dx}{p}\int_{x_{k-1/2}}^{x} q\phi\,d\xi\right),$$

$$(f^h)_k = \frac{1}{\Delta x_k}\int_{x_{k-1/2}}^{x_{k+1/2}} f\,dx,$$

(here and below, for any continuous function u on $[0, 1]$ we adopt the symbol $(u)_h$ to denote the $(n-1)$-dimensional vector from F_h with the components $u(x_k)$);

$$(\Theta^h)_k = -\frac{1}{\Delta x_k}\left(\frac{1}{\int_{x_k}^{x_{k+1}}\frac{dx}{p}}\int_{x_k}^{x_{k+1}}\frac{dx}{p}\int_{x_{k+1/2}}^{x} f\,d\xi\right.$$

$$\left. -\frac{1}{\int_{x_{k-1}}^{x_k}\frac{dx}{p}}\int_{x_{k-1}}^{x_k}\frac{dx}{p}\int_{x_{k-1/2}}^{x} f\,d\xi\right).$$

Let us estimate the norms $\|\xi^h\|_{F_h}, \|\eta^h\|_{F_h}, \|\Theta^h\|_{F_h}$. To this end assume that $q, f \in Q^{(2)}(0, 1)$ and $p \in Q^{(3)}(0, 1)$, where $Q^{(s)}(0, 1)$ is the space of piece-wise continuously differentiable functions (up to and including the order s), the possible discontinuities being those of the first kind at points $0 < y_1 < y_2 < \cdots < y_m < 1$. We will assume everywhere in what follows that the set $\{y_l\}_{l=1}^m$ belongs to the set of net points $\{x_k, k = 1, \ldots, n-1\}$. This assumption will be needed in analyzing the approximation error.

From the assumption made it follows that the solution ϕ of Problem (1.1) will be continuous, while on each of the segments $[y_l, y_{l+1}]$, $l = 1, \ldots, m-1$, the solution will have a fourth derivative, that is $\phi \in Q^{(4)}(0, 1)$. Let us now investigate the behavior of the components of ξ^h, η^h and θ^h under the assumption that $h \ll 1$, where $h = \max_{0 \le k \le n-1} |x_{k+1} - x_k|$. Expanding into the Taylor series in the vicinity of the net points, it is not difficult to show that the components of these vectors are majorized in modulus by the corresponding components of the vector ω^h, where

$$(\omega^h)_k = \begin{cases} Nh, & \text{if } x_k \text{ is one of the points } y_l, l = 1, 2, \ldots, m, \\ M(|\Delta x_{k+1/2} - \Delta x_{k-1/2}| + h^2) & \text{otherwise;} \end{cases}$$

M, N are positive constants. Here we have introduced the notation $\Delta x_{k+1/2} = x_{k+1} - x_k$. For proof let us assume that in the domain of definition of the solution there is a point of discontinuity of the coefficients. Denote it by $x = x_l$. Keeping in mind (1.10), we can write

$$\|\omega^h\|_{F_h}^2 = \sum_{\substack{k=1 \\ k \neq l}}^{n-1} (\omega^h)_k^2 \Delta x_k + (\omega^h)_l^2 \Delta x_l.$$

Using the above local estimates for $(\omega^h)_k$, we obtain

$$\|\omega^h\|_{F_h} \leq Ch^{3/2},$$

where $C \leq M + N$.

Hence we have the following estimate for the norms of approximation errors of ξ^h, η^h, and Θ^h:

$$\max(\|\xi^h\|_{F_h}, \|\eta^h\|_{F_h}, \|\Theta^h\|_{F_h}) \leq Ch^{3/2} \qquad (1.14)$$

(C being a constant independent of h), provided one of the two conditions below is satisfied: either the net is uniform on each of the intervals $[0, y_1], [y_1, y_2], \ldots, [y_m, 1]$; or the net is quasiuniform, that is, the inequality $|\Delta x_{k+1/2} - \Delta x_{k-1/2}| \leq ch^2$ as $h \to 0$ is violated only finitely many times (c a positive constant). We could use a number of other conditions of this kind, but the two mentioned are most often met in practice.

Let us note that if the order of smoothness of any of the functions p, q, and f is decreased by one, the following estimate is obtained

$$\max(\|\xi^h\|_{F_h}, \|\eta^h\|_{F_h}, \|\Theta^h\|_{F_h}) \leq C_1 h.$$

The difference scheme (1.11), which we have considered, is rarely used in practice the way it stands, since the explicit integration of the functions p, q, and f can become very difficult. Therefore, as a rule, instead of (1.11) we use its simplified version:

$$(A^h \phi^h)_k = -\frac{1}{\Delta x_k}\left\{ p_{k+1/2} \frac{\phi_{k+1} - \phi_k}{\Delta x_{k+1/2}} - p_{k-1/2} \frac{\phi_k - \phi_{k-1}}{\Delta x_{k-1/2}} - (q\Delta x)_k \phi_k \right\},$$

$$(f^h)_k = \frac{1}{\Delta x_k}(f\Delta x)_k = f_k \qquad (k = 1, 2, \ldots, n-1),$$

$$\phi_0 = \phi_n = 0.$$

It turns out that all the conclusions we have made, with regard to the size of the approximation error, still hold, provided all the corresponding assumptions on smoothness of the parameters also remain unchanged.

We will now turn to the convergence properties of (1.11) and (1.12), keeping the smoothness assumptions on p, q, and f. Let us use the values of ϕ at the net points rather than $(\phi^h)_k$. Correspondingly, we introduce the vector $(\phi)_h$ with the components $\phi(x_k)$, $k = 1, \ldots, n - 1$. Simple manipulations lead to

$$A^h[\phi^h - (\phi)_h] = -(\xi^h - \eta^h + \Theta^h). \tag{1.15}$$

We then define the error vector $\varepsilon^h = \phi^h - (\phi)_h$ and take the scalar product with ε^h on both sides of (1.15). Using next the Schwartz inequality and the triangle inequality we obtain

$$(A^h\varepsilon^h, \varepsilon^h) \leq (\|\xi^h\|_{F_h} + \|\eta^h\|_{F_h} + \|\Theta^h\|_{F_h})\|\varepsilon^h\|_{F_h}, \tag{1.16}$$

where the scalar product is taken in the sense

$$(\phi, \psi) = \sum_{n=1}^{n-1} \Delta x_k \phi_k \psi_k \qquad (\phi, \psi \in F_h).$$

Let us investigate the left-hand side of (1.16) in more detail. Since $p(x) \geq p_0 > 0$ by assumption, then for any $\psi \in F_h$

$$(A^h\psi, \psi) = \sum_{k=1}^{n} \frac{(\psi_k - \psi_{k-1})^2}{\int_{x_{k-1}}^{x_k} \frac{dx}{p(x)}} + \sum_{k=1}^{n-1} \left[\psi_k^2 \int_{x_{k-1/2}}^{x_{k+1/2}} q\, dx\right]$$

$$\geq p_0 \sum_{k=1}^{n} \frac{(\psi_k - \psi_{k-1})^2}{\Delta x_{k-1/2}} = p_0(L^h\psi, \psi) > 0, \tag{1.17}$$

where ψ is a nonzero vector with the components satisfying $\psi_0 = \psi_n = 0$, and L^h is the difference operator

$$(L^h\psi)_k = -\frac{1}{\Delta x_k}\left(\frac{\psi_{k+1} - \psi_k}{\Delta x_{k+1/2}} - \frac{\psi_k - \psi_{k-1}}{\Delta x_{k-1/2}}\right) \qquad (k = 1, \ldots, n-1),$$

$$\psi_0 = \psi_n = 0.$$

Since L^h is selfadjoint and positive-definite [that it is selfadjoint is easy to establish, while positive-definiteness follows from (1.17)], we have finally

$$(A^h\psi, \psi) \geq p_0 \lambda_1^h(\psi, \psi),$$

where λ_1^h is the smallest eigenvalue of the problem

$$L^h\psi^h = \lambda^h\psi^h. \tag{1.18}$$

Method of Constructing Difference Equations 41

This problem is a difference analog of the simplest Sturm–Liouville problem

$$-\frac{d^2u}{dx^2} = \lambda u \qquad (0 < x < 1),$$
$$u(0) = u(1) = 0,$$ (1.19)

on a nonuniform net; the eigenvalues of the latter problem are $\lambda_k = \pi^2 k^2$, $k = 1, 2, \ldots$. From Tichonov and Samarskii, [4], it follows that for any $k \geq 1$

$$|\lambda_k^h - \lambda_k|_{h \to 0} \to 0,$$

that is, the eigenvalues of the difference Problem (1.18) converge to the corresponding eigenvalues of Problem (1.19).

This fact is established quite easily in the case of a uniform net. Indeed, the operator L^h is then expressed as the matrix

$$L^h = \frac{1}{h^2} \begin{Vmatrix} 2 & -1 & 0 & \cdots & 0 & 0 \\ -1 & 2 & -1 & \cdots & 0 & 0 \\ \cdots & \cdots & \cdots & \cdots & \cdots & \cdots \\ 0 & 0 & 0 & \cdots & -1 & 2 \end{Vmatrix},$$

the eigenvalues of which are given by

$$\lambda_k^h = \frac{4}{h^2} \sin^2 \frac{\pi h k}{2} \qquad (k = 1, 2, \ldots, n-1).$$

From here it is clear that for any $k \geq 1$

$$\lambda_k^h \to k^2 \pi^2$$

as $h \to 0$, and in particular $\lambda_1^h = (4/h^2) \sin^2 (\pi h/2) \to \pi^2$ as $h \to 0$.

Therefore coming back to (1.16) and using the estimates already obtained, we have

$$\|\varepsilon^h\|_{F_h} \leq \frac{1}{p_0 \lambda_1^h} (\|\xi^h\|_{F_h} + \|\eta^h\|_{F_h} + \|\Theta^h\|_{F_h}). \qquad (1.20)$$

Further estimates can be derived quite easily if one knows the behavior of the approximation errors for ξ^h, η^h, and Θ^h. For instance, if the inequality (1.14) holds true, then

$$\|\varepsilon^h\|_{F_h} \leq M h^{3/2},$$

where $M \geq 3C/p_0 \lambda_1^h$ is a positive constant. (Let us note that this estimate is coarse; using more delicate arguments, Tichonov and Samarskii [4] have shown that $\|\varepsilon^h\|_{F_h} \leq M h^2$.)

In conclusion let us note that although the technique we have described here is of considerable practical importance, it will only be used for comparison with the new approaches to the construction of difference analogs of differential equations.

2.2. Variational Methods in Mathematical Physics

In this section we are going to give a number of facts concerning variational problem formulations in mathematical physics and describe some fundamental methods of their solution. Many of the results have been taken from the well-known monograph by Michlin [1], to which the reader is also referred for proofs of the theorems formulated below.

For illustration purposes we are going to use the elliptic equation

$$Lu \equiv -\sum_{i,j=1}^{2} \frac{\partial}{\partial x_i} A_{ij}(x) \frac{\partial u}{\partial x_j} + \sum_{i=1}^{2} B_i(x) \frac{\partial u}{\partial x_i} + q(x)u = f, \quad (2.1)$$

$$x = (x_1, x_2) \in D,$$

on a bounded region D with the boundary condition of the form

$$u = 0, \quad x \in \partial D \quad (2.2)$$

(the first boundary problem).

The basic assumptions regarding this problem will be as follows. The operator

$$L_0 = -\sum_{i,j=1}^{2} \frac{\partial}{\partial x_i} A_{ij}(x) \frac{\partial}{\partial x_j} \quad (2.3)$$

is selfadjoint in the sense of Lagrange and is nonsingular, that is, for any nonzero vector $\xi = (\xi_1, \xi_2)$ one has

$$\inf_{x \in D} \sum_{i,j=1}^{2} A_{ij}(x)\xi_i\xi_j \geq \mu_0 \sum_{i=1}^{2} \xi_i^2 \quad (2.4)$$

for some positive constant μ_0. Further we will assume that the function $q(x)$ is nonnegative in D and that the solution of (2.1), (2.2) exists and is unique. The reader may wonder about the smoothness properties of the parameters, whether the solution is taken in the classical or in the generalized sense, etc.

First of all we will assume that the solution is taken in the classical sense and moreover that it belongs to the Sobolev space $\overset{\circ}{W}_2^1$ consisting of functions from $L_2(D)$ whose generalized derivatives are square-integrable over D and which are zero on the boundary ∂D. The norm in $\overset{\circ}{W}_2^1$ is given by (see also Section 1.1)

$$\|u\|_{\overset{\circ}{W}_2^1} = \left\{ \int_D u^2 dD + \int_D \left[\left(\frac{\partial u}{\partial x_1} \right)^2 + \left(\frac{\partial u}{\partial x_2} \right)^2 \right] dD \right\}^{1/2}. \quad (2.5)$$

Properties of the input data (for instance smoothness of the coefficients, smoothness properties of the boundary) will be assumed so as to guarantee that the solution belongs to the above space. Second, if it is necessary for the discussion of a concrete problem, we will take it for granted that all the additional smoothness requirements regarding the coefficients or the solutions are satisfied. The above assumptions allow for easier fulfillment of the objectives of the present chapter, namely, to study the principles for building the net-analogs for partial differential equations.

2.2.1. The Ritz Method

Let us assume in addition that B_i, $i = 1, 2$, are identically zero in D. Then it is not difficult to show that

$$(L\phi, \psi) \equiv \int_D \psi L\phi \, dD = \int_D \left\{ \sum_{i,j=1}^{2} A_{i,j}(x) \frac{\partial \phi}{\partial x_i} \frac{\partial \psi}{\partial x_j} + q(x)\phi\psi \right\} dD. \quad (2.6)$$

for all functions ϕ and ψ from the domain of the operator L. In correspondence to the original Problems (2.1) and (2.2), let us now formulate the variational problem of finding an element $u \in \Phi(L)$ on which the quadratic functional

$$J(u) = \int_D uLu \, dD - 2 \int_D fu \, dD$$

assumes its minimum ($\Phi(L)$ denotes the domain of L). The question of what that element is, is answered by the following theorem.

Theorem 1. *Assume that the function $u \in \Phi(L)$ solves Problem (2.1), (2.2); then it minimizes the quadratic functional $J(u)$. Conversely, if*

$$J(u) = \inf_{v \in \Phi(J)} J(v), \quad (2.7)$$

where $\Phi(J)$ is the domain of definition of the functional $J(u)$,† and if also $u \in \Phi(L)$, then u is a solution of Problem (2.1), (2.2).

This theorem has many applications and can be easily generalized to equations with selfadjoint and positive definite operators.

Because of our uniqueness assumption on Problem (2.1), (2.2), it is natural to replace this problem by (2.7) and study the methods of its solution.

The best known method for solving (2.7) is the Ritz method. We will describe it as applied to solving the operator equation

$$Lu = f \quad (2.8)$$

in a Hilbert space F with the inner product (u, v), in which the operator L is selfadjoint and positive definite. By Theorem 1 we conclude that solving (2.8) is equivalent to minimization of the functional

$$J(u) = (Lu, u) - 2(u, f). \quad (2.9)$$

over the Hilbert space F.

---

† Clearly, $\Phi(L) \subseteq \Phi(J)$.

Introduce a sequence of finite-dimensional subspaces $F_h \subseteq F$ corresponding to the sequence of parameters $h_1, h_2, \ldots, h_k, \ldots$ with $h_k \to 0$ as $k \to \infty$. We say that the sequence F_h is complete in F if for any $u \in F$ and $\varepsilon > 0$ there is a $\hat{h} = \hat{h}(u, \varepsilon)$ such that

$$\inf_{w \in F_h} \|u - w\| < \varepsilon \tag{2.10}$$

for all $\hat{h} > h$. In other words, completeness indicates that any element of F can be approximated with an arbitrary accuracy by elements from F_h, starting with some $h = \hat{h} > 0$.

In this framework the Ritz method is formulated as follows: *find an element $u^h \in F_h$ which minimizes $J(u)$ in the space F_h.*

Theorem 2. *Let L be a positive definite operator mapping from F_h, and let $\{F_h\}$ be a complete sequence in F. Then the sequence of the Ritz approximations $\{u^h\}$ converges in F to the solution u of the Equation (2.8).*

In the case when the basis for F_h is assumed to be known and consists of functions $\{\phi_i^h\}_{i=1}^{N_h}$, the problem of finding $u^h \in F_h$ is equivalent to that of finding the coefficients $\{\alpha_i\}_{i=1}^{N_h}$ in the expansion

$$u^h = \sum_{i=1}^{N_h} \alpha_i \phi_i^h, \tag{2.11}$$

by minimizing the functional $J(u)$. As usual, substituting (2.11) into the functional $J(u)$ and letting the derivatives $\partial J(u^h)/\partial \alpha_i$ be zero for $i = 1, \ldots, N_h$, we obtain a system of algebraic equations

$$A\alpha = g, \tag{2.12}$$

where α and g are N_h-vectors, while

$$g_i = (f, \phi_i^h), \tag{2.13}$$

and $A = (a_{ij})$ is the Gramm matrix for the system of vectors $\{\phi_i\}$ in the inner product

$$a_{ij} = (L\phi_i, \phi_j)(1 \leq i, j \leq N_h). \tag{2.14}$$

Since $a_{ij} = a_{ji}$, the matrix A is symmetric, and by the inequality

$$(A\xi, \xi) = \left(L\left(\sum_{i=1}^{N_h} \xi_i \phi_i^h\right), \sum_{i=1}^{N_h} \xi_i \phi_i^h\right) > 0 \tag{2.15}$$

for $\xi \neq 0$, it is also positive definite.

Boundary conditions are called *natural* if the solution of the variational problem is the limit of a sequence of elements, which do not satisfy the boundary conditions of the differential problem (for example the Neumann conditions). If every element of this sequence satisfies the boundary conditions, then they are called *principal* (for example the Dirichlet conditions). Lions [1, 2] and Babuška [5] have introduced the so called *penalty method*, which enables one to reduce the problem with the principal boundary conditions to the problem with the natural boundary conditions. For instance, for the Dirichlet problem we minimize in this case the functional

$$J_\varepsilon(u_\varepsilon) = \int_D (u_\varepsilon L u_\varepsilon)\, dD - 2 \int_D f u_\varepsilon\, dD + \frac{1}{\varepsilon} \int_{\partial D} u_\varepsilon^2\, ds,$$

where ε is a sufficiently small parameter.

In the case of the difference Dirichlet problem for the Poisson equation (see Babuška [5]) one can choose $\varepsilon = h^\sigma$, where $\sigma > 0$, and for $f \in W_2^k$ we have the estimate $\|u - u_\varepsilon\|_{W_2^1} \leq ch^\mu \|f\|_{W_2^k}$, where

$$\mu = \min\left(k + 1, k + \frac{3}{2} - \frac{\sigma}{2}, \frac{\sigma}{2}, k - 1, k - \frac{1}{2} - \frac{\sigma}{2}\right).$$

From here we can compute σ for a given k.

2.2.2. The Galerkin Method

The main deficiency of the Ritz method is the fact that it is applicable only for equations with selfadjoint and positive definite operators. Another variational method, the so-called Galerkin method (or Bubnov–Galerkin method) is free from this constraint. We will describe this method with an example of the equation

$$Lu = f$$

in the Hilbert space F ($\Phi(L)$ is dense in F). As in the previous subsection, we introduce a sequence of finite dimensional spaces F_h, $h = h_1, h_2, \ldots$, with the corresponding basis systems $\{\phi_i^h\}_{i=1}^{N_h}$. The Galerkin approximations are sought in the form

$$u^h = \sum_{i=1}^{N_h} \alpha_i \phi_i^h \tag{2.16}$$

with the residual $Lu^h - f$ orthogonal to all the vectors from F_h. As a result we obtain the system

$$A\alpha = g, \tag{2.17}$$

where

$$a_{ij} = (L\phi_j^h, \phi_i^h), \quad g_i = (f, \phi_i^h) \quad (i, j = 1, \ldots, N_h). \tag{2.18}$$

After evaluating the coefficients $\{\alpha_i\}$ one easily finds the approximate solution by Formula (2.16). It is not difficult to see that for a selfadjoint and positive operator L, System (2.17) coincides with System (2.12) obtained by the Ritz method.

Next let us investigate the convergence of the approximate solutions (2.16). In contrast to the Ritz method, here the problem is considerably more difficult. Because of this, we only consider a particular case when $L = L_0 + K$, where L_0 is a symmetric and positive definite operator and $\Phi(K) \subseteq \Phi(L_0)$. We have the following:

Theorem 3. *If u is a unique solution of the problem $Lu = f$ in F, if the sequence $\{F_h\}$ is complete in F (in the sense of the definition in Section 2.2.1), and if furthermore the operator $L_0^{-1}K$ is compact in F, then the sequential approximations u^h, obtained by the Galerkin method, converge in F to the exact solution u.*

Mikhlin [1] has shown that if we take for F the space $\overset{0}{W}{}_2^1$ (in the case of our present example (2.1), (2.2)), then the assumptions of Theorem 3 are satisfied. Note also, that in the Galerkin method the functions $\{B_i(x)\}_{i=1}^2$ are no longer assumed to be identically zero.

2.2.3. The Least-Squares Method

The least-squares method has been widely used for solving boundary value problems of mathematical physics. Again, let us use the operator equation

$$Lu = f \qquad (2.19)$$

in the Hilbert space F. The method has the following scheme:

Let F_h be finite-dimensional subspaces with the corresponding basis vectors $\phi_1^h, \ldots, \phi_{N_h}^h$, and let $F_h \subseteq \Phi(L)$. (Note that instead of L we may consider its extension $\hat{L}$ such that $\Phi(L)$ is a complete space.)

Now the approximate solutions (2.16) are constructed by the least-squares method, using the equations

$$\frac{\partial}{\partial \alpha_i} \|Lu - f\| = 0 \qquad (i = 1, \ldots, N_h). \qquad (2.20)$$

There results a linear system of Equations (2.17) with the corresponding matrix $A = (a_{ij})$ and the vector $g = (g_i)$, where

$$a_{ij} = (L\phi_i^h, L\phi_j^h), \; g_i = (f, L\phi_i^h) \qquad (1 \leq i, j \leq N_h). \qquad (2.21)$$

It is clear that the matrix A is symmetric and positive definite, provided the operator L is nonsingular in $\Phi(L)$.

We now state the sufficiency conditions for the convergence of the least-squares method.

Theorem 4. *The sequential least-square approximations u^h converge in F to the exact solution u of the Equation (2.19), provided the solution is unique, and assuming also that the sequence LF_h is complete in $\Phi(L)$ and that the inverse L^{-1} exists and is bounded.*

Let us clear up the second requirement in the statement of the theorem. The completeness of the subspaces LF_h means (see Section 2.2.1) that for any $u \in \Phi(L)$ and $\varepsilon > 0$ there is a $\hat{h} = \hat{h}(u, \varepsilon) > 0$ such that

$$\inf_{w \in F_h} \|Lu - Lw\| < \varepsilon \qquad (2.22)$$

for all F_h with $h < \hat{h}$. Note that the expression LF_h makes sense, since $F_h \subset \Phi(L)$ by assumption. Further, the hypothesis of the theorem and the uniqueness of the solution u of (2.19) clearly imply that $Lu_h \to Lu$ and $u^h \to u$ as $h \to 0$.

The boundary-value problem for the limiting solution is now much more complicated than in the two previous methods. We will briefly discuss two possible approaches to this problem.

The first and most straightforward route is to require that all the functions from F_h satisfy the boundary conditions. The practical implementation of this approach however becomes extremely complicated even in the case of mixed boundary problem for the elliptic Equation (2.1).

The second possibility is the *method of weights* as proposed by Bramble and Schatz [5] and others. Briefly, the idea is as follows: Corresponding to the $2m$-order partial differential equation

$$Lu = f \quad \text{in} \quad D \tag{2.23}$$

with the boundary conditions

$$l_i u = f_i \quad \text{on} \quad \partial D \quad (i = 1, \ldots, m), \tag{2.24}$$

let us form the functional

$$J_h(u) = \|Au - f\|^2 + \sum_{i=1}^{m} c_i(h) \|l_i u - f_i\|^2, \tag{2.25}$$

where $\{c_i(h)\}_{i=1}^{m}$ are positive functions of h which characterize the sequence of the subspaces F_h. For the difference analog of the selfadjoint Problem (2.23), (2.24) we have $c_i(h) = h^{-2(2m - m_i - 1/2)}$ on the smooth solutions, where m_i is the order of the highest derivative in the operator l_i. The approximations u^h are now thought of as the solutions of the variational problems

$$\inf_{u \in F_h} J_h(u) = J(u^h),$$

using the least-square method.

The functions u^h converge to u as $h \to 0$, while both (2.23) and (2.24) are asymptotically satisfied. At the same time the functions from the subspaces F_h do not necessarily satisfy the boundary conditions.

2.3. Difference Schemes for Equations with Discontinuous Coefficients Based on Variational Principles

Starting with the works of Lions [1, 2], Oganesyan and Rukhovec [5], Aubin [5], Birkhoff, Schultz and Varga [5], Babuška [5], and others, there has emerged in the recent years a sweeping scientific flow of new methods of constructing difference equations on the basis of variational principles. This direction has been enriched by a number of interesting ideas, of which the most important has been the idea of introducing the test functions with finite support, i.e. the functions which vanish everywhere except relatively small sets (say, of the order of the mesh size). As it turns out, it is often convenient to seek the solution of a given problem as a linear combination of functions with finite support and to choose subsequently the coefficients so as to minimize a suitable functional related to the variational principle. This methodology has been used for various classes of problems and has led to a very effective algorithm of constructing difference schemes. We will illustrate this algorithm on a one-dimensional diffusion problem.

2.3.1. Simple Difference Equations for a Diffusion Based on the Ritz Method

As pointed out earlier the Ritz method can only be used for problems with selfadjoint operators. In order to find an approximate solution of the selfadjoint boundary-value problem [Equations (1.1) and (1.2)], let us introduce the following variational functional:

$$J = \int_0^1 \left[p\left(\frac{d\phi}{dx}\right)^2 + q\phi^2 - 2f\phi \right] dx. \tag{3.1}$$

As we have seen in Section 2.2, the minimum of the functional (3.1) is achieved on the solution of Equations (1.1) and (1.2) from Section 2.1. Hence we will construct an approximate solution on the net D_h so as to choose its unspecified parameters in correspondence to the minimizations of (3.1). With this in mind, let us seek the approximate solution in the form of a continuous, piece-wise linear function belonging to $\overset{\circ}{W}{}_2^1$:

$$\phi^h(x) = \frac{x_{k+1} - x}{\Delta x_{k+1/2}} \phi_k + \frac{x - x_k}{\Delta x_{k+1/2}} \phi_{k+1} \qquad (x_k \leqslant x \leqslant x_{k+1}). \tag{3.2}$$

We first introduce two linear functions $\omega_1(x), \omega_2(x)$ as follows:

$$\omega_1(x) = \frac{x - x_k}{\Delta x_{k+1/2}}, \qquad \omega_2(x) = \frac{x_{k+1} - x}{\Delta x_{k+1/2}}. \tag{3.3}$$

The interpolation Formula (3.2) can then be rewritten as

$$\phi^h(x) = \omega_1(x)\phi_{k+1} + \omega_2(x)\phi_k, \tag{3.4}$$

where $\Delta x_{k+1/2} = x_{k+1} - x_k$, and ϕ_k designates the values of the approximate solution of the problem at the net points. These values are not *a priori* known however; they are to be found by minimizing the functional $J(\phi)$. The functions $\omega_1(x)$ and $\omega_2(x)$ depend on their interval of definition in general. Therefore we perhaps should use the notation $\omega_{1,k+1/2}$ and $\omega_{2,k+1/2}$. But, in what follows, this correspondence will be ultimately apparent at any transformation stage; therefore the additional indices of the functions ω_1, ω_2 will be omitted for the sake of simplicity. Note that in agreement with the *a priori* information about the solution of Equations (1.1) and (1.2) we construct the approximate solution $\phi^h(x)$ so as to be continuous on the interval $0 \leqslant x \leqslant 1$. The coefficients $p(x) \geqq p_0 > 0$, $q(x) \geqq 0$, and $f(x)$ are

Difference Schemes for Equations with Discontinuous Coefficients

assumed to be piece-wise continuous functions with possible discontinuities of the first kind. We will assume that the points of discontinuity coincide with the points x_k. For convenience, the functional $J(\phi^h)$ to be minimized is taken in the form

$$J(\phi^h) = \sum_{k=0}^{n-1} \int_{x_k}^{x_{k+1}} \left[p\left(\frac{d\phi}{dx}\right)^2 + q\phi^2 - 2f\phi \right] dx. \tag{3.5}$$

Substituting (3.4) into (3.5), and noting the relations

$$\frac{d\omega_1}{dx} = \frac{1}{\Delta x_{k+1/2}}, \quad \frac{d\omega_2}{dx} = -\frac{1}{\Delta x_{k+1/2}}, \quad x_k < x < x_{k+1},$$

we obtain

$$J(\phi^h) = \sum_{k=0}^{n-1} \int_{x_k}^{x_{k+1}} \left[p \frac{\phi_{k+1}^2 - 2\phi_{k+1}\phi_k + \phi_k^2}{\Delta x_{k+1/2}^2} \right.$$
$$+ q(\omega_1^2 \phi_{k+1}^2 + 2\omega_1\omega_2 \phi_{k+1}\phi_k + \omega_2^2 \phi_k^2)$$
$$\left. - 2f(\omega_1 \phi_{k+1} + \omega_2 \phi_k) \right] dx. \tag{3.6}$$

The values of ϕ_k, $k = 1, 2, \ldots, n-1$, will be chosen so as to minimize the functional $J(\phi^h)$. The conditions for minimum are as follows:

$$\frac{\partial J}{\partial \phi_k} = 0, \quad \frac{\partial^2 J}{\partial \phi_k^2} > 0 \quad (k = 1, \ldots, n-1). \tag{3.7}$$

Differentiating (3.6) with respect to ϕ_k and putting the result equal to zero, we get

$$\frac{p_{k-1/2}}{\Delta x_{k-1/2}}(\phi_k - \phi_{k-1}) - \frac{p_{k+1/2}}{\Delta x_{k+1/2}}(\phi_{k+1} - \phi_k) + q_{k+1/2}^{1,2}\phi_{k+1}$$
$$+ (q_{k+1/2}^{2,2} + q_{k-1/2}^{1,1})\phi_k + q_{k-1/2}^{1,2}\phi_{k-1} = F_k, \tag{3.8}$$

where

$$p_{k+1/2} = \frac{1}{\Delta x_{k+1/2}} \int_{x_k}^{x_{k+1}} p\, dx; \quad q_{k+1/2}^{i,j} = \int_{x_k}^{x_{k+1}} \omega_i \omega_j q\, dx;$$

$$F_k = \int_{x_{k-1}}^{x_k} f\omega_1 dx + \int_{x_k}^{x_{k+1}} f\omega_2 dx.$$

To specify the boundary problem completely, we complement the difference equations of (3.8) by the boundary conditions

$$\phi_0 = 0, \quad \phi_n = 0.$$

Equation (3.8) along with the boundary conditions can be rewritten as a *three-point difference problem*

$$a_k\phi_{k-1} - b_k\phi_k + c_k\phi_{k+1} = -F_k,$$
$$\phi_0 = 0, \quad \phi_n = 0, \tag{3.9}$$

where

$$a_k = \frac{p_{k-1/2}}{\Delta x_{k-1/2}} - q_{k-1/2}^{1,2}; \qquad c_k = \frac{p_{k+1/2}}{\Delta x_{k+1/2}} - q_{k+1/2}^{1,2};$$

$$b_k = \frac{p_{k+1/2}}{\Delta x_{k+1/2}} + \frac{p_{k-1/2}}{\Delta x_{k-1/2}} + (q_{k+1/2}^{1,1} + q_{k-1/2}^{2,2}). \qquad (3.10)$$

In particular, consider the case when the functions p and q are piecewise constant on each of the intervals $x_k \leq x \leq x_{k+1}$. Then it is not difficult to obtain

$$a_k = \frac{p_{k-1/2}}{\Delta x_{k-1/2}} - \frac{\Delta x_{k-1/2}}{6} q_{k-1/2}; \qquad c_k = \frac{p_{k+1/2}}{\Delta x_{k+1/2}} - \frac{\Delta x_{k+1/2}}{6} q_{k+1/2};$$

$$b_k = \frac{p_{k+1/2}}{\Delta x_{k+1/2}} + \frac{p_{k-1/2}}{\Delta x_{k-1/2}} + \frac{1}{3}(\Delta x_{k+1/2} q_{k+1/2} + \Delta x_{k-1/2} q_{k-1/2}). \qquad (3.11)$$

Here

$$p_{k+1/2} = p(x_{k+1/2}); \qquad q_{k+1/2} = q(x_{k+1/2}).$$

Assume that the solution of Problem (3.9) has been found. Then the continuous solution $\phi^h(x)$ is restored with the help of relations (3.4). The significant property of the variational approach is the fact that we—*a priori*—construct the approximate solution by means of an interpolation polynomial, and the solution is found in the form of a continuous, piece-wise linear function, which is by construction the best approximation of the exact solution in the class of continuous, piece-wise linear functions.

Of course, the variational construction method can be used to obtain approximate solutions of any order of accuracy. For this, it is necessary to choose more accurate interpolation formulae than those from (3.4). Here one can formally exploit the interpolation formulae of Lagrange.

Next let us turn to the analysis of approximating the difference equations obtained. For this we rewrite (3.8) as

$$A^h \phi = -\frac{1}{\Delta x_k} \left\{ p_{k+1/2} \frac{\phi_{k+1} - \phi_k}{\Delta x_{k+1/2}} - p_{k-1/2} \frac{\phi_k - \phi_{k-1}}{\Delta x_{k-1/2}} \right.$$
$$\left. - [q_{k+1/2}^{1,2} \phi_{k+1} + (q_{k+1/2}^{2,2} + q_{k-1/2}^{1,1})\phi_k + q_{k-1/2}^{1,2} \phi_{k-1}] \right\} = f_k^h, \qquad (3.12)$$

where

$$f_k^h = \frac{1}{\Delta x_k} F_k. \qquad (3.13)$$

Here $\Delta x_k = (\Delta x_{k+1/2} + \Delta x_{k-1})/2$. Since the solution $\phi(x)$, and all the coefficients p, q, and f are smooth on the intervals $x_k \leq x \leq x_{k+1}$, we can

Difference Schemes for Equations with Discontinuous Coefficients

show, using the Taylor series expansion, that

$$\|(A\phi)_h - A^h(\phi)_h\| \leq Mh^\alpha,$$
$$\|(f)_h - f^h\| \leq Nh^\alpha,$$

where the quantity α is determined by the properties of the net (see Section 2.1). Further analysis is identical to that of Section 2.1 and is therefore omitted.

2.3.2. Constructions of Simple Difference Schemes Based on the Galerkin (Finite Elements) Method

Consider Problem (2.1), (2.2) on the interval $0 \leq x \leq 1$. Cover this region by the system of intervals $x_{k-1} \leq x \leq x_k$, and for each k introduce the following function from $\overset{0}{W}{}_2^1$:

$$\omega_k(x) = \begin{cases} 0 & (0 \leq x \leq x_{k-1}), \\ \omega_1(x) & (x_{k-1} \leq x \leq x_k), \\ \omega_2(x) & (x_k \leq x \leq x_{k+1}), \\ 0 & (x_{k+1} \leq x \leq x_n = 1). \end{cases} \quad (3.14)$$

We can see (Fig. 2.1) that the functions $\omega_k(x)$ ($k = 1, \ldots, n - 1$) are defined on the whole interval $0 \leq x \leq 1$.

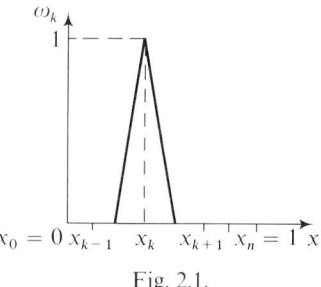

Fig. 2.1.

They are continuous and zero everywhere except for the interval $x_{k-1} \leq x \leq x_{k+1}$, on which they consist of two linear segments with the maximum attained at the point $x = x_k$. Such functions possess a sort of the completeness property; namely, the system is complete in the sense that any continuous, piece-wise linear function $\phi(x)$ with possible corners at the net points x_k can be represented as a linear combination of these functions:

$$\phi(x) = \sum_k \phi_k \omega_k(x), \quad (3.15)$$

where the "Fourier coefficients" ϕ_k are actually the values of ϕ at the points x_k.

Let us also note that the functions $\omega_k(x)$ have certain orthogonality properties, though not in the usual sense. Indeed, consider the scalar product

$$(g, h) = \int_0^1 gh\,dx,$$

then for each $\omega_k(x)$

$$\int_0^1 \omega_k(x)\omega_n(x)\,dx = \begin{cases} 0 & (n \leq k-2), \\ \frac{1}{6}\Delta x_{k-1/2} & (n = k-1), \\ \frac{1}{3}(\Delta x_{k-1/2} + \Delta x_{k+1/2}) & (n = k), \\ \frac{1}{6}\Delta x_{k+1/2} & (n = k+1), \\ 0 & (n \geq k+2). \end{cases} \quad (3.16)$$

From (3.16) we have that $\omega_k(x)$ are orthogonal to all $\omega_n(x)$ except ω_{k-1}, ω_k, and ω_{k+1}. This is a specific feature of the basis chosen. Having defined the functions $\omega_k(x)$, we are now going to exploit them for obtaining the equations in finite differences. To this end consider again the diffusion equation

$$-\frac{d}{dx}p\frac{d\phi}{dx} + q\phi = f \quad \text{in} \quad D, \quad (3.17)$$

$$\phi(0) = 0, \quad \phi(1) = 0.$$

With reference to the Galerkin method, let us take the scalar product of ω_n and Equation (3.17). We obtain

$$\int_0^1 \left(-\frac{d}{dx}p\frac{d\phi}{dx} + q\phi - f\right)\omega_n(x)\,dx = 0. \quad (3.18)$$

Rewrite the functional in this equation to get

$$\int_0^1 \left[p\frac{d\phi}{dx}\cdot\frac{d\omega_n}{dx} + (q\phi - f)\omega_n\right]dx = 0. \quad (3.19)$$

(We have used integration by parts and the fact that $\omega_n(0) = \omega_n(1) = 0$.) Equation (3.19) can be further rewritten as

$$\sum_k \int_{x_{k-1}}^{x_{k+1}} \left[p\frac{d\phi}{dx}\cdot\frac{d\omega_n}{dx} + (q\phi - f)\omega_n\right]dx = 0. \quad (3.20)$$

Taking into account the form of the function ω_n, it is not difficult to obtain

$$\int_{x_{k-1}}^{x_k} p\frac{d\phi}{dx}\cdot\frac{d\omega_n}{dx}\,dx = \begin{cases} 0, & \text{if } n \neq k, \\ \dfrac{p_{k-1/2}}{\Delta x_{k-1/2}}(\phi_k - \phi_{k-1}), & \text{if } n = k; \end{cases}$$

$$\int_{x_k}^{x_{k+1}} p\frac{d\phi}{dx}\cdot\frac{d\omega_n}{dx}\,dx = \begin{cases} 0, & \text{if } n \neq k, \\ -\dfrac{p_{k+1/2}}{\Delta x_{k+1/2}}(\phi_{k+1} - \phi_k), & \text{if } n = k \end{cases} \quad (3.21)$$

and

$$\int_{x_{k-1}}^{x_k} q\phi \omega_k \, dx = q_{k-1/2}^{1,2} \phi_{k-1} + q_{k-1/2}^{1,1} \phi_k,$$

$$\int_{x_k}^{x_{k+1}} q\phi \omega_k \, dx = q_{k+1/2}^{2,2} \phi_k + q_{k+1/2}^{1,2} \phi_{k+1}.$$
(3.22)

The notation here corresponds to that of (3.8).

Substitute next (3.21) and (3.22) into (3.19). There results

$$\frac{p_{k-1/2}}{\Delta x_{k-1/2}}(\phi_k - \phi_{k-1}) - \frac{p_{k+1/2}}{\Delta x_{k+1/2}}(\phi_{k+1} - \phi_k) + q_{k-1/2}^{1,2}\phi_{k-1}$$

$$+ (q_{k-1/2}^{1,1} + q_{k+1/2}^{2,2})\phi_k + q_{k+1/2}^{1,2}\phi_{k+1} = F_k, \qquad (3.23)$$

where

$$F_k = \int_{x_{k-1}}^{x_k} f\omega_1 \, dx + \int_{x_k}^{x_{k+1}} f\omega_2 \, dx = \int_{x_{k-1}}^{x_{k+1}} f\omega_k \, dx. \qquad (3.24)$$

In order to have a complete picture, one only has to add the boundary conditions

$$\phi_0 = 0, \qquad \phi_n = 0. \qquad (3.25)$$

The formulation is complete.

An examination of the difference equations (3.23) and (3.8) shows that they are identical. In other words, the Ritz and the Galerkin methods on the basis of finite functions give the same difference analogs for the selfadjoint problems. We point out, however, that the Galerkin method can be applied regardless whether the problem is selfadjoint or not; thus, it has a broader range of applicability.

If the equations have smooth coefficients, variational principles yield difference approximations with higher-order accuracy.

2.4. General Approach to Variational-Difference Schemes for One-Dimensional Equations and Construction of Subspaces

In this section we will consider a more general approach for obtaining high-order-accuracy difference approximation schemes by variational methods. Such approximations have been considered by Aubin [5], Babuška [5], Strang and Fix [5], and Bramble and Schatz [5]. We will only discuss a method due to Varga [1].

The exposition of this section is primarily intended for the benefit of those who wish to get a deeper insight into the methods of constructing difference schemes of improved accuracy. For this purpose we bring in some additional tools from functional analysis.

Of fundamental importance for variational-difference schemes are the piece-wise polynomial functions. We will demonstrate this with an example of a one-dimensional diffusion equation

$$Lu \equiv -\frac{d}{dx}p(x)\frac{du}{dx} + r(x)\frac{du}{dx} + q(x)u = f, \quad 0 \leq x \leq 1 \quad (4.1)$$

with the boundary conditions

$$u(0) = u(1) = 0. \quad (4.2)$$

We will assume that on the segment [0, 1] the function $p(x)$ is positive and the function $q(x)$ is nonnegative and that there is a unique solution of the problem of (4.1), (4.2). We also assume that the coefficients in the right-hand side of (4.1), (4.2), as well as its solution, are as smooth as needed in every specific situation.

Consider an interval $[a, b]$ on the real line, and let it be covered by a uniform net with the mesh size h. Thus, the net points satisfy $a = x_0 < x_1 = a + h < \cdots < x_N = a + Nh < x_{n+1} = b$, where $h = (b - a)/(N + 1)$. (N is a positive integer.) Denote by $H_N^m(a, b)$ the set of functions $\omega(x)$ with the property that they coincide with polynomials of order $2m + 1$ on each of the intervals $[x_k, x_{k+1}] \subset [a, b]$, and moreover

$$\omega^{(i)}(x_k) = d_{k,i},$$

for any $0 \leq i \leq m$ and $0 \leq k \leq N + 1$, where $d_{k,i}$ are given numbers and $\omega^{(i)}(x)$ denotes the ith derivative of $\omega(x)$ [$\omega^{(0)}(x) = \omega(x)$].

Let us also assume that $\omega(a) = \omega(b) = 0$, i.e., $d_{0,0} = d_{M+1,0} = 0$. Thus it follows that the functions $\omega(x) \in H_N^m(a, n)$ are piece-wise polynomial and m-times continuously differentiable on $[a, b]$, the dimensionality of the space $H_N^m(a, b)$ being $m(N + 2) + N$. Varga [1] has shown that the system of $m(N + 2) + N$ functions $\{\omega_{k,l}\}$, defined below, forms a basis for the space $H_N^m(a, b)$:

$$\omega_{k,0}(x_j) = \delta_{k,j}, \quad \omega_{k,0}^{(i)}(x_j) = 0 \quad (4.3)$$
$$(k = 1, \ldots, N; \; j = 0, 1, \ldots, N + 1; \; i = 1, \ldots, m),$$

$$\omega_{k,l}^{(l)}(x_j) = \delta_{k,j}, \quad \omega_{k,l}^{(i)}(x_j) = 0 \quad (4.4)$$
$$(k = 0, 1, \ldots, N + 1; \; j = 0, 1, \ldots, N + 1;$$
$$i = 0, 1, \ldots, l - 1, l + 1, \ldots, m; \; l = 1, 2, \ldots, m),$$

where $\delta_{k,j}$ is the Kronecker symbol.

It is not difficult to see that the total of N functions are defined by Equation (4.3), while Equation (4.4) defines $m(N + 2)$ functions. Also, since any function $\omega(x) \in H_N^m(a, b)$ is defined by means of $\{d_{k,0}\}_{k=1}^N$ and $d_{k,l}$ ($k = 0, 1, \ldots, N + 1$; $l = 1, \ldots, m$), we have to have the representation

$$\omega(x) = \sum_{k=1}^{N} d_{k,0}\omega_{k,0}(x) + \sum_{k=0}^{N+1}\sum_{l=1}^{m} d_{k,l}\omega_{k,l}(x). \quad (4.5)$$

Let us consider the functions $\{\omega_{k,l}\}$ in more detail. Suppose $m = 1$. Then the space $H_N^1(a, b)$ is of dimension $2N + 2$, and the basis functions are defined by

$$\omega_{k,0}(x_j) = \delta_{kj}, \quad \omega_{k,0}^{(1)}(x_j) = 0 \quad (k = 1, \ldots, N; j = 0, 1, \ldots, N + 1), \quad (4.6)$$

$$\omega_{k,1}(x_j) = 0, \quad \omega_{k,1}^{(1)}(x_j) = \delta_{kj} \quad (k, j = 0, 1, \ldots, N + 1). \quad (4.7)$$

By definition, the functions $\omega_{k,l}$ are seen to be different from zero only on the intervals $(x_k - h, x_k + h)$. In order to find the function $\omega_{k,0}(x)$ we have to solve the following two problems:

The first problem is to construct a polynomial

$$\omega_{k,0}(x) = \sum_{j=0}^{3} C_j^{0,-}(x_k - x)^j \quad (4.8)$$

on the interval $[x_k - h, x_k]$, such that $\omega_{k,0}(x_k - h) = 0$, $\omega_{k,0}^{(1)}(x_k - h) = 0$, $\omega_{k,0}(x_k) = 1$, $\omega_{k,0}^{(1)}(x_k) = 0$. From this we obtain a system of four equations for the unknowns $\{C_j^{0,-}\}_{j=0}^{3}$:

$$\sum_{j=0}^{3} C_j^{0,-}(h)^j = 0, \quad \sum_{j=1}^{3} jC_j^{0,-}(h)^{j-1} = 0, \quad C_1^{0,-} = 0, \quad C_0^{0,-} = 1; \quad (4.9)$$

therefore

$$C_0^{0,-} = 1, \quad C_2^{0,-} = -\frac{3}{h^2}, \quad C_1^{0,-} = 0, \quad C_3^{0,-} = \frac{2}{h^3}. \quad (4.10)$$

For the interval $[x_k, x_{k+1}]$ we have similarly

$$\omega_{k,0}(x) = \sum_{j=0}^{3} C_j^{0,+}(x_k - x)^j, \quad (4.11)$$

where (taking into account the conditions $\omega_{k,0}(x_k + h) = 0$, $\omega_{k,0}(x_k) = 1$, $\omega_{k,0}^{(1)}(x_k + h) = 0$, $\omega_{k,0}^{(1)}(x_k) = 0$) the coefficients $\{C_j^{0,+}\}_{j=0}^{3}$ are given by

$$C_0^{0,+} = 1, \quad C_2^{0,+} = -\frac{3}{h^2}, \quad C_1^{0,+} = 0, \quad C_3^{0,+} = -\frac{2}{h^3}. \quad (4.12)$$

Finally we have

$$\omega_{k,0}(x) = \begin{cases} \sum_{j=0}^{3} C_j^{0,-}(x_k - x)^j, & \text{if } x \in [x_k - h, x_k], \\ \sum_{j=0}^{3} C_j^{0,+}(x_k - x)^j, & \text{if } x \in [x_k, x_k + h], \\ 0 & \text{for other } x \in [a, b]. \end{cases} \quad (4.13)$$

The graph of the function $\omega_{2,0}(x)$ (for $a = 0$, $b = 2$, and $N = 3$) is shown in Fig. 2.2.

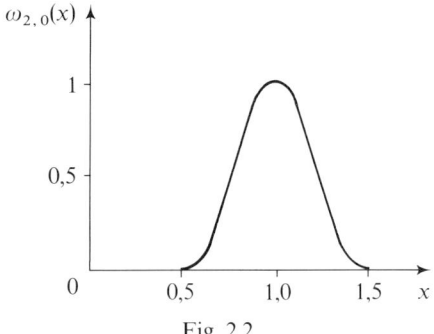

Fig. 2.2.

Let us now use Equation (4.7) to derive a formula for the functions $\omega_{k,1}(x)$. In analogy to the above it can be proved that for any $0 \leq k \leq N + 1$ we have the following expression:

$$\omega_{k,1}(x) = \begin{cases} \sum_{j=0}^{3} C_j^{1,-}(x_k - x)^j, & \text{if } x \in [x_k - h, x_h] \cap [a, b], \\ \sum_{j=0}^{3} C_j^{1,+}(x_k - x)^j, & \text{if } x \in [x_k, x_h + h] \cap [a, b], \\ 0 & \text{otherwise,} \end{cases} \quad (4.14)$$

where

$$C_0^{1,-} = 0, \qquad C_2^{1,-} = \frac{2}{h},$$

$$C_1^{1,-} = -1, \qquad C_3^{1,-} = -\frac{1}{h^2},$$

$$C_0^{1,+} = 0, \qquad C_2^{1,+} = -\frac{2}{h}, \quad (4.15)$$

$$C_1^{1,+} = -1, \qquad C_3^{1,+} = -\frac{1}{h^2}.$$

Figure 2.3 gives the graph of $\omega_{2,1}(x)$ for the case $a = 0$, $b = 2$, and $N = 3$.

The formulas for $\{\omega_{k,0}\}$ become especially simple when $m = 0$. Then $H_N^0(a, b)$ consists of piece-wise linear functions $\{\omega_k\}_{k=1}^N$, such that

$$\omega_k(x_j) = \delta_{k,j} \qquad (k, j = 1, 2, \ldots, N). \quad (4.16)$$

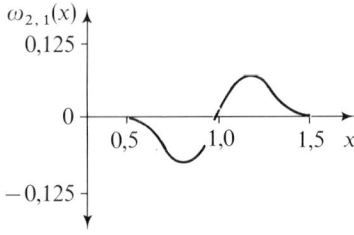

Fig. 2.3.

Note that the functions $\{\omega_k(x)\}$ have been studied in the preceding section (see Fig. 2.1).

We will now describe yet another method of constructing the space F_h. As before, let us take the interval $[a, b]$ on the real line and cover it by the net such that its net points $a = x_0 < x_1 < \cdots < x_N < x_{N+1} = b$ satisfy $x_k = a + kh$, $k = 0, 1, \ldots, N + 1$, $h = (b - a)/(N + 1)$. Denote by $M_N^m(a, b)$ a set of functions $g(x)$ satisfying the following two properties. First, $g(x)$ is an mth-order polynomial on each of the intervals $[x_k, x_{k+1}]$; second, for any $0 \le k \le N$ and $0 \le j \le m$ we have

$$g(x_{k,j}) = d_{k,j},$$

where $x_{k,j} = x_k + (h/m)j$ and $d_{k,j}$ are given numbers; in addition $g(a) = g(b) = 0$, that is, $d_{0,0} = d_{N,m} = 0$. Hence it follows that the function $g \in M_N^m(a, b)$ is a piece-wise polynomial function belonging to $\overset{\circ}{W}_2^1$; in other words, the function $g(x)$ is continuous, with possible discontinuities in the first derivative at the points $\{x_k\}_{k=1}^N$. Let us now take a look at the explicit construction of $g(x)$, given $\{d_{k,j}\}$.

Let us choose an arbitrary k, $0 \le k \le N$, and construct the function $g(x)$ on $[x_k, x_{k+1}]$ (denote it by $g_k(x)$). By a well-known result from the theory of approximations, there exists a unique mth order polynomial, passing through the $(m + 1)$ points $d_{k,0}, \ldots, d_{k,m}$. It coincides with the Lagrange polynomial

$$g_k(x) = \sum_{i=0}^{m} d_{k,i} \prod_{\substack{l=0 \\ l \ne i}}^{m} \frac{(x_{k,l} - x)}{(x_{k,l} - x_{k,i})}. \tag{4.17}$$

For the case $m = 1$ we obtain a usual linear function

$$g_k(x) = d_k \frac{(x_{k+1} - x)}{(x_{k+1} - x_k)} + d_{k+1} \frac{(x_k - x)}{(x_k - x_{k+1})}. \tag{4.18}$$

It follows from this, in particular, that the space $M_N^1(a, b)$ is the same as $H_N^0(a, b)$, and therefore the basis obtained for H_N^0 can be used for $M_N^1(a, b)$. We will not consider the problem of a basis for $M_N^m(a, b)$ for large m.

Thus, we have constructed two kinds of sets of subspaces F_h for the space $F = \overset{\circ}{W}_2^1$, each set being complete in $\overset{\circ}{W}_2^1$.

2.5. Variational-Difference Schemes for Two-Dimensional Equations of Elliptic Type

2.5.1. The Ritz Method

In this section we are going to describe how to obtain the variational-difference schemes for Problem (2.1), (2.2) under the following additional assumptions:

$$B_i(x) = 0 \quad \text{in} \quad D \quad (i = 1, 2);$$
$$q(x) = 0 \quad \text{in} \quad D;$$

further, for any vector $\xi = (\xi_1, \xi_2)'$

$$\mu_0 \sum_{i=1}^{2} \xi_i^2 \leqslant \inf_{x \in D} \sum_{i,j=1}^{2} A_{ij}(x)\xi_i\xi_j \leqslant \sup_{x \in D} \sum_{i,j=1}^{2} A_{ij}(x)\xi_i\xi_j \leqslant \mu_1 \sum_{i=1}^{2} \xi_i^2 \quad (5.1)$$

with some positive constants $\mu_0 \leq \mu_1$; the boundary ∂D of D is supposed to be piece-wise linear. In principle, the last three assumptions could be weakened; we will not do that for the sake of simplicity. Our problem is thus to find the solution of the equation

$$-\sum_{i,j=1}^{2} \frac{\partial}{\partial x_i} A_{ij}(x) \frac{\partial u}{\partial x_j} = f \quad \text{in} \quad D \quad (5.2)$$

with the boundary conditions

$$u = 0 \quad \text{on} \quad \partial D. \quad (5.3)$$

As we have seen in Section 2.2, this is equivalent to finding a function which minimizes the quadratic functional

$$J(u) = \int_D \left[\sum_{i,j=1}^{2} A_{ij}(x) \frac{\partial u}{\partial x_i} \frac{\partial u}{\partial x_j} \right] dD - 2 \int_D uf \, dD \quad (5.4)$$

over the space $\mathring{W}_2^1(D)$.

To solve the latter, we will use the Ritz method with a special kind of subspaces F_h. To this end, let us triangularize the region D, that is, cover it by a triangular net D_h (Fig. 2.4). In each of the triangles we then consider an

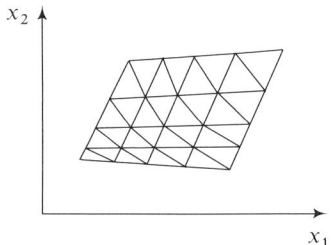

Fig. 2.4.

mth order polynomial in the variables x_1, x_2, of the following form

$$g(x_1, x_2) = \sum_{i=0}^{m} \sum_{j_1 + j_2 = i} c_{j_1, j_2} x_1^{j_1} x_2^{j_2}. \quad (5.5)$$

In every triangle the coefficients of the polynomials are chosen in such a way, that the overall function becomes an element of $\mathring{W}_2^1(D)$, that is, the function must be continuous, and zero on the boundary ∂D.

The method has been introduced by Courant [5] for the case $m = 1$, and also considered by Oganesyan [5] and others. The cases $m = 2, 3, 5$ have been solved in detail by Zlámal [5].

We will illustrate the method with the example $m = 1$, i.e. a piece-wise linear function
$$g(x_1, x_2) = c_{0,0} + c_{1,0}x_1 + c_{0,1}x_2. \tag{5.6}$$
For a specific triangle the coefficients of this function are defined by means of given values $u(p_1)$, $u(p_2)$, and $u(p_3)$ at the vertices p_1, p_2, and p_3 of the triangle. Performing this procedure in each triangle from the decomposition of the region D, while taking $u(p) = 0$ for $p \in \partial D$, we find that the resulting function is continuous in D and zero on the boundary ∂D. Its continuity on the common boundaries of the neighboring triangles (for example, when this common boundary coincides with the line segment joining the triangle vertices p_1 and p_2) follows from the fact that for every triangle the function $g(x_1, x_2)$ is linear on this boundary, with the values $u(p_1)$ and $u(p_2)$ at p_1 and p_2 respectively.

For the quadratic interpolation ($m = 2$)
$$g(x_1, x_2) = c_{0,0} + c_{1,0}x_2 + c_{0,1}x_2 + c_{2,0}x_1^2 + c_{1,1}x_1x_2 + c_{0,2}x_2^2, \tag{5.7}$$
a convenient way of making sure that the resulting function belongs to the space $\overset{0}{W}{}^1_2$ is as follows: The function u^h is specified at the triangle vertices p_1, p_2, and p_3, and at the points $p_{1,2}, p_{2,3}$, and $p_{3,1}$, which divide the segments $[p_1, p_2]$, $[p_2, p_3]$, and $[p_3, p_1]$ in half (see Fig. 2.5).

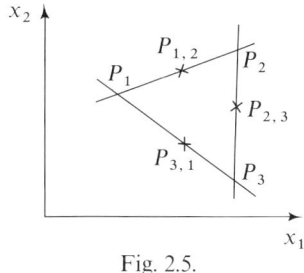

Fig. 2.5.

The continuity of u^h then follows from the simple fact that this function is uniquely defined on any triangle segment. For instance, $u^h(x_1, x_2)$ is uniquely defined on $[p_1, p_2]$ by specifying $u(p_1)$, $u(p_{1,2})$, and $u(p_2)$.

If now the solution of the problem is sought in the form (5.7) in every triangular region Δ, then the difference equations are found in a routine way, using the Ritz method for the variational functional. Zlámal has shown that if the solution $\hat{u}$ of Equations (5.2) and (5.3) is from the class $C^{(3)}(D)$, if the third derivatives of $\hat{u}(x)$ are bounded by M in modulus, and if in addition the minimal angle in the triangles which decompose the region D is bounded by a number $v_0 > 0$ from below, then the Ritz approximation error for the quadratic interpolation (5.7) can be estimated by
$$\|u^h - \hat{u}\|_{\overset{0}{W}{}^1_2} \leqslant Ch^2$$
where $C = C_1 M / \sin v_0$ and C_1 are independent of the particular triangle.

If the smoothness assumptions on the solution of (5.2), (5.3) are taken in a weaker form ($\hat{u} \in C^{(2)}(D)$), then it is easy to prove that

$$\|u^h - \hat{u}\|_{\mathring{W}_2^1} \leq Ch, \quad C > 0,$$

for the case of piece-wise linear approximations.

This concludes the discussion of quadratic approximations on triangles; we will now concentrate on piece-wise linear approximations. In order to define uniquely a piece-wise linear function $g(x) \in \mathring{W}_2^1(D)$ in each triangle, it suffices (by what we have said earlier) to specify the values of $g(x)$ at the vertices of the triangles. For that, let us index all the inner vertices, using the notation $\{p_k\}_{k=1}^{N_h}$, and let us also denote by $D_{h,k}$ the union of all triangles with p_k the common vertex. Then the basis for the space F_h can be taken as the system of functions $\{\omega_k(x)\}_{k=1}^{N_h}$ defined by

1. $\omega_k(p_j) = \delta_{k,j}$, where $\delta_{k,j}$ is the Kronecker symbol, and
2. $\omega_k(x)$ is linear on each of the triangles, i.e. is represented by Expression (5.6).

Thus ω_k can be viewed geometrically as a pyramid with the vertex at the point p_k and zero outside the region $D_{h,k}$ (Figure 2.6).

More specifically, let us assume that D is actually the unit square $\{x_1, x_2 : 0 < x_1, x_2 < 1\}$. Let us cover D with the usual uniform orthogonal net with the mesh size $h = 1/(N+1)$, where N is a positive integer, and triangularize next the region D in the manner shown in Fig. 2.7. The basis

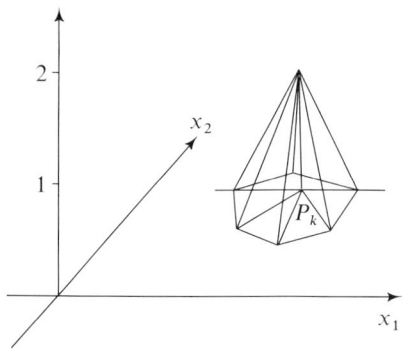

Fig. 2.6.

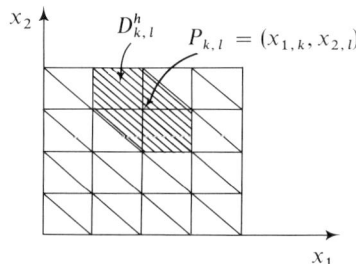

Fig. 2.7.

Variational-Difference Schemes for Two-Dimensional Equations

functions $\{\omega_k(x)\}_{k,l=1}^{N^2}$ will be denoted in this case by $\{\omega_{k,l}(x)\}_{k,l=1}^{N}$ (note $N_h = N^2$).

Consider the matrix A from the system of linear equations

$$A\alpha = g \qquad (5.8)$$

corresponding to the Ritz method. Here $\alpha = (\alpha_1, \ldots, \alpha_{N^2})'$ is a vector composed of the coefficients $\{\alpha_{N(k-1)+l} = \alpha_{k,l}\}_{k,l=1}^{N}$ from the decomposition

$$u^h(x) = \sum_{k,l=1}^{N} \alpha_{k,l} \omega_{k,l}(x), \qquad (5.9)$$

$g = (g_1, \ldots, g_{N^2})'$ is a vector with the components

$$g_{N(k-1)+l} = g_{k,l} = \int_{D_{k,l}} f\omega_{k,l}(x)\, dD \qquad (k, l = 1, \ldots, N), \qquad (5.10)$$

and for the elements of A we have the formulas

$$a_{N(k-1)+l, N(i-1)+j} = \int_D \sum_{s,t=1}^{2} A_{s,t}(x) \frac{\partial \omega_{k,l}}{\partial x_s} \frac{\partial \omega_{i,j}}{\partial x_t}\, dD \qquad (5.11)$$

$$(k, l, i, j = 1, \ldots, N).$$

Let us use the notation $a_{k,l}^{i,j} = a_{N(k-1)+l, N(i-1)+j}$.

Considering the form of the functions $\{\omega_{k,l}(x)\}_{k,l=1}^{N}$ (see Figs. 2.6 and 2.7), it is not difficult to show that

$$a_{k,l}^{i,j} = 0$$

provided at least one of the two following inequalities holds true:

$$|i - k| > 1, \qquad |j - l| > 1 \qquad (k, l, i, j = 1, \ldots, N).$$

From this, one can show directly that A is a three-diagonal, block matrix of the form

$$A = \begin{Vmatrix} A_{11} & A_{12} & 0 & \cdots & 0 & 0 \\ A_{21} & A_{22} & A_{23} & \cdots & 0 & 0 \\ \multicolumn{6}{c}{\dotfill} \\ 0 & 0 & 0 & \cdots & A_{N,N-1} & A_{NN} \end{Vmatrix}, \qquad (5.12)$$

where $A_{kk} = A_{k,k}^*, A_{k,k+1} = A_{k+1,k}^*$ ($k = 1, \ldots, N$), and each of the matrices $A_{k,l}$ is itself a three-diagonal matrix of order N. Further analysis shows that the matrices $\{A_{k,k-1}\}_{k=2}^{N}$ are in fact two-diagonal matrices:

$$A_{k,k-1} = \begin{Vmatrix} a_{k,1}^{k-1,1} & a_{k,1}^{k-1,2} & 0 & \cdots & 0 & 0 \\ 0 & a_{k,2}^{k-1,2} & a_{k,2}^{k-1,3} & \cdots & 0 & 0 \\ \multicolumn{6}{c}{\dotfill} \\ 0 & 0 & 0 & \cdots & 0 & a_{k,N}^{k-1,N} \end{Vmatrix}. \qquad (5.13)$$

$$(k = 2, \ldots, N).$$

Let us compute the elements $\{a_{k,l}^{i,j}\}$ of the matrix A for the particular case of (5,2), (5.3):

$$-\frac{\partial}{\partial x}p(x,y)\frac{\partial u}{\partial x} + \frac{\partial}{\partial y}q(x,y)\frac{\partial u}{\partial y} = f \quad \text{in} \quad D, \tag{5.14}$$

$$u = 0 \quad \text{on} \quad \partial D.$$

In order to do that, we first represent $D_{k,l}^h$ as the union of the six triangles $\{D_{k,l,m}^h\}_{m=1}^6$, the indexing of which is clear from Fig. 2.8. By a straightforward

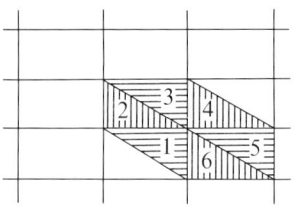

Fig. 2.8.

computation we obtain

$$\omega_{k,l}(x,y) = \begin{cases} 1 - \frac{1}{h}(x_k - x) - \frac{1}{h}(y_l - y), & \text{if } x,y \in D_{k,l,1}^h, \\ 1 - \frac{1}{h}(x_k - x), & \text{if } x,y \in D_{k,l,2}^h, \\ 1 + \frac{1}{h}(y_l - y), & \text{if } x,y \in D_{k,l,3}^h, \\ 1 + \frac{1}{h}(x_k - x) + \frac{1}{h}(y_l - y), & \text{if } x,y \in D_{k,l,4}^h, \\ 1 + \frac{1}{h}(x_k - x), & \text{if } x,y \in D_{k,l,5}^h, \\ 1 - \frac{1}{h}(y_l - y), & \text{if } x,y \in D_{k,l,6}^h. \end{cases} \tag{5.15}$$

Since the matrix A is symmetric, and since the matrices $\{A_{k,k}\}_{k=1}^N$ are three-diagonal and the matrices $\{A_{k,k-1}\}_{k=2}^N$, $\{A_{k,k+1}\}_{k=1}^{N-1}$ are two-diagonal, it is enough to compute the elements

$$a_{k,l}^{k,l}, a_{k,l}^{k,l-1}, a_{k,l}^{k-1,l}, a_{k,l}^{k-1,l+1} \qquad (1 \leq k, l \leq N).$$

Variational-Difference Schemes for Two-Dimensional Equations

From (5.11) and (5.15) we have (for simplicity we put $D_i = D^h_{k,l,i}$)

$$a_{k,l}^{k,l} = \int_{D^h_{k,l}} \left[p(x,y) \left(\frac{\partial \omega_{k,l}}{\partial x} \right)^2 + q(x,y) \left(\frac{\partial \omega_{k,l}}{\partial y} \right)^2 \right] dxdy$$

$$= \frac{1}{h^2} \left[\int_{D_1 \cup D_2} p(x,y)\, dxdy + \int_{D_4 \cup D_5} p(x,y)\, dxdy \right.$$

$$\left. + \int_{D_5 \cup D_4} q(x,y)\, dxdy + \int_{D_1 \cup D_6} q(x,y)\, dxdy \right];$$

(5.16)

$$a_{k,l}^{k,l-1} = \int_{D^h_{k,l} \cup D^h_{k,l-1}} \left[p(x,y) \frac{\partial \omega_{k,l}}{\partial x} \cdot \frac{\partial \omega_{k,l-1}}{\partial x} + q(x,y) \frac{\partial \omega_{k,l}}{\partial y} \right.$$

$$\left. \times \frac{\partial \omega_{k,l-1}}{\partial y} \right] dxdy = -\frac{1}{h^2} \left[\int_{D_1 \cup D_6} q(x,y)\, dxdy \right];$$

$$a_{k,l}^{k-1,l} = -\frac{1}{h^2} \left[\int_{D_1 \cup D_2} p(x,y)\, dxdy \right]; \qquad a_{k,l}^{k-1,l+1} = 0.$$

From this one can immediately show that the matrices $\{A_{k,k-1}\}_{k=2}^N$ are diagonal. Moreover, if we introduce a vector u with the components $u_{k,l} = \alpha_{k,l}$, where α is a vector from the system (5.8), then it is interesting to note that the system $Au = g$ has the representation

$$(A_1 + A_2)u = g, \tag{5.17}$$

where

$$(A_1 u)_{k,l} = -\tilde{P}_{k-1/2,l} u_{k-1,l} + (\tilde{P}_{k-1/2,l} + \tilde{P}_{k+1/2,l}) u_{k,l} - \tilde{P}_{k+1/2,l} u_{k+1,l} \tag{5.18}$$

and

$$(A_2 u)_{k,l} = -\tilde{Q}_{k,l-1/2} u_{k,l-1} + (\tilde{Q}_{k,l-1/2} + \tilde{Q}_{k,l+1/2}) u_{k,l} - \tilde{Q}_{k,l+1/2} u_{k,l+1}.$$

Here we have used the notation

$$\tilde{P}_{k\pm 1/2,l} = \frac{1}{h^2} \int_{D^h_{k,l} \cap D^h_{k\pm 1,l}} p(x,y)\, dxdy,$$

$$\tilde{Q}_{k,l\pm 1/2} = \frac{1}{h^2} \int_{D^h_{k,l} \cap D_{k,l\pm 1}} q(x,y)\, dxdy.$$

(5.19)

From (5.17) to (5.19) it is easy to see that the variational-difference scheme we have derived by the Ritz method is essentially identical to standard difference schemes, as far as the structure and distribution of the nonzero elements and their form is concerned. In particular, for constant $p(x,y)$ and $q(x,y)$, both difference- and variational-difference analogs of the differential operator are exactly the same. This fact allows one to solve System (5.17) by effective iteration methods, such as (among others) the splitting method and the sequential over-relaxation method.

2.5.2. The Galerkin Method

Since the basic features of solving two-dimensional problems have been already demonstrated with an example of the Ritz method, we will not dwell on the detailed constructions of the Galerkin variational-difference schemes. Let us only remark that for the nonzero coefficients $\{B_i(x)\}_{i=1}^2$ in (2.1), (2.2), the matrix $\tilde{A}$ of the Galerkin method with the basis functions (5.15) differs from the matrix A of the system of (5.17) by a certain matrix B ($B = \tilde{A} - A$). The elements $b_{k,l}^{i,j}$ of the latter matrix are given by

$$b_{k,l}^{i,j} = \int_{D_{k,l}^h \cap D_{i,j}^h} \left[B_1(x,y) \frac{\partial \omega_{ij}}{\partial x} \omega_{k,l} + B_2(x,y) \frac{\partial \omega_{ij}}{\partial y} \omega_{k,l} \right] dxdy \qquad (5.20)$$

$$(k, l, i, j = 1, \ldots, N).$$

From this it can be seen that the matrices $\{A_{k,k-1}\}_{k=2}^N$ in the Galerkin method may not turn out to be diagonal even for the equation

$$\sum_{i=1}^{2} \left[-\frac{\partial^2 u}{\partial x_i^2} + B_i(x)u \right] = f \quad \text{in} \quad D, \qquad (5.21)$$

$$u = 0 \quad \text{on} \quad \partial D.$$

A more complicated matrix structure of the algebraic system calls for simpler subspaces F_h than those in the Ritz method. The same conclusion follows if we try to construct the Galerkin variational-difference schemes of more than first-order approximation.

Using the results from Section 2.4, the subspaces F_h can be constructed fairly easily for the regions which can be represented as finite unions of rectangles. Below we describe the structure of these spaces F_h with an example of a piece-wise linear approximation and construct a variational-difference scheme for a particular case.

Assume the region D is a union of r rectangles $\{D_i\}_{i=1}^r$, the sides of which are parallel to the coordinate lines. Let $\tilde{D}$ be the smallest rectangle containing the region D (see Fig. 2.9, where $\tilde{D} = \{(x, y) : a \leq x \leq b, c \leq y \leq d\}$. On the intervals $[a, b]$ and $[c, d]$ let us choose the nets $a = x_0 < x_1 < \cdots < x_{N+1} = b$ and $c = y_0 < y_1 < \cdots < y_{M+1} = d$ respectively, in such a way, that

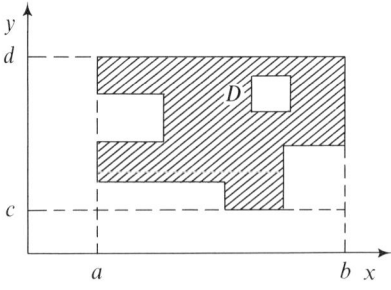

Fig. 2.9.

any boundary segment of any rectangle from the decomposition of D belongs to one of the lines

$$x \equiv x_k, \quad y \equiv y_l \qquad (k = 0, 1, \ldots, N+1; l = 0, 1, \ldots, M+1). \quad (5.22)$$

Finally, denote by D^h the net region consisting of the points (x_k, y_l) belonging to D $(k = 0, 1, \ldots, N; l = 0, 1, \ldots, M)$.

Let us now turn to constructing the subspace $F_h \in \mathring{W}_2^1(D)$. Introduce the functions

$$\omega_{x,k}(x) = \begin{cases} \dfrac{x - x_{k-1}}{x_k - x_{k-1}}, & \text{if} \quad x \in [x_{k-1}, x_k], \\ \dfrac{x - x_{k+1}}{x_k - x_{k+1}}, & \text{if} \quad x \in [x_k, x_{k+1}], \\ 0 & \text{otherwise} \end{cases} \quad (5.23)$$

$(k = 1, \ldots, N)$,

$$\omega_{y,l}(y) = \begin{cases} \dfrac{y - y_{l-1}}{y_l - y_{l-1}}, & \text{if} \quad y \in [y_{l-1}, y_l], \\ \dfrac{y - y_{l+1}}{y_l - y_{l+1}}, & \text{if} \quad y \in [y_l, y_{l+1}], \\ 0 & \text{otherwise} \end{cases} \quad (5.24)$$

$(l = 1, \ldots, M)$,

and the system of functions

$$\omega_{k,l}(x, y) = \omega_{x,k}(x)\omega_{y,l}(y), \qquad (x_k, y_l) \in D^h \quad (5.25)$$

Let us take for F_h the linear hull of the functions $\{\omega_{k,l}\}$. Since the system $\{\omega_{k,l}\}$ is linearly independent, it can be clearly taken for the basis of the space F_h.

Next we will derive the linear equations corresponding to the Galerkin method in the subspace F_h with the basis (5.25) for the following simple problem:

$$-\frac{\partial^2 u}{\partial x^2} - \frac{\partial^2 u}{\partial y^2} + \frac{\partial u}{\partial x} = f \quad \text{in} \quad D \quad (5.26)$$

$$u = 0 \quad \text{on} \quad \partial D.$$

In agreement with Section 2.2, if the approximation u^h is taken as

$$u^h = \sum_{(x_i, y_j) \in D^h} u_{i,j} \omega_{i,j}(x, y), \quad (5.27)$$

then the linear system is of the form

$$Au = g, \quad (5.28)$$

where the components of u and g are $\{u_{k,l}\}$ from (5.27), and $g_{k,l} = \int_D f\omega_{k,l} dD$ respectively; the elements of A are given by

$$a_{k,l}^{i,j} = \int_D \left[\frac{\partial \omega_{i,j}}{\partial x} \cdot \frac{\partial \omega_{k,l}}{\partial x} + \frac{\partial \omega_{i,j}}{\partial y} \cdot \frac{\partial \omega_{k,l}}{\partial y} + \frac{\partial \omega_{i,j}}{\partial x} \omega_{k,l} \right] dD$$

$$= \int_D \left[\omega_j \omega_l \frac{\partial \omega_i}{\partial x} \cdot \frac{\partial \omega_k}{\partial x} + \omega_i \omega_k \frac{\partial \omega_j}{\partial y} \cdot \frac{\partial \omega_l}{\partial y} + \omega_j \frac{\partial \omega_i}{\partial x} \omega_k \omega_l \right] dD. \quad (5.29)$$

Similarly as in the previous section we can easily see, that $a_{k,l}^{i,j} = 0$ if either one of the inequalities below is satisfied:

$$|i - k| > 1, |j - l| > 1.$$

Hence it follows that A is a three-diagonal block matrix of the form (5.12). We now give the final formulas for $a_{k,l}^{i,j}$, assuming for simplicity that the net is uniform with the mesh size h:

$$a_{k,l}^{k,l} = \frac{1}{h^2} \int_{x_{k-1}}^{x_{k+1}} dx \int_{y_{l-1}}^{y_{l+1}} dy [\omega_l^2(y) + \omega_k^2(x)] = \frac{8}{3},$$

$$a_{k,l}^{k,l-1} = \frac{1}{h^2} \int_{x_{k-1}}^{x_{k+1}} dx \int_{y_{l-1}}^{y_l} dy [\omega_{l-1}(y)\omega_l(y) - \omega_k^2(x)] = -\frac{1}{3}, \quad (5.30a)$$

$$a_{k,l}^{k,l+1} = -\frac{1}{3},$$

$$a_{k,l}^{k-1,l} = \int_{x_{k-1}}^{x_k} dx \int_{y_{l-1}}^{y_{l+1}} dy \left[-\frac{\omega_l^2(y)}{h^2} + \frac{\omega_{k-1}(x)\omega_k(x)}{h^3} \right.$$

$$\left. - \frac{1}{h} \omega_k(x)\omega_l^2(y) \right] = -\frac{1}{3} - \frac{h}{3}, \quad (5.30b)$$

$$a_{k,l}^{k-1,l-1} = \int_{x_{k-1}}^{x_k} dx \int_{y_{l-1}}^{y_l} dy \left[-\frac{1}{h^2} \omega_l(y)\omega_{l-1}(y) \right.$$

$$\left. - \frac{1}{h^2} \omega_k(x)\omega_{k-1}(x) - \omega_k(x)\omega_l(y)\omega_{l-1}(y) \right] = -\frac{1}{3} - \frac{h}{12},$$

$$a_{k,l}^{k-1,l+1} = \int_{x_{k-1}}^{x_k} dx \int_{y_l}^{y_{l+1}} dy \left[-\frac{\omega_{l+1}(y)\omega_l(y)}{h^2} \right.$$

$$\left. - \frac{\omega_{l+1}(x)\omega_{k+1}(x)}{h^2} - \frac{\omega_k(x)\omega_l(y)\omega_{l+1}(y)}{h} \right] = -\frac{1}{3} - \frac{h}{12}, \quad (5.30c)$$

$$a_{k,l}^{k+1,l} = -\frac{1}{3} + \frac{h}{3}, \quad a_{k,l}^{k+1,l-1} = -\frac{1}{3} + \frac{h}{12},$$

$$a_{k,l}^{k+1,l+1} = -\frac{1}{3} + \frac{h}{12}, \quad (x_k, y_l) \in D^h.$$

Thus, it turns out that in the case $f(x, y) = f = \text{const.}$ our variational-difference scheme is equivalent to a somewhat unusual difference scheme

$$\frac{8}{3}u_{k,l} - \frac{1}{3}u_{k,l-1} - \frac{1}{3}u_{k,l+1} - \left(\frac{1}{3} + \frac{h}{3}\right)u_{k-1,l} - \left(\frac{1}{3} + \frac{h}{12}\right)u_{k-1,l-1}$$

$$- \left(\frac{1}{3} + \frac{h}{12}\right)u_{k-1,l+1} - \left(\frac{1}{3} - \frac{h}{3}\right)u_{k+1,l} - \left(\frac{1}{3} - \frac{h}{12}\right)u_{k+1,l-1}$$

$$- \left(\frac{1}{3} - \frac{h}{12}\right)u_{k+1,l+1} = h^2 f. \tag{5.31}$$

This difference scheme can be easily obtained using three-point approximations of the second derivatives, provided the differential part in (5.26) is replaced in advance by the approximate expression on the net points D_h:

$$\left.\frac{\partial^2 u}{\partial x^2} + \frac{\partial^2 u}{\partial y^2} - \frac{\partial u}{\partial x}\right|_{\substack{x=x_k\\y=y_l}} \approx \frac{1}{6}\left[\sum_{i=k-1}^{k+1} \beta_{k-1}\left(\frac{\partial^2 u}{\partial y^2}\right)_{\substack{x=x_i\\y=y_l}}\right.$$

$$\left. + \sum_{j=l-1}^{l+1} \beta_{l-j}\left(\frac{\partial^2 u}{\partial x^2} - \frac{\partial u}{\partial x}\right)_{\substack{x=x_k\\y=y_j}}\right], \tag{5.32}$$

where

$$\beta_{-1} = \beta_1 = 1, \beta_0 = 4, \quad \text{and} \quad (x_k, y_l) \in D^h.$$

2.6. Variational Methods for Multidimensional Problems

We will briefly discuss some possibilities for constructing the variational-difference schemes for problems with more than two independent variables.

2.6.1. Methods of Choosing the Subspaces

Consider a bounded region D in the space of the variables x, y, z. Suppose the boundary ∂D is piece-wise linear. In this case there is a well-known method of constructing the subspaces $F_h \subset \mathring{W}_2^1(D)$; it is as follows: First, the region D is triangularized, i.e. D is covered by a nonintersecting finite system of three-edged pyramids Δ_k, such that $D = \bigcup_{k=1}^{N} \Delta_k$. Denote by h_k the length of the longest edge of the pyramid Δ_k. In analogy to the one-dimensional and two-dimensional cases (see Sections 2.4 and 2.5.1) we now construct the sequence of spaces F_h of piece-wise polynomial functions, where the index $h = \max_{1 \leq k \leq N} h_k$. Let us illustrate this procedure with an example of piece-wise linear approximations and with the region $D = \{x, y, z: 0 \leq x \leq 1, 0 \leq y \leq 1, 0 \leq z \leq 1\}$.

Let us decompose the interval $[0, 1]$ by a uniform grid $0 = \xi_0 < \xi_1 < \cdots < \xi_{n+1} = 1$, $\xi_k = \Delta \xi \times k = k/(n + 1)$. Consider *elementary cubes*, with edges of equal length $\Delta \xi$, and with the vertices at the points $[x = \xi_k, y = \xi_l, z = \xi_m]$, $k, l, m = 0, 1, \ldots, n + 1$. This system of cubes covers D, and hence it is enough to triangularize the elementary cubes. There are many ways of doing this; one of them is shown in Fig. 2.10.

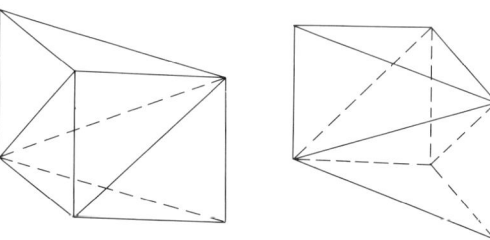

Fig. 2.10.

Here every elementary cube is taken as the union of six pyramids, the maximum edge length being $h = \sqrt{3} \Delta \xi = \sqrt{3}/(n + 1)$. Thus the whole of D is covered by a total of $N = 6(n + 1)^3$ pyramids. The space F_h is now as follows: In each pyramid Δ_k consider a polynomial of the form

$$g_k(x, y, z) = \sum_{t=0}^{m} \sum_{i_1 + i_2 + i_3 = t} C_{i_1, i_2, i_3}^k x^{i_1} y^{i_2} z^{i_3} \qquad (6.1)$$

such that the function

$$g(x, y, z) = \begin{cases} g_1(x, y, z), & \text{if } (x, y, z) \in \Delta_1, \\ \vdots \\ g_N(x, y, z), & \text{if } (x, y, z) \in \Delta_N \end{cases} \qquad (6.2)$$

belongs to the space $\overset{\circ}{W}{}_2^1(D)$; that is, its first derivatives are square integrable, the function itself being zero on the boundary. For this it is enough that the piece-wise polynomial function g is continuous and is zero on ∂D.

In the case of piece-wise linear functions

$$g_k(x, y, z) = C_{0,0,0}^k + C_{1,0,0}^k x + C_{0,1,0}^k y + C_{0,0,1}^k z \qquad (6.3)$$

the continuity of $g(x, y, z)$ is assured if, for instance, the coefficients of the functions g_k are specified by the function values of g at the vertices of the corresponding pyramid Δ_k. In other words, it is enough to require that g_k assumes strictly specified values at the vertices of the pyramid Δ_k.

The approach just described is sufficiently universal and can be applied to problems with more general boundaries. It leads, however, to very complex algorithms even for the simplest boundaries. Below we give another method of constructing the subspaces F_h which, in our opinion, is preferable for the multidimensional regions of a special kind.

Consider a region D in the p-dimensional Euclidean space, and assume D can be represented as a finite union of rectangles D_v. Let $\tilde{D} = \{x_i: a_i \leq x_i \leq b_i, i = 1, \ldots, p\}$ be the smallest rectangle containing D (it is assumed

Variational Methods for Multidimensional Problems

that the edges of D_y are parallel to the coordinate lines). For such i, $1 \leq i \leq p$, consider a subdivision of the interval $[a_i, b_i]$ of the x_i axis

$$a_i = x_{i,0} < x_{i,1} < \cdots < x_{i,N_i+1} = b_i \qquad (i = 1, \ldots, p),$$

and define the net D^h as the set of points $x_k = (x_{i,k_1}, x_{2,k_2}, \ldots, x_{p,k_p})$ belonging to D ($1 \leq k_i \leq N_i$ for $i = 1, \ldots, p$).

On each one-dimensional net introduce the family of one-dimensional basis functions

$$g_{i,k_i}(x_i) = \begin{cases} \dfrac{x - x_{i,k_i-1}}{x_{i,k_i} - x_{i,k_i-1}}, & \text{if } x_i \in [x_{i,k_i-1}, x_{i,k_i}], \\ \dfrac{x - x_{i,k_i+1}}{x_{i,k_i} - x_{i,k_i+1}}, & \text{if } x_i \in [x_{i,k_i}, x_{i,k_i+1}], \\ 0 & \text{otherwise,} \end{cases} \qquad (6.4)$$

$$(k_i = 1, \ldots, N_i; 1 \leqslant i \leqslant p).$$

Then the functions

$$g_k(x) = \prod_{i=1}^{p} g_{i,k_i}(x_i), \; (x_k \in D^h, \; k = (k_1, k_2, \ldots, k_p)) \qquad (6.5)$$

can be taken for the basis of the space of piece-wise linear functions $F_h \subset \overset{0}{W}{}^1_2(D)$. The case when $p = 2$ was considered in detail in the previous section.

2.6.2. Coordinate-by-Coordinate Methods for Variational-Difference Schemes

The approaches we have used above to construct the variational-difference schemes and the corresponding subspaces F_h become very complicated in practical implementations. In particular, this involves the choice of the basis functions (the first method in the previous section), the computation of the nonzero elements of the matrix characterizing the system, etc. These difficulties may be overcome by combining the variational approach and the ordinary difference approach into one single method. The system of net equations in this case will not be of the strictly difference or variational-difference types any more; in many cases, however, the mixture will allow one to relax the assumptions of one approach at the expense of borrowing from the other approach. Let us illustrate this with an example of a p-dimensional equation of elliptic type.

Consider the equation

$$-\sum_{i=1}^{p} \frac{\partial}{\partial x_i} P_i(x) \frac{\partial u}{\partial x_i} = f \qquad (6.6)$$

in a bounded region D with the boundary ∂D. Suppose

$$u = 0 \quad \text{on} \quad \partial D \tag{6.7}$$

(the first boundary-value problem). If $p \geq 2$, and the region D is not one of those considered above or for which the sequence of subspaces F_h can be easily constructed, we propose the following method. It combines the difference and the variational approaches.

Define in the region D a rectangular net D^h by intersecting the region D with a set of hyperplanes parallel to the coordinate planes. Denote by $x_k = (x_{1,k_1}, x_{2,k_2}, \ldots, x_{p,k_p})$ the net points of D^h. The net equations are obtained from the approximations of the Problem (6.6), (6.7) at every net point of D^h, while the approximations themselves are in turn derived with the help of the one-dimensional variational-difference schemes. The procedure in this case is as follows:

Assume we need to approximate Problem (6.6), (6.7) at $x_k \in D^h$. The first thing to do is to rewrite the equation in the form

$$\sum_{i=1}^{p} \left[-\frac{\partial}{\partial x_i} P_i(x) \frac{\partial u}{\partial x_i} - f_i \right] = 0, \tag{6.8}$$

where $\sum_{i=1}^{p} f_i = f$. For each member of the sum (6.8) we then construct the variational difference scheme along the corresponding coordinate line x_i, $1 \leq i \leq p$. The next step is to add up the one-dimensional schemes. In other words, the net analog of (6.8) has the form

$$\sum_{i=1}^{p} (L_i^{(h)} u^h - f_i^h) = 0, \tag{6.9}$$

where $(L_i^{(h)} u^h - f_i^h)$ is the variational-difference analog of

$$-\frac{\partial}{\partial x_i} P_i(x) \frac{\partial u}{\partial x_i} - f_i$$

along the segments of the x_i-coordinate line. Variational-difference schemes for the one-dimensional operators have been considered in detail in Sections 2.3 and 2.4.

Let us illustrate the proposed approach with an example of piece-wise linear approximations, when the region D is the p-dimensional unit box with a uniform net D_h ($h = 1/(N + 1)$). From Section 2.4 and Equation (6.9) we have that the system of net equations for Equation (6.8) is of the form

$$\frac{1}{h^2} \sum_{i=1}^{p} [-P_{k_i - 1/2} u_{k_i - 1}^h + (P_{k_i - 1/2} + P_{k_i + 1/2}) u_{k_i}^h - P_{k_i + 1/2} u_{k_i + 1}^h] - f_{k_i}^h = 0,$$

$$x_k \in D_h. \tag{6.10}$$

Here we have used the notation ($1 \leq i \leq p$):

$$u^h_{k_i+\alpha} = u^h(x_{k_1}, \ldots, x_{k_i-1}, x_{k_i+\alpha}, x_{k_i+1}, \ldots, x_{k_p})$$

$$(\alpha = 0, \pm 1);$$

$$P_{k_i+\alpha} = \frac{1}{h} \int_{x_{k_i}+\alpha-1/2}^{x_{k_i}+\alpha+1/2} P(x_{k_i}, \ldots, x_{k_i-1}, x_i, x_{k_i+1}, \ldots, x_{k_p}) \, dx \quad (6.11)$$

$$(\alpha = \pm 1/2).$$

2.7. Interpolation of Solutions of Difference Equations by Means of Splines

The interpolation problem for a function of a continuous argument, given the discrete data, is closely related to the problem of constructing the variational-difference schemes and to continuous representations of solutions of difference problems. For, as a rule, in order to obtain the difference equations, one has to discretize the operator and solution of the problem by taking suitable projections. At the same time the solution of the differential problem represents usually the approximate solution of the original problem on a discrete set of points. Let us assume that the difference problem has been solved and that we know the approximate solution of the problem. The procedure now envolves the interpolation from the results obtained onto the whole region on which the solution of the original problem is sought. The interpretation of this kind requires that certain conditions are taken into account: namely, if the solution of the difference equations has been obtained with certain accuracy, then the accuracy of the interpolation must be at least the same. If we possess additional information regarding the error of the approximate solution, then the interpolation may be implemented not from the exact data, but from those involving possible error at the net points. *A priori* information about the smoothness of solutions may then also allow one to improve the accuracy of the approximate solution in some cases, using one or another difference method. Apart from this, the interpolation problem also has merits of its own.

As a rule, the interpolation algorithms for the exact data, specified on a discrete set of points, are based on Lagrange interpolation polynomials. Also, the interpolated function $\phi(x)$ is *a priori* assumed to be suitably differentiable.

Another problem closely related to interpolation arises when the function values of ϕ at the net points x_k are known only up to a certain error, the maximum magnitude of which is *a priori* known at each of the points. In this case we want to construct a curve which would be in some sense the best approximation of the function, the latter being specified at the net points up to random errors.

The theory of interpolation has been recently enriched by what is being referred to as *spline interpolation methods*. By a spline we usually mean a function on the region D which in a vicinity of any point from the interior of D can be represented as a polynomial of certain degree, say m. Also, this function must be continuous in D along with its first $m - 1$ derivatives, the mth derivatives being square integrable. The most often used spline polynomials in practice seem to be those of the third order.

The spline interpolation methods are discussed in detail in the papers by Holladey [14], Shoenberg [14], Ryabenkii [3], Anselon and Laurent [14], and Zavyalov [14]. Generalizations are found in Birkhof, Schultz, and Varga [5], Varga [1], and others.

2.7.1. Interpolation of Functions of One Variable

Consider the net $a = x_0 < x_1 < \cdots < x_n = b$ on an interval $[a, b]$ of the real line. Assume we are given the values $\{f_k\}_{k=0}^n$ of a function $f(x)$, with $[a, b]$ as its domain of definition. The piece-wise cubic interpolation problem for this case is formulated as follows. Find a function $g(x)$ on $[a, b]$ satisfying the four conditions below:

1. $g(x) \in C^{(2)}(a, b)$, that is, g is continuous along with its derivatives up to and including the second order;
2. On each of the intervals $[x_{k-1}, x_k]$, $g(x)$ is identical to a cubic polynomial of the form

$$g(x) \equiv g_k(x) = \sum_{l=0}^{3} a_l^{(k)}(x_k - x)^l \quad (k = 1, \ldots, n); \tag{7.1}$$

3. At the net points x_k ($k = 0, 1, \ldots, n$),

$$g(x_k) = f_k, \quad k = 0, 1, \ldots, n; \quad \text{and} \tag{7.2}$$

4. The boundary conditions

$$g''(a) = g''(b) = 0. \tag{7.3}$$

are satisfied by $g(x)$.

The advantages of this type of interpolation will be understood later, after we establish certain natural extremal property of the function $g(x)$.

We will now prove that the piece-wise cubic interpolation function $g(x)$ is uniquely defined by the above requirements 1–4.

First we exploit the continuity of $g(x)$ and its first two derivatives at the net points. The following $3(n - 1)$ equations are thus obtained:

$$g_{k+1}(x_k) = g_k(x_k); \quad g'_{k+1}(x_k) = g'_k(x_k); \quad g''_{k+1}(x_k) = g''_k(x_k) \tag{7.4}$$
$$(k = 1, \ldots, n - 1).$$

From Equation (7.2) and Representation (7.1) we have

$$a_0^{(k)} = f_k \quad (k = 1, \ldots, n). \tag{7.5}$$

Interpolation of Solutions of Difference Equations by Means of Splines

Combining this with Equations (7.4) we obtain the total of $(3n - 2)$ equations

$$f_{k-1} = a_3^{(k)}h_k^3 + a_2^{(k)}h_k^2 + a_1^{(k)}h_k + f_k \quad (k = 1, \ldots, n)$$
$$a_1^{(k-1)} = 3a_3^{(k)}h_k^2 + 2a_2^{(k)}h_k + a_1^{(k)}, \quad (7.6)$$
$$a_2^{(k-1)} = 3a_3^{(k)}h_k + a_2^{(k)} \quad (k = 2, \ldots, n),$$

for $3n$ unknowns $\{a_1^{(k)}, a_2^{(k)}, a_3^{(k)}\}_{k=1}^n$, where we have denoted $h_k = x_k - x_{k-1}$, $k = 1, \ldots, n$. The additional two equations needed follow from (7.3) and (7.1):

$$3a_3^{(1)}h_1 + a_2^{(1)} = 0, \quad a_2^{(n)} = 0. \quad (7.7)$$

After simple manipulations on (7.6), (7.7) we obtain

$$a_3^{(k)} = \frac{a_2^{(k-1)} - a_2^{(k)}}{3h_k} \quad (k = 1, \ldots, n), \quad (7.8)$$

where we have introduced $a_2^{(0)} = 0$ for notational convenience, and

$$a_1^{(k)} = -\frac{h_k}{3}(a_2^{(k-1)} + 2a_2^{(k)}) + \frac{f_{k-1} - f_k}{h_k} \quad (k = 1, \ldots, n) \quad (7.9)$$

Next, let us use (7.8), (7.9) in the second set of equations in (7.6) [continuity conditions on the derivatives of $g(x)$]. We obtain

$$[h_k a_2^{(k-1)} + 2(h_k + h_{k+1})a_2^{(k)} + h_{k+1}a_2^{(k+1)}] = F_k, \quad (7.10)$$

where

$$F_k = 3\left[\frac{f_{k-1} - f_k}{h_k} - \frac{f_k - f_{k+1}}{h_{k+1}}\right] \quad (k = 1, \ldots, n-1). \quad (7.11)$$

Using the assumption $a_2^{(0)} = 0$ and the condition $a_2^{(n)} = 0$ from (7.7), we can rewrite the system of (7.10) and (7.11) in the vector-matrix form

$$A\phi = F; \quad (7.12)$$

here

$$A = \begin{Vmatrix} 2(h_1 + h_2) & h_2 & 0 & \cdots & 0 & 0 \\ h_2 & 2(h_2 + h_3) & h_3 & \cdots & 0 & 0 \\ \cdots & \cdots & \cdots & \cdots & \cdots & \cdots \\ 0 & 0 & 0 & \cdots & h_{n-1} & 2(h_{n-1} + h_n) \end{Vmatrix}$$

is a three-diagonal matrix of order $(n - 1)$, and $\phi = (a_2^{(1)}, \ldots, a_2^{(n-1)})'$, $F = (F_1, \ldots, F_{n-1})'$ are $(n - 1)$-dimensional vectors.

Since the matrix A is symmetric with a strong diagonal predominance, it is easy to see that $(A\phi, \phi) > 0$, hence positivity. Hence it follows, that the coefficients $\{a_2^{(k)}\}_{k=1}^n$, and for that matter the coefficients $\{a_0^{(k)}, a_1^{(k)}, a_3^{(k)}\}_{k=1}^n$ [see Equations (7.5), (7.8), and (7.9)], are uniquely defined. Thus, the interpolation problem as formulated always has a unique solution.

The usual standard method (see Section 3.8) gives a very convenient numerical procedure for solving (7.12); the formulas become as follows:

$$\beta_k \equiv \begin{cases} \dfrac{-h_2}{2(h_1 + h_2)}, & \text{if } k = 1, \\ \dfrac{-h_{k+1}}{2(h_k + h_{k+1}) + h_k \beta_{k-1}}, & \text{if } 1 < k < n, \end{cases}$$

$$z_k = \begin{cases} \dfrac{F_1}{2(h_1 + h_2)}, & \text{if } k = 1, \\ \dfrac{F_k - h_k z_{k-1}}{2(h_k + h_{k+1}) + h_k \beta_{k-1}}, & \text{if } 1 < k < n, \end{cases} \quad (7.13)$$

$$a_2^{(k)} = \begin{cases} z_{n-1}, & \text{if } k = n - 1, \\ \beta_k a_2^{(k+1)} + z_k, & \text{if } 1 \leqslant k < n - 1. \end{cases}$$

Having computed $\{a_2^{(k)}\}_{k=1}^n$, the coefficients $\{a_0^{(k)}, a_1^{(k)}, a_3^{(k)}\}_{k=1}^n$ are obtained by means of Equations (7.5), (7.9), and (7.8) correspondingly.

It remains to clarify why the piece-wise cubic interpolation is so effective in practice. We can easily show that the usual piece-wise linear interpolation

$$g(x) = \begin{cases} g_1(x), & \text{if } x \in [x_0, x_1], \\ g_2(x), & \text{if } x \in [x_1, x_2], \\ \vdots \\ g_n(x), & \text{if } x \in [x_{n-1}, x_n] \end{cases} \quad (7.14)$$

of the net function $f(x_k) = f_k$, $k = 0, \ldots, n$, satisfies the variational problem

$$\int_a^b \left(\frac{du}{dx}\right)^2 dx = \min_{W_2^1(a,b)},$$
$$u(x_k) = f_k \quad (k = 0, 1, \ldots, n). \quad (7.15)$$

Above we have used the notation

$$g_k(x) = f_{k-1} \frac{x_k - x}{x_k - x_{k-1}} + f_k \frac{x - x_{k-1}}{x_k - x_{k-1}}. \quad (7.16)$$

In other words, the function $g(x)$ from (7.14), (7.16) is a solution of (7.15) in the class of functions with square-integrable first generalized derivatives. These functions are usually called *splines*. (The term has its origins in the theory of elasticity, where the above problem corresponds to the principle of minimal potential energy.)

The above remark suggests that in order to obtain smoother interpolation results one should consider variational problems with higher derivatives.

Interpolation of Solutions of Difference Equations by Means of Splines

The following problem serves as a particular example:

$$\Phi(u) \equiv \int_a^b \left(\frac{d^2 u}{dx^2}\right)^2 dx = \min_{W_2^2(a,b)}, \tag{7.17}$$

$$u(x_k) = f_k \quad (k = 0, 1, \ldots, n),$$

where $W_2^2(a, b)$ is the space of functions whose second generalized derivatives are square integrable. (Note that the functions from $W_2^2(a, b)$ are continuous, along with their first derivatives.) We will prove, that the piece-wise cubic interpolation function $g(x)$ constructed above is a solution of (7.17). To this end, consider

$$\Phi(u - g) = \int_a^b \left[\frac{d^2}{dx^2}(u - g)\right]^2 dx. \tag{7.18}$$

Integrating by parts and using the properties of the space $W_2^2(a, b)$ we obtain

$$\Phi(u - g) = \Phi(u) - \Phi(g) - 2\left[\left(\frac{du}{dx} - \frac{dg}{dx}\right)\frac{d^2 g}{dx^3}\Big|_{x=a}^{x=b}\right.$$

$$\left. - \int_a^b \left(\frac{du}{dx} - \frac{dg}{dx}\right)\frac{d^3 g}{dx^3} dx\right] = \Phi(u) - \Phi(g) \tag{7.19}$$

$$- 12 \sum_{k=1}^n a_3^k (u - g)\Big|_{x=x_{k-1}}^{x=x_k} = \Phi(u) - \Phi(g),$$

or, equivalently,

$$\Phi(g) = \Phi(u) - \Phi(u - g) \leqslant \Phi(u) \tag{7.20}$$

for an arbitrary function $u \in W_2^2(a, b)$. Therefore, the function $g(x)$ is a unique solution of Problem (7.17) (the uniqueness is easily established by using the equality $\Phi(u) - \Phi(g) = \Phi(u - g)$).

In summary we may conclude that the piece-wise cubic interpolation is not only smooth ($g(x)$ is twice continuously differentiable) but is also highly accurate, since it minimizes the integral of the squares of the second derivatives.

More precisely let $f(x)$ be defined and continuous on (a, b), and suppose we know its values at the points $a = x_0 < x_1 < \cdots < x_n = b$: $f_k = f(x_k)$, $k = 0, \ldots, n$. Using these values, let us construct a piece-wise cubic interpolation $g(x)$ satisfying the requirements 1–4. Then the interpolation error $\phi(x) = g(x) - f(x)$ may be estimated by the following inequality:

$$\max_{a \leqslant x \leqslant b} |\phi(x)| \leqslant Ch^\alpha, \tag{7.21}$$

where α and C are nonnegative and independent of the net, and

$$h = \max_{1 \leqslant k \leqslant n} |x_k - x_{k-1}|. \tag{7.22}$$

The constant α can be obtained sufficiently accurately, provided we have more information about the behavior of $f(x)$, that is the existence of the first derivatives. If the function $f(x)$ is twice continuously differentiable, then $\alpha = 2$.

2.7.2. Piece-Wise Cubic Interpolation with Smoothing

The interpolation problem we are going to consider in this section is fundamentally different from the ordinary interpolation considered above.

Let $f(x)$ be a function of a single variable defined on an interval $[a, b]$ of the real line. Assume further, that at the net points $a = x_0 < x_1 < \cdots < x_n = b$ we know the approximate values $\{\tilde{f}_k\}_{k=0}^n$ of the function f. Assuming a random nature of the errors $\varepsilon_k = f(x_k) - \tilde{f}_k$, $k = 0, 1, \ldots, n$, the ordinary interpolation may lead to large approximation errors if the errors ε_k are large at some net points. In such a case it is advantageous to use the interpolation with smoothing. We will consider one of its variants.

The *interpolation-smoothing* function $g(x)$ will be required to satisfy the following conditions:

1. The function $g(x)$ along with its first and second derivatives are continuous on the interval $[a, b]$;
2. On each subinterval $[x_{k-1}, x_k]$, the function $g(x)$ coincides with the third-order polynomial (cubic polynomial) of the form

$$g(x) \equiv g_k(x) = \sum_{l=0}^{3} a_l^{(k)} (x_k - x)^l \qquad (1 \leq k \leq n); \quad \text{and} \qquad (7.23)$$

3. The boundary conditions

$$g''(a) = g''(b) = 0 \qquad (7.24)$$

are satisfied by $g(x)$.

Comparing the above requirements with those from the previous section, we see that the first two conditions are identical with the conditions 1 and 2 from the Section 2.7.1, whereas the third condition of (7.24) coincides with the fourth condition of (7.3).

Similarly, as in the case of ordinary interpolation, we construct a system of equations for determining the $4n$ coefficients of the polynomial $g_k(x)$. $3(n - 1)$ equations are obtained from the conditions 1 and 2:

$$g_k(x_k) = g_{k+1}(x_k), \qquad g_k'(x_k) = g_{k+1}'(x_k), \qquad g_k''(x_k) = g_{k+1}''(x_k), \quad (7.25)$$

and two more from the condition 3:

$$g_1''(a) = 0, \qquad g_n''(b) = 0. \qquad (7.26)$$

A substitution of (7.23) into (7.25), (7.26) yields $(3n - 1)$ equations

$$a_0^{(k)} = \sum_{l=0}^{3} a_l^{(k+1)} h_{k+1}^l,$$

$$a_1^{(k)} = \sum_{l=1}^{3} l a_l^{(k+1)} h_{k+1}^{l-1}, \qquad (7.27)$$

$$2a_2^{(k)} = \sum_{l=2}^{3} l(l-1) a_l^{(k+1)} h_{k+1}^{l-2}$$

$(k = 1, \ldots, n-1)$, $3a_3^{(1)} h_1 + a_2^{(1)} = 0$, $a_2^{(n)} = 0$,

where $h_k = x_k - x_{k-1}$ $(k = 1, \ldots, n)$. Hence it is seen that in order to uniquely determine the coefficients $\{a_j^{(k)}\}$ we need in addition $(n + 1)$ more independent equations.

These additional equations are obtained by considering the variational formulation for finding the interpolation-smoothing functions; namely, we require that $g(x)$ solves the problem

$$\Phi_1(u) \equiv \int_a^b \left(\frac{d^2 u}{dx^2}\right)^2 dx + \sum_{k=0}^{n} p_k [u(x_k) - \tilde{f}_k]^2 = \max_{W_2^2(a,b)} \qquad (7.28)$$

where $\{p_k\}_{k=0}^n$ are positive numbers. In other words, the function $g(x)$ must satisfy the relation

$$\Phi_1(g) = \min_{u \in W_2^2(a,b)} \Phi_1(u). \qquad (7.29)$$

The quantities $\{p_k\}_{k=0}^n$ can be used to control the smoothing process as dependent on the errors $\{\varepsilon_k\}_{k=0}^n$ or on its distributions, provided such distributions are known at least approximately.

Later we will also need the following functional:

$$\Phi_0(u) \equiv \int_a^b \left(\frac{d^2 u}{dx^2}\right)^2 dx + \sum_{k=0}^{n} p_k u^2(x_k).$$

We have the relation

$$\Phi_0(u - g) = \int_a^b (u'' - g'')^2 dx + \sum_{k=0}^{n} p_k [u(x_k) - g(x_k)]^2$$

$$= \sum_{k=1}^{n} \int_{x_{k-1}}^{x_k} (u'' - g'')^2 dx + \sum_{k=0}^{n} p_k [(u(x_k) - \tilde{f}_k) - (g(x_k) - \tilde{f}_k)]^2.$$

The right-hand side above can be further rewritten using the identity

$$p(\alpha - \beta)^2 \equiv p\alpha^2 - p\beta^2 - 2p(\alpha - \beta)\beta.$$

Specifically, taking
$$p = 1, \qquad \alpha = u'', \qquad \beta = g'',$$
respectively
$$p = p_k, \qquad \alpha = u(x_k) - \tilde{f}_k, \qquad \beta = g(x_k) - \tilde{f}_k,$$
we obtain
$$\Phi_0(u - g) = \int_a^b (u'')^2 dx - \int_a^b (g'')^2 dx - 12 \sum_{k=1}^n a_3^{(k)}\{[u(x_k)$$
$$- g(x_k)] - [u(x_{k-1}) - g(x_{k-1})]\} + \sum_{k=0}^n p_k[u(x_k) - \tilde{f}_k]^2$$
$$- \sum_{k=0}^n p_k[g(x_k) - \tilde{f}_k]^2 - 2\sum_{k=0}^n p_k[u(x_k) - g(x_k)][g(x_k) - \tilde{f}_k]$$
$$= \Phi_1(u) - \Phi_1(g) - 12 \sum_{k=1}^n a_3^{(k)}[u(x_k) - g(x_k)]$$
$$+ 12 \sum_{k=0}^{n-1} a_3^{(k+1)}[u(x_k) - g(x_k)]$$
$$- 2 \sum_{k=0}^n p_k[g(x_k) - \tilde{f}_k][u(x_k) - g(x_k)]$$
$$= \Phi_1(u) - \Phi_1(g) - 2 \sum_{k=0}^n C_k[u(x_k) - g(x_k)], \qquad (7.30)$$

where
$$C_k = \begin{cases} -6a_3^{(1)} + p_0[g(x_0) - \tilde{f}_0], & k = 0, \\ -6[a_3^{(k+1)} - a_3^{(k)}] + p_k[g(x_k) - \tilde{f}_k], & 0 < k < n, \\ 6a_3^{(n)} + p_n[g(x_n) - \tilde{f}_n], & k = n. \end{cases} \qquad (7.31)$$

Formula (7.30) shows that
$$0 \leq \Phi_0(u - g) = \Phi_1(u) - \Phi_1(g),$$
or equivalently
$$\Phi_1(g) = \Phi_1(u) - \Phi_0(u - g) \leq \Phi_1(u), \qquad (7.32)$$
which holds for any $u \in W_2^2(a, b)$, provided all the coefficients $\{C_k\}_{k=0}^n$ are equal to zero:
$$C_k = 0 \qquad (k = 0, 1, \ldots, n). \qquad (7.33)$$
We have thus obtained the $(n+1)$ equations needed for completing the system of (7.27).

Interpolation of Solutions of Difference Equations by Means of Splines

The final system with $4n$ equations and $4n$ unknowns is as follows:

$$a_0^{(k)} = a_3^{(k+1)}h_{k+1}^3 + a_2^{(k+1)}h_{k+1}^2 + a_1^{(k+1)}h_{k+1} + a_0^{(k+1)}, \quad (7.34a)$$

$$a_1^{(k)} = 3a_3^{(k+1)}h_{k+1}^2 + 2a_2^{(k+1)}h_{k+1} + a_1^{(k+1)}, \quad (7.34b)$$

$$a_2^{(k)} = 3a_3^{(k+1)}h_{k+1} + a_2^{(k+1)}, \quad (7.34c)$$

$$a_3^{(k+1)} - a_3^{(k)} = \frac{p_k}{6}[a_0^{(k)} - \tilde{f}_k] \quad (k = 1, \ldots, n-1); \quad (7.34d)$$

$$3a_3^{(1)}h_1 + a_2^{(1)} = 0, \quad (7.35a)$$

$$a_2^{(n)} = 0, \quad (7.35b)$$

$$a_3^{(1)} = \frac{p_0}{6}[a_3^{(1)}h_1^3 + a_2^{(1)}h_1^2 + a_1^{(1)}h_1 + a_0^{(1)} - \tilde{f}_0]; \quad (7.35c)$$

$$a_3^{(n)} = -\frac{p_n}{6}[a_0^{(n)} - \tilde{f}_n]. \quad (7.35d)$$

Let us rewrite this system by eliminating some of the unknowns. First, using (7.34c) and (7.35a), we have

$$a_3^{(k)} = \frac{a_2^{(k-1)} - a_2^{(k)}}{3h_k} \quad (k = 1, \ldots, n), \quad (7.36)$$

where we introduce $a_2^{(0)} = 0$ for notational purposes. Further, from (7.34d) and (7.35d)

$$a_0^{(k)} = -6 \frac{a_3^{(k)} - a_3^{(k+1)}}{p_k} + \tilde{f}_k \quad (k = 1, \ldots, n), \quad (7.37)$$

where, similarly, $a_3^{(n+1)} = 0$ has been introduced. Equations (7.36) and (7.37) are equivalent to

$$a_0^{(k)} = -\frac{2}{p_k}\left[\frac{a_2^{(k-1)} - a_2^{(k)}}{h_k} - \frac{a_2^{(k)} - a_2^{(k+1)}}{h_{k+1}}\right] + \tilde{f}_k$$

$$(k = 1, \ldots, n-1), \quad (7.38)$$

$$a_0^{(n)} = -\frac{2}{p_n}\left[\frac{a_2^{(n-1)} - a_2^{(n)}}{h_n}\right] + \tilde{f}_n.$$

The first $(n-1)$ equations above can be further rewritten in matrix form (recall $a_2^{(0)} = 0$, $a_2^{(n)} = 0$):

$$\psi = B\phi + \tilde{f}. \quad (7.39)$$

Here the symbols ψ, ϕ, $\tilde{f}$ denote the $(n-1)$-vectors with the components $\{a_0^{(k)}\}_{k=1}^{n-1}$, $\{a_2^{(k)}\}_{k=1}^{n-1}$, and $\{\tilde{f}_k\}_{k=1}^{n-1}$ respectively, and

$$B = PH, \quad (7.40)$$

where

$$P = 2 \begin{Vmatrix} \dfrac{1}{p_1} & 0 & 0 & \cdots & 0 & 0 \\ 0 & \dfrac{1}{p_2} & 0 & \cdots & 0 & 0 \\ \multicolumn{6}{c}{\dotfill} \\ 0 & 0 & 0 & \cdots & 0 & \dfrac{1}{p_{n-1}} \end{Vmatrix} \qquad (7.41)$$

is a diagonal matrix and

$$H = \begin{Vmatrix} \left(\dfrac{1}{h_1}+\dfrac{1}{h_2}\right) & -\dfrac{1}{h_2} & 0 & \cdots & 0 & 0 \\ -\dfrac{1}{h_2} & \left(\dfrac{1}{h_2}+\dfrac{1}{h_3}\right) & -\dfrac{1}{h_3} & \cdots & 0 & 0 \\ \multicolumn{6}{c}{\dotfill} \\ 0 & 0 & 0 & \cdots & -\dfrac{1}{h_{n-1}} & \left(\dfrac{1}{h_{n-1}}+\dfrac{1}{h_n}\right) \end{Vmatrix}$$
(7.42)

is a tridiagonal symmetric matrix.

Continuing the elimination process, Equations (7.34a) and (7.35c), with the help of Equations (7.36) and (7.37) yield

$$a_1^{(k)} = \frac{a_0^{(k-1)} - a_0^{(k)}}{h_k} - \frac{h_k}{3}(2a_2^{(k)} + a_2^{(k-1)}) \qquad (k = 2, \ldots, n),$$

$$a_1^{(1)} = -\frac{1}{h_1} a_0^{(1)} - \left(\frac{2}{p_0 h_1^2} + \frac{2}{3} h_1\right) a_2^{(1)} + \frac{\tilde{f}_0}{h_1}.$$
(7.43)

Next, consider the equations

$$h_{k+1}(a_2^{(k)} + a_2^{(k+1)}) + a_1^{(k+1)} - a_1^{(k)} = 0 \qquad (k = 1, \ldots, n-1), \quad (7.44)$$

which result from (7.34b) by substituting $\{a_3^{(k)}\}_{k=1}^n$ from (7.36). Using Formulas (7.34) [7.43] for $\{a_1^{(k)}\}_{k=1}^n$ in (7.44), we get

$$\frac{1}{3}[h_k a_2^{(k-1)} + 2(h_k + h_{k+1})a_2^{(k)} + h_{k+1} a_2^{(k+1)}] + \frac{a_0^{(k)} - a_0^{(k+1)}}{h_{k+1}}$$

$$- \frac{a_0^{(k-1)} - a_0^{(k)}}{h_k} = 0 \qquad (k = 2, \ldots, n-1), \qquad (7.45)$$

$$\frac{1}{3}\left[2\left(\frac{3}{p_0 h_1^2} + h_1 + h_2\right) a_2^{(1)} + h_2 a_2^{(2)}\right] + \frac{a_0^{(1)} - a_0^{(2)}}{h_2} + \frac{a_0^{(1)}}{h_1} = \frac{\tilde{f}_0}{h_1}$$

Since $a_2^{(n)} = 0$ and

$$a_0^{(n)} = -\frac{2}{p_n}\left[\frac{a_2^{(n-1)} - a_2^{(n)}}{h_n}\right] + \tilde{f}_n,$$

the system of (7.45) becomes

$$\tfrac{1}{3}S\phi + H\psi = w, \qquad (7.46)$$

where

$$S = \begin{Vmatrix} 2(\alpha + h_1 + h_2) & h_2 & 0 & \cdots & 0 & 0 \\ h_2 & 2(h_2 + h_3) & h_3 & \cdots & 0 & 0 \\ \hdotsfor{6} \\ 0 & 0 & 0 & \cdots & h_{n-1} & 2(\beta + h_{n-1} + h^n) \end{Vmatrix} \qquad (7.47)$$

is a symmetric three-dimensional $(n-1) \times (n-1)$ matrix $[\alpha = 3/(p_0 h_1^2)$, $\beta = 3/(p_n h_n^2)]$, the matrix H is as defined in (7.42), w is an $(n-1)$-vector with the components

$$w_k = \begin{cases} \dfrac{\tilde{f}_0}{h_1}, & \text{if } k = 1, \\ 0, & \text{if } 1 < k < n-1, \\ \dfrac{\tilde{f}_n}{h_n}, & \text{if } k = n-1, \end{cases}$$

and the vectors ψ, ϕ are as defined earlier. Finally, the application of (7.39) leads to

$$\tilde{A}\phi = \tilde{F}, \qquad (7.48)$$

where

$$\tilde{A} = \tfrac{1}{3}S + HB = \tfrac{1}{3}S + HPH \qquad (7.49)$$

and

$$\tilde{F} = w - H\tilde{f}. \qquad (7.50)$$

The vector $\tilde{F}$ has the components

$$\tilde{F}_k = \frac{\tilde{f}_{k-1} - \tilde{f}_k}{h_k} - \frac{\tilde{f}_k - \tilde{f}_{k+1}}{h_{k+1}} \qquad (k = 1, 2, \ldots, n-1). \qquad (7.51)$$

It is not difficult to see that the matrix $\tilde{A}$ can be written as a sum of two symmetric and positive definite matrices HPH and $S/3$. Positivity of S follows by direct computations, since S is a Jacobi matrix with diagonal predominance. Thus $\tilde{A}$ is symmetric, and nonsingular (positive definite). Hence the coefficients $\{a_2^{(k)}\}_{k=1}^n$ are uniquely defined. Since the remaining coefficients $\{a_0^{(k)}, a_1^{(k)}, a_3^{(k)}\}_{k=1}^n$ are also determined uniquely by $\{a_2^{(k)}\}_{k=1}^n$ and by Relations (7.33), (7.36), and (7.43), we conclude that there exists a unique piece-wise cubic interpolation-smoothing function $g(x)$.

In order to illustrate the method for solving (7.48), let us consider the particular case $p_k = p$ $(k = 0, 1, \ldots, n)$; the system now becomes

$$\tilde{A}\phi \equiv \left(\frac{1}{3}S + \frac{2}{p}H^2\right)\phi - \tilde{F}. \tag{7.52}$$

First compute the elements $\{t_{k,l}\}_{k,l=1}^{n-1}$ of the five-diagonal matrix $T = H^2$:

$$t_{kk} = \begin{cases} \left(\dfrac{1}{h_1} + \dfrac{1}{h_2}\right)^2 + \dfrac{1}{h_2^2}, & \text{if } k = 1, \\[2mm] \left(\dfrac{1}{h_k} + \dfrac{1}{h_{k+1}}\right)^2 + \dfrac{1}{h_k^2} + \dfrac{1}{h_{k+1}^2}, & \text{if } 1 < k < n-1, \\[2mm] \left(\dfrac{1}{h_{n-1}} + \dfrac{1}{h_n}\right)^2 + \dfrac{1}{h_{n-1}^2}, & \text{if } k = n-1 \end{cases}$$

$$(k = 1, \ldots, n-1);$$

$$t_{k,k+1} = -\left(\frac{1}{h_k} + \frac{2}{h_{k+1}} + \frac{1}{h_{k+2}}\right)\frac{1}{h_{k+1}} \quad (k = 1, \ldots, n-2);$$

$$t_{k,k+2} = \frac{1}{h_{k+1}h_{k+2}} \quad (k = 1, \ldots, n-3);$$

$$t_{k,k-1} = t_{k-1,k} \quad (k = 2, \ldots, n-1);$$

$$t_{k,k-2} = t_{k-2,k} \quad (k = 3, \ldots, n-1).$$

All the remaining elements of the matrix T equal to zero, i.e. $t_{k,l} = 0$ if $|k - l| > 2$, $(k, l = 1, \ldots, n-1)$. If $\tilde{A}$ is now defined by

$$\tilde{A} = \begin{Vmatrix} b_1 & c_1 & d_1 & 0 & \cdots & 0 & 0 \\ c_1 & b_2 & c_2 & d_2 & \cdots & 0 & 0 \\ d_1 & c_2 & b_3 & c_3 & \cdots & 0 & 0 \\ \multicolumn{7}{c}{\dotfill} \\ 0 & 0 & 0 & 0 & \cdots & b_{n-2} & c_{n-2} \\ 0 & 0 & 0 & 0 & \cdots & c_{n-2} & b_{n-1} \end{Vmatrix}, \tag{7.54}$$

Interpolation of Solutions of Difference Equations by Means of Splines

then its entries are given by the formulas

$$b_k = \begin{cases} \dfrac{2}{3}\left(\dfrac{3}{ph_1^2} + h_1 + h_2\right) + \dfrac{2}{p}t_{1,1}, & \text{if } k = 1, \\[1em] \dfrac{2}{3}(h_k + h_{k+1}) + \dfrac{2}{p}t_{k,k}, & \text{if } 1 < k < n - 1, \\[1em] \dfrac{2}{3}\left(\dfrac{3}{ph_n^2} + h_{n-1} + h_n\right) + \dfrac{2}{p}t_{n-1,n-1}, & \text{if } k = n - 1, \end{cases} \quad (7.55)$$

$$c_k = \frac{h_{k+1}}{3} + \frac{2}{p}t_{k,k+1} \quad (k = 1, \ldots, n - 2);$$

$$d_k = \frac{2}{p}t_{k,k+2} \quad (k = 1, \ldots, n - 3).$$

Equation (7.52) can be solved by either direct or iterative methods. In most cases it is convenient to use the well-developed direct methods for solving algebraic equations, based on the Gauss elimination method (see Section 3.8).

Note in passing that if the matrices in question are such that except for some diagonals all the elements are zero, and if the coefficients entering the equations satisfy some additional requirements, then in a sense the best realization of the Gauss elimination method is the so-called *factorization method*. Its basic ideas, as well as the corresponding numerical scheme, are found in Section 3.8. In our present case we have a matrix with zero entries except for five diagonals.

Thus, we have described a method for finding the interpolating-smoothing and piece-wise cubic function $g(x)$ of one variable. The analytic formulation along with the elements of variational formulation have been discussed. We have also proved the existence and uniqueness of $g(x)$ and explained in detail the algorithms of its construction.

Let us briefly consider the problem of choosing the variables $\{p_k\}_{k=0}^n$ appearing in our procedure; except for positivity there have been no other requirements regarding these variables. Suppose that at some points of the net, say $x_{k_1}, x_{k_2}, \ldots, x_{k_m}$, the values of f are known exactly, i.e. $f(x_{k_i}) = \tilde{f}_{k_i}$ ($i = 1, \ldots, m$); then putting $1/p_{k_i} = 0$, we obtain from (7.38)

$$a_0^{(k_i)} = \tilde{f}_{k_i}$$

and hence

$$g(x_{k_i}) = \tilde{f}_{k_i} \quad (i = 1, \ldots, m).$$

In particular, for $1/p_{k_i} = 0$ ($k = 0, 1, \ldots, n$), we arrive at a piece-wise cubic interpolation problem in which System (7.48) is fully equivalent to System (7.12).

It follows from what we have said that for more accurate values of f at the points x_k we need larger values of p_k ($0 \le k \le n$). This observation can be exploited for compensating the nonuniformity in the distributions of the errors $\varepsilon_k = f(x_k) - \tilde{f}_k$ ($k = 0, 1, \ldots, n$).

More complete recommendations regarding the quantities $\{p_k\}_{k=0}^n$ can only be given either after having made some experimental test computations or by analyzing first the dependence of the solution of (7.48) on its right-hand side and on the variables $\{p_k\}_{k=0}^n$ themselves.

2.7.3. Interpolation of Functions of Two Variables

A two-dimensional interpolation problem by piece-wise bicubic functions has been considered by Alberg, Nilson and Walsh [14], Zavyalov [14], and others. We will restrict ourselves to a brief consideration of the following example.

Let $D = \{x, y : a \le x \le b, c \le y \le d\}$ be a rectangular region. Define in this region a net $D_h = \{x_k, y_l : a = x_0 < x_1 < \cdots < x_n = b, c = y_0 < y_1 < \cdots < y_m = d\}$. Let f be a function on D, the function values of which have been specified at the net points. The piece-wise bicubic interpolation problem is to find a function $g(x, y)$ with the following properties:

1. $g(x, y) \in C^{(2)}(D)$, that is twice continuously differentiable;
2. In each cell of the net, the function $g(x, y)$ coincides with a bicubic polynomial of the form

$$g(x, y) \equiv g_{k,l}(x, y) = \sum_{i,j=0}^{3} a_{i,j}^{(k,l)}(x_k - x)^i(y_l - y)^j \qquad (7.56)$$

$$(k = 1, \ldots, n; l = 1, \ldots, m);$$

3. For all $0 \le k \le n$ and $0 \le l \le m$

$$g(x_k, y_l) = f_{k,l}; \quad \text{and} \qquad (7.57)$$

4. $g(x, y)$ satisfies the boundary conditions

$$\frac{\partial^2}{\partial v^2} g = 0, \qquad (7.58)$$

where v is the normal to the boundary ∂D.

There is a unique solution to the above problem, and it is moreover the unique solution to the following variational problem at the same time:

$$\int_D \left[\left(\frac{\partial^2 u}{\partial x^2}\right)^2 + \left(\frac{\partial^2 u}{\partial y^2}\right)^2 \right] dD = \min_{W_2^2(a,b)}, \qquad (7.59)$$

$$u(x_k, y_l) = f_{k,l} \quad (k = 0, 1, \ldots, n; l = 0, 1, \ldots, m).$$

In other words, Conditions 1–4 are necessary conditions for the solution of Problem (7.59).

Let us pause briefly and consider the problem of constructing the interpolating function $g(x, y)$. In principle, the derivation of $g(x, y)$ does not differ from the one-dimensional case, and eventually it is reduced to solving a system of order nm. (For more detail we refer the reader to Zavyalov [14].) In a number of significant practical cases, however, it turns out that solving the problem in general may be useful. For a simple example, consider the case when $g(x, y)$ is to be evaluated at a relatively small set of points. We will show that this case can be easily handled using the one-dimensional interpolation methods described in Section 2.7.1.

From the representation of (7.56) it follows that the function $g(x, y)$ is piece-wise cubic and twice continuously differentiable along any line $x \equiv \text{const}$ which intersect the region D. In addition, it satisfies the boundary conditions

$$\left.\frac{\partial^2 g}{\partial y^2}\right|_{y=c} = \left.\frac{\partial^2 g}{\partial y^2}\right|_{y=d} = 0. \tag{7.60}$$

Hence we conclude that in order to find $g(x_k, y)$ for any x_k ($0 \leq k \leq n$), it is enough to solve the one-dimensional piece-wise cubic interpolation problem below:

1. $g(x_k, y)$ is twice continuously differentiable with respect to y;
2. On each interval $[y_l, y_{l-1}]$ the function $g(x_k, y)$ has the form

$$g(x_k, y) \equiv g_{k,l}(y) = \sum_{j=0}^{3} a_j^{(k,l)}(y_l - y)^j \tag{7.61}$$

$$(l = 1, \ldots, m);$$

3.
$$g(x_k, y_l) = f_{k,l}, \quad 0 \leq l \leq m; \quad \text{and} \tag{7.62}$$

4. $g(x_k, y)$ satisfies the boundary condition (7.60).

This formulation is fully equivalent to the piece-wise cubic interpolation problem, for which the solution algorithms were described in detail in Section 2.7.1. Clearly, along the lines $y \equiv y_l$, where y_l ($0 \leq l \leq m$) correspond to the net points of D_h, the function $g(x, y)$ can be evaluated in a similar fashion.

With the above remarks in mind, we suggest the following algorithm for evaluating the interpolation function at a given point $(\tilde{x}, \tilde{y}) \in D$: To be specific, let $n \leq m$. As a first step, we solve n one-dimensional interpolation problems along the lines $x \equiv x_k$ and hence determine the function at $(x_k, \tilde{y})$, $k = 0, 1, \ldots, n$. Second, we solve one piece-wise cubic interpolation problem along $y \equiv \tilde{y}$, thereby evaluating $g(x, y)$ at the required point $(\tilde{x}, \tilde{y})$. In this way, the total of $(n + 1)$ one-dimensional interpolations are required for evaluating the function $g(x, y)$ at a single point in D: n of them along the y-coordinate line, and one along the x-coordinate line. If the function g is to be evaluated at $N > 1$ different points of D, one can show that no more than $(m + N)$ one-dimensional interpolations need be executed.

The above described algorithm can be easily generalized to regions expressible as finite unions of rectangles, and to multidimensional regions of a similar type.

Chapter 3

Methods for Solving Stationary Problems of Mathematical Physics

Numerical methods for stationary problems of mathematical physics constitute a more or less selfcontained subject, even though many stationary problems can be treated as asymptotic cases ($t \to \infty$) for nonstationary problems. In solving stationary problems by asymptotic methods, we do not pay any attention to the transient behavior, since it is of no interest whatsoever. In the nonstationary case, however, the transient behavior has a physical meaning. Generally speaking, this is the point where the differences between the two classes of problems start and end. Let us illustrate this with an example.

Consider the problem

$$A\phi = f,$$

where

$$A > 0, \phi \in \Phi \quad \text{and} \quad f \in F.$$

Replace this problem by the nonstationary one:

$$\frac{\partial \psi}{\partial t} + A\psi = f,$$

$$\psi = 0 \quad \text{for} \quad t = 0.$$

(In what follows, we will not indicate the domains of the operators.) Solutions of these problems will be sought in the form

$$\phi = \sum_n \phi_n u_n, \qquad \psi = \sum_n \psi_n u_n, \qquad f = \sum_n f_n u_n,$$

$$Au_n = \lambda_n u_n, \qquad A^* u_n^* = \lambda_n u_n^*,$$

where $\phi_n = (\phi, u_n^*)$, $\psi_n = (\psi, u_n^*)$, $f_n = (f, u_n^*)$. In a standard manner we obtain the following relations for the Fourier coefficients:

$$\lambda_n \phi_n = f_n,$$

and

$$\frac{d\psi_n}{dt} + \lambda_n \psi_n = f_n, \qquad \psi_n(0) = 0.$$

Solving these equations, we obtain

$$\phi = \sum_n \frac{f_n}{\lambda_n} u_n, \qquad \psi = \sum_n \frac{f_n}{\lambda_n} (1 - e^{-\lambda_n t}) u_n.$$

Suppose that the spectrum of the operator A is real, and that in fact $\lambda_n > 0$ ($n = 1, 2, \ldots$). It follows that

$$\lim_{t \to \infty} \psi = \phi.$$

Clearly, the nonstationary problem for ψ can be solved using difference methods with respect to t. For example,

$$\frac{\psi^{j+1} - \psi^j}{\tau} + A\psi^j = f,$$

or

$$\psi^{j+1} = \psi^j - \tau(A\psi^j - f).$$

As for the stationary problem, the solution is given by

$$\lim_{j \to \infty} \psi^j = \phi,$$

assuming certain relation between τ and $\beta(A)$.

The parameter τ may or may not depend on j. In any case, from the point of view of solving the stationary problem, it is convenient to interpret the index j as denoting the iteration rather than time. The above type of simple, clear-cut relation between the stationary and nonstationary problems does not necessarily exist if the structure of the spectrum of A is arbitrary.

A survey of iterative methods of linear algebra can be found in the monographs by Faddeev and Faddeeva [8], Forsythe and Moler [8], Wilkinson [8], Householder [3], Voevodin [8], Bakhvalov [8], Marchuk and Lebedev [17], Varga [3], and others.

3.1. Some Iterative Methods and their Optimization

We will assume that the operator A coincides with a matrix, and hence that the original problem has already been reduced to a system of linear algebraic equations. Thus let

$$A\phi = f, \tag{1.1}$$

where A is a matrix, ϕ and f are vectors, $A > 0$, and the spectrum $\lambda(A) > 0$.

In order to solve Equation (1.1), let us use the following iterative process:

$$\phi^{j+1} = \phi^j - \tau(A\phi^j - f), \tag{1.2}$$

where τ is an arbitrary parameter. Let us introduce the notation

$$\xi^j = A\phi^j - f.$$

ξ^j is called a *residual* of the iterative process. Apply the operator A to both sides of (1.2) and subtract the vector f from the equation thereby obtained. As a result we get the following iterations for the residual:

$$\xi^{j+1} = T\xi^j, \tag{1.3}$$

where $T = E - \tau A$ is the *transition* operator. Suppose we know the bounds of the positive spectrum of A, i.e. the maximal and minimal eigenvalues corresponding to the spectral problems

$$Au_n = \lambda_n u_n, \; A^* u_n^* = \lambda_n u_n^*, \tag{1.4}$$

namely

$$\alpha \leqslant \lambda_n \leqslant \beta. \tag{1.5}$$

Assume further that (1.4) defines a bi-orthogonal basis of eigenvectors $\{u_n\}$ and $\{u_n^*\}$. A simple method of choosing the parameter τ is as follows. Write the Fourier series for the residual ξ in terms of the complete system of eigenvectors of (1.4):

$$\xi = \sum_n \xi_n u_n, \tag{1.6}$$

where

$$\xi_n = (\xi, u_n^*).$$

Substitute the series (1.6) into (1.3); the usual procedure gives

$$\xi_n^{j+1} = T_n \xi_n^j, \tag{1.7}$$

where

$$T_n = 1 - \tau \lambda_n.$$

In order that the successive approximations (1.2) converge, it is enough that for all $\lambda_n \in [\alpha, \beta]$

$$|1 - \tau \lambda_n| < 1.$$

Hence

$$\tau < \frac{2}{\beta}. \tag{1.8}$$

Before we proceed with the problem of choosing the free parameter τ, let us note the following terminology:

The iterative process of (1.2) is called *stationary* if the parameter τ does not depend on a particular iteration. If $\tau = \tau_j$ is changing from one iteration to another, then the process is called *nonstationary*

Our main objective now is to choose the parameter τ so as to optimize the rate of convergence of the iterative process. When looking for effective iterative schemes, we will make certain *a priori* assumptions. Since any optimization method is tied down to specific *a priori* information about the operator A, we could hardly talk about optimization of the iterative method in the absence of any knowledge regarding this operator. The information which is usually of crucial significance for the optimization of stationary iterations includes the positivity of the operator A, the spectral bounds, and selfadjointness. The key (and at the same time difficult to obtain) information is that regarding the spectral bounds. The spectrum is real if in addition the problem is selfadjoint. In applications we often have to deal with matrices of a more general structure, and hence it is always necessary to have a complete system of eigenvalues. As a rule, completeness is very difficult to establish. For optimization purposes it is usually assumed *a priori*. In the absence of completeness our optimal algorithms result into nonoptimal but still highly effective schemes. Methods for finding spectral bounds have been considered in Section 1.1.

3.1.1. The Simple Iterative Method

Assume at first that $\lambda(A) > 0$ and that only the upper spectral bound is available: $\beta = \max_n \{\lambda_n\}$. Consider the basic problem

$$A\phi = f \qquad (1.9)$$

and two auxiliary problems

$$Au = \lambda u, \qquad A^*u^* = \lambda u^*. \qquad (1.10)$$

Consider the iterative process described in (1.2):

$$\phi^{j+1} = \phi^j - \tau(A\phi^j - f), \qquad \phi^0 = 0, \qquad (1.11)$$

or, equivalently,

$$\phi^{j+1} = \phi^j - \tau\xi^j, \qquad \xi^j = A\phi^j - f. \qquad (1.12)$$

We require that the Fourier coefficient from the residual expansion which corresponds to the largest eigenvalue $\max_n \{\lambda_n\} = \lambda_m = \beta$ becomes zero in one iteration. Hence we get the following formula for choosing τ:

$$\tau = \frac{1}{\beta}. \qquad (1.13)$$

Naturally, for such a τ

$$\xi_m^{j+1} = 0.$$

Having chosen τ, let us now estimate the reduction in the residual components after one iteration. To this end consider T_n from (1.7):

$$T_n = 1 - \tau\lambda_n = 1 - \frac{\lambda_n}{\beta}. \qquad (1.14)$$

Clearly, the maximal value of T_n corresponds to the minimal eigenvalue $\min_n \{\lambda_n\} = \alpha$. Thus

$$q = \max_n \{T_n\} = 1 - \frac{\alpha}{\beta}. \qquad (1.15)$$

Let us see how the iteration process evolves. The first iteration allows the residual component, which corresponds to the largest eigenvalue, to become zero by taking a special choice of $\tau = 1/\beta$. The remaining coefficients ξ_n^{j+1} decrease in absolute value by the multiplicative factor of T_n. The least affected is the coefficient ξ_1^{j+1}, which corresponds to the minimal eigenvalue, in which case $T_1 = 1 - \alpha/\beta$. Repeating this process leads us to the asymptotic case, for $j \gg 1$ all the coefficients ξ_n^j ($n = 2, 3, \ldots, m - 1$) become negligibly small compared to ξ_1^j, and hence we get approximately

$$\xi^j = q\xi^{j-1}, \tag{1.16}$$

where

$$\xi^j = \xi_1^j u_1.$$

Let us introduce the norm

$$\|\xi^j\|^2 = (\xi^j, \xi^j); \tag{1.17}$$

then it follows from (1.16) that

$$\|\xi^j\| = q\|\xi^{j-1}\|. \tag{1.18}$$

The performance of the iterative process is usually judged by the following asymptotic quantity ($j \to \infty$):

$$s = -\ln q = -\ln \frac{\|\xi^j\|}{\|\xi^{j-1}\|}. \tag{1.19}$$

Generally speaking, q depends on the iteration. In many cases, however, it is possible to find an estimate of q as $j \to \infty$, which does not depend on j, and it is this quantity that is used in the definition of an *asymptotic rate of convergence* of an iterative process. By the recursive relation (1.18) we obtain

$$\|\xi^j\| = q^j\|\xi^0\|,$$

where q^j is the jth power of the asymptotic quantity q. This equality can be written equivalently as

$$\|\xi^j\| = e^{j \ln q}\|\xi^0\|.$$

Therefore

$$\|\xi^j\| = e^{-js}\|\xi^0\|.$$

In this manner, the asymptotic rate of convergence s characterizes the rate of the exponential residual decrease.

Recalling (1.15), it can be shown, that

$$q = 1 - (1/p)$$

where $p = \beta/\alpha$.

The quantity p is usually called a *conditioning number of the matrix A*. If $p = \beta/\alpha \gg 1$, the asymptotic estimate becomes

$$s = 1 - q = (1/p). \tag{1.20}$$

3.1.2. The Displacement Method

Let us now turn to a more effective optimization method, assuming an *a priori* knowledge of both upper and lower spectral bounds:

$$\alpha \leq \lambda_n \leq \beta, \quad n = 1, \ldots$$

Consider the following first-order polynomial in λ:

$$P(\lambda) = 1 - \tau\lambda, \quad P(0) = 1.$$

Let us find τ_0 which minimizes the absolute value of the polynomial $P(\lambda)$, that is,

$$\max_\lambda |P(\lambda)| = \min_\tau.$$

Since $P(\lambda) = 1 - \tau\lambda$ is linear in λ, its maximal value is attained at one of the end points of the interval $[\alpha, \beta]$. Hence the best τ_0 is determined from the equation

$$1 - \tau_0\alpha = -(1 - \tau_0\beta),$$

so that

$$\tau_0 = 2/(\alpha + \beta). \tag{1.21}$$

It can be easily seen, that in this case

$$\min_\tau \max_\lambda |P(\lambda)| = |1 - \tau_0\alpha| = |1 - \tau_0\beta| = \frac{\beta - \alpha}{\beta + \alpha} = q,$$

and hence

$$T_n \leq \frac{\beta - \alpha}{\beta + \alpha}.$$

The successive approximation method with τ in the form of (1.21) is stationary. The expression for q can be conveniently written as

$$q = 1 - 2/(1 + p).$$

If $\beta/\alpha = p \gg 1$, then $q = 1 - (2/p)$, and the asymptotic rate of convergence is given by

$$s = (2/p). \tag{1.22}$$

The optimization method with the choice of the parameter as in (1.21) is usually called a *displacement method*. It can be seen by comparing Equations (1.22) and (1.20) that the asymptotic rate of convergence of this method is double that of the previous section, when only the largest eigenvalue was supposed to be known.

3.1.3. The Chebyshev Acceleration Method

A more general iterative process (sometimes called the Richardson method) adjusts its parameter τ at each consecutive iteration, so as to maximize the rate of decrease of each component of the residual

$$\xi^j = (E - \tau_j A)\xi^{j-1}, \cdot \tag{1.23}$$

Some Iterative Methods and their Optimization

(E is the identity operator). This method belongs to the family of non-stationary processes. With the help of (1.23) we obtain

$$\xi^j = \prod_{i=1}^{j}(E - \tau_i A)\xi^0$$

or

$$\xi^j = P_j(A)\xi^0,$$

where

$$P_j(A) = \prod_{i=1}^{j}(E - \tau_i A).$$

Expanding the residual into a Fourier series with respect to the eigenfunctions of the operator A, we have the following relations for the Fourier coefficient ξ_n^j of the residual ξ^j:

$$\xi_n^j = P_j(\lambda_n)\xi_n^0; \qquad \xi_n^j = (\xi^j, u_n^*),$$

where

$$P_j(\lambda) = \prod_{i=1}^{j}(1 - \tau_i \lambda). \qquad (1.24)$$

Suppose that the spectrum of A is real, that its bounds are known:

$$0 < \alpha \leq \lambda_n \leq \beta,$$

and that the eigenvectors of A form a basis. Then the condition of the fastest decrease of all the coefficients ξ_n^j implies the following formulation: find a polynomial $P_j(\lambda)$ such that $P_j(0) = 1$ and that

$$\max_{\alpha \leq \lambda \leq \beta} |P_j(\lambda)| \qquad (1.25)$$

attains its minimum by suitably choosing τ_i. This problem has been solved by A. A. Markov using Chebyshev polynomials $T_j(y)$; the solution is as follows:

$$P_j(\lambda) = \frac{T_j\left(\dfrac{\beta + \alpha - 2\lambda}{\beta - \alpha}\right)}{T_j\left(\dfrac{\beta + \alpha}{\beta - \alpha}\right)}, \qquad (1.26)$$

where

$$T_j(y) = \frac{(y + \sqrt{y^2 - 1})^j + (y - \sqrt{y^2 - 1})^j}{2};$$

the linear function $y = (\beta + \alpha - 2\lambda)/(\beta - \alpha)$ maps the interval $\alpha \leq \lambda \leq \beta$ into the interval $-1 \leq y \leq 1$ (with the ordering reversed). It is easy to show that $T_j(y) = \cos(j \arccos y)$ for all $|y| \leq 1$.

Chebyshev polynomials $T_j(\lambda)$ are known to tend in the best (uniform) manner to the null function $\theta(\lambda) \equiv 0$ on the interval $-1 \leq \lambda \leq 1$. At the same time the number of roots of the polynomial depends on its degree j. Outside $[-1, 1]$ the Chebyshev polynomials quickly become large in absolute value, but this increase occurs outside the spectral region. Therefore, we can see that the optimization using Chebyshev polynomials is very sensitive to the spectral boundaries of A. The sequence of the quantities τ is found as follows: First we find all the roots of the polynomials $P_j(\lambda)$. It is well known that for a fixed j the roots of the Chebyshev polynomials $T_j(y)$ are given by the formula

$$y_i = \cos\frac{(2i-1)\pi}{2j} \quad (i = 1, 2, \ldots, j). \tag{1.27}$$

With this it is not difficult to establish a formula for choosing λ_i; we need only solve the equation

$$\frac{\beta + \alpha - 2\lambda_i}{\beta - \alpha} = y_i,$$

for λ_i—the roots of $P_j(\lambda)$.

We have thus found the best polynomial approximation $P_j(\lambda)$ of the null function. Our problem now is to make this polynomial identical to the one in (1.26) by suitably choosing the parameters τ_i in the expression

$$\prod_{i=1}^{j}(1 - \tau_i \lambda).$$

For this it is enough to require that $P_j(\lambda)$ and $\prod_{i=1}^{j}(1 - \tau_i \lambda)$ have identical roots and scaling. Since we already know the zeros of $P_j(\lambda)$, the choice of $\tau_i = 1/\lambda_i$ yields the desired coincidence (note that for $\lambda = 0$ the two polynomials have the same value).

Now, since we have a decomposition of $P_j(\lambda)$ into linear multiplicative factors [see Equation (1.24)], the parameter values τ_i are determined from the condition that the above-mentioned factors are zero. Hence

$$\tau_i = \frac{1}{\lambda_i} = \frac{2}{(\beta + \alpha) - (\beta - \alpha)\cos\dfrac{(2i-1)\pi}{2j}} \quad (i = 1, 2, \ldots, j). \tag{1.28}$$

Let us turn to estimating the asymptotic rate of convergence of the iterative process of Equation (1.23). For this we need to estimate

$$q^j = \max_{\alpha \leq \lambda \leq \beta} |P_j(\lambda)|. \tag{1.29}$$

Since the maximum of $P_j(\lambda)$ is attained at the boundary of $[\alpha, \beta]$, and since Chebyshev polynomials have the property

$$T_j(\pm 1) = (\pm 1)^j,$$

Some Iterative Methods and their Optimization

we have that

$$q^j = \frac{1}{|T_j(r)|} = \frac{2}{(r+\sqrt{r^2-1})^j + (r-\sqrt{r^2-1})^j} \leq \frac{2}{(r+\sqrt{r^2-1})^j},$$

as $j \to \infty$, where

$$r = \frac{\beta + \alpha}{\beta - \alpha}.$$

Taking $\beta \gg \alpha$, we obtain an asymptotic relation

$$r = 1 + \frac{2}{p}$$

and hence

$$s = -\ln q = -\frac{1}{j} \ln 2 + \ln(r + \sqrt{r^2 - 1}).$$

As $j \to \infty$,

$$s = \ln(r + \sqrt{r^2 - 1})$$

or

$$s = \ln\left(1 + \frac{2}{p} + \sqrt{\left(1 + \frac{2}{p}\right)^2 - 1}\right) = \ln\left(1 + \frac{2}{\sqrt{p}}\right)$$

$$+ O\left(\frac{1}{p}\right) = \frac{2}{\sqrt{p}} + O\left(\frac{1}{p}\right).$$

Thus if $j \to \infty$ and $p \gg 1$, the asymptotic rate of convergence of the method of (1.23), (1.28) is given by the formula

$$s = \frac{2}{\sqrt{p}}. \tag{1.30}$$

Hence we conclude that the present method of (1.23), (1.28) has very good convergence performance for the case of ill-conditioned systems with $p \gg 1$. Since this method requires the advance knowledge of the upper and lower spectral boundaries, it is very effective when used repeatedly for solving problems with the same operator and different input data; in such a case the values α and β have to be found only once.

Thus, the foremost task for an implementation of the method is finding the spectral boundaries $\alpha(A)$, $\beta(A)$ and constructing the sequence of optimal parameter values

$$\tau_i = \frac{2}{\beta(A) + \alpha(A) - [\beta(A) - \alpha(A)]\cos\dfrac{(2i-1)\pi}{2j_0}} \qquad (i = 1, 2, \ldots, j_0).$$

Here $j = j_0$ is the number of steps in the cycle.

For a very large $p = \beta/\alpha$ the method may become unstable. This essentially well-known fact has stirred some pessimism as to the effectiveness of the method for a broader class of problems. Another special feature is the necessity of the preliminary computation of both spectral boundaries. This latter problem has been dealt with in Section 1.1.2. The stability problem on the other hand has been solved by Lebedev and Finogenov [9], who have shown that for ordered τ_i's the method is stable with respect to the roundoff errors (see also Samarskii [3]).

Let r be the least integer such that $j_0 < 2^r$ and $1 \leq k \leq 2^r$. Let us represent $k - 1$ in the binary system: $k = \gamma_1 \gamma_2 \ldots \gamma_r$, where γ_σ is zero or one. Let $æ(k) = \gamma_r \ldots \gamma_2 \gamma_1 + 1$. We will say that τ_i comes before τ_σ if $æ(i) < æ(\sigma)$. If $j_0 = 2^r$, it means that τ_i has a new ordering number $\kappa(i)$. This second ordering procedure is conveniently defined by a recursive formula.

Let $(\sigma_1, \sigma_2, \ldots, \sigma_{j_0/2})(1 \leq \sigma_i \leq j_0/2, \sigma_i \neq \sigma_k$ for $i \neq k, i, k = 1, 2, \ldots, j_0/2)$ be an ordering of τ_i determined by this procedure for $j_0/2$. Then for $j_0 = 2^r$ we require that the τ_i ordering be as follows:

$$(k_1, k_2, \ldots, k_{j_0}) = (\sigma_1, j_0 + 1 - \sigma_1, \sigma_2, j_0 + 1 - \sigma_2, \ldots, j_0 + 1 - \sigma_{j_0/2}).$$

The two alternate ways of ordering as described above seem to "uniformly shuffle" the transition operators $(E - \tau_i A)$ of large and small norms. One has to emphasize however, that if the cycle length j_0 is not large ($j_0 \leq 10$), then, as a rule, the computational procedure is stable and does not require any further special procedure for choosing the parameters τ_i.

In conclusion let us note that for $j_0 = 1$, that is for one-step Chebyshev acceleration of the successive approximation method, the optimal parameter τ is obtained using Equation (1.28), from which for $i = 1, j = 1$ one obtains

$$\tau = \frac{2}{\beta + \alpha},$$

The asymptotic rate of convergence in this case follows directly using the formulas

$$q^1 = \frac{1}{|T_1(r)|} = \frac{1}{r} = \frac{\beta - \alpha}{\beta + \alpha}$$

and is given by

$$s = \frac{2}{p}.$$

It is easy to see that the expressions for τ and s are identical to those corresponding to the displacement method considered earlier. This latter method is thus a special case of the more general multistage method of the Chebyshev iterative process.

3.1.4. The Over-Relaxation Method

Young and Frankel have devised an iterative method which has aquired popularity in many applications. It is called the *over-relaxation* method. The idea is as follows:

Consider a linear algebraic system

$$A\phi = f \tag{1.31}$$

with A a block matrix of three-dimensional type:

$$A = \begin{Vmatrix} E & -T_1 & 0 & 0 & \cdots & 0 & 0 \\ -S_2 & E & -T_2 & 0 & \cdots & 0 & 0 \\ \cdots & \cdots & \cdots & \cdots & \cdots & \cdots & \cdots \\ 0 & 0 & 0 & 0 & \cdots & -S_k & E \end{Vmatrix}.$$

Note that this kind of structure arises in many problems of mathematical physics of the elliptic type when writing the corresponding equations in finite differences.

Write the matrix A in the form

$$A = E - S - T,$$

where

$$T = \begin{Vmatrix} 0 & T_1 & 0 & \cdots & 0 & 0 \\ 0 & 0 & T_2 & \cdots & 0 & 0 \\ \cdots & \cdots & \cdots & \cdots & \cdots & \cdots \\ 0 & 0 & 0 & \cdots & 0 & 0 \end{Vmatrix}, \quad S = \begin{Vmatrix} 0 & 0 & 0 & \cdots & 0 & 0 \\ S_2 & 0 & 0 & \cdots & 0 & 0 \\ \cdots & \cdots & \cdots & \cdots & \cdots & \cdots \\ 0 & 0 & 0 & \cdots & S_k & 0 \end{Vmatrix},$$

and E is the identity matrix. Equation (1.31) can now be written in the form

$$\phi = B\phi + f,$$

where

$$B = S + T.$$

We will assume that the iterative process

$$\phi^{n+1} = B\phi^n + f \tag{1.32}$$

converges. This means that the largest (in magnitude) eigenvalue is smaller than unity, i.e. $\beta(B) < 1$. This particular condition is of key importance for the applicability of the over-relaxation method.

System (1.31) can be written in the form

$$\Phi_p - S_p \Phi_{p-1} - T_p \Phi_{p+1} = F_p \quad (p = 1, 2, \ldots, k) \tag{1.33}$$

under the assumption that

$$S_1 = 0, \quad T_k = 0,$$

where Φ_p is the vector component of ϕ, and F_p is the vector component of f.

To solve Equation (1.33), we use the sequential approximation method in the form of a relaxation process

$$\Phi_p^{j+1} = \Phi_p^j - \tau(\Phi_p^j - S_p\Phi_{p-1}^{j+1} - T_p\Phi_{p+1}^j - F_p) \quad (1.34)$$

$$(p = 1, 2, \ldots, k).$$

Let us solve (1.34) starting with $p = 1$ and continuing in the order of increasing p; for a given p, the vector Φ_{p-1}^{j+1} in the right-hand side of (1.34) is already known from the previous computational stage. Note that in some cases the process (1.34) is conveniently written as a two-stage algorithm

$$\Phi_p^{j+1/2} = S_p\Phi_{p-1}^{j+1} + T_p\Phi_{p+1}^j + F_p,$$
$$\Phi_p^{j+1} = \tau\Phi_p^{j+1/2} + (1-\tau)\Phi_p^j, \quad (1.35)$$

where Φ_{p-1}^{j+1} from the right of the first formula (1.35) is assumed to be known for a given p.

Consider the homogeneous equation (1.34):

$$\Phi_p^{j+1} = \Phi_p^j - \tau(\Phi_p^j - S_p\Phi_{p-1}^{j+1} - T_p\Phi_{p+1}^j). \quad (1.36)$$

Its solution will be sought in the form

$$\Phi_p^j = \lambda^j(G)V_p, \quad (1.37)$$

where $\lambda^j(G)$ is the jth power of $\lambda(G)$—the eigenvalue of the operator

$$G = (E - \tau S)^{-1}[(1-\tau)E + \tau T].$$

The constant $\lambda(G)$ is found as follows. Substitute (1.37) into (1.36), and cancel out $\lambda^j(G)$:

$$\lambda(G)V_p = V_p - \tau(V_p - S_p\lambda(G)V_{p-1} - T_pV_{p+1}). \quad (1.38)$$

Consider the spectral problems

$$Aw = \lambda(A)w; \quad Bw = \lambda(B)w. \quad (1.39)$$

Obviously, the two have the same eigenvectors, since $A = E - B$.

Assume, that $\{w^n\}$ is a complete system of eigenvectors. Then (1.38) can be solved in the form

$$V_p = \lambda^{p/2}(G)w_p. \quad (1.40)$$

Substituting (1.40) into (1.38), and using (1.39), we obtain

$$[(\lambda(G) + \tau - 1]w_p = \tau\sqrt{\lambda(G)}\lambda(B)w_p.$$

Hence

$$\lambda(G) - \tau\lambda(B)\sqrt{\lambda(G)} + \tau - 1 = 0, \quad (1.41)$$

that is, we have obtained a quadratic equation for $\sqrt{\lambda(G)}$ and hence

$$\sqrt{\lambda(G)} = \frac{\tau\lambda(B)}{2} \pm \sqrt{\frac{\tau^2\lambda^2(B)}{4} - \tau + 1}. \quad (1.42)$$

Some Iterative Methods and their Optimization   99

If, in addition, we also want that for $|\lambda(B)| < 1$ the quantity $|\lambda(G)|$ be smaller than one (and this condition guarantees convergence of the successive approximations method), then we require

$$|\tau - 1| < 1.$$

Therefore we will assume

$$0 < \tau < 2.$$

Conider the maximum value of the magnitude of $\sqrt{\lambda(G)}$:

$$\max|\sqrt{\lambda(G)}| = \frac{\tau|\beta(B)|}{2} + \sqrt{\left|\frac{\tau^2\beta^2(B)}{4} - \tau + 1\right|}.$$

It is not difficult to show that the minimum of the quantity $\max\{|\sqrt{\lambda(G)}|\}$ is attained by choosing τ so as to satisfy the relation

$$\frac{\tau^2\beta^2(B)}{4} = \tau - 1.$$

Thus the optimal parameter is given by

$$\tau_0 = \frac{2}{1 + \sqrt{1 - \beta^2(B)}}, \qquad (1.43)$$

since this choice minimizes $\max\{|\sqrt{\lambda(G)}|\} = \beta(G)$ for all eigenvalues $\lambda(B)$. With this choice of τ we also have

$$\beta(G) = \left(\frac{\beta(B)}{1 + \sqrt{1 - \beta^2(B)}}\right)^2 = \tau_0 - 1. \qquad (1.44)$$

Let us find next the asymptotic rate of convergence for the over-relaxation method. Recall that $|\lambda(B)| < 1$. Moreover, for ill-conditioned systems of algebraic equations we usually have $|\beta(B)| \simeq 1$, which also explains the slow convergence of (1.32).

Let us estimate the asymptotic rate of convergence of the over-relaxation method for ill-conditioned matrices. First of all, we will show that for the convergent iterative processes of (1.31) we always have

$$0 < \alpha(A) \leqslant \lambda(A) \leqslant \beta(A) < 2.$$

Indeed, from the structure of the matrix $A = E - B$, and from the natural linear relation $\lambda(A) = 1 - \lambda(B)$, we have that

$$|\beta(B)| = \max\{|1 - \beta(A)|, |1 - \alpha(A)|\} < 1.$$

Suppose now that the matrix A which has resulted by solving a problem of mathematical physics is ill-conditioned. In such a case one usually has $\alpha(A) \simeq 0$, $\beta(A) \simeq 2$ and $\alpha(A)/\beta(A) \ll 1$. It means that for a given B the spectrum of A is mapped into the interval $[-1, 1]$. Moreover, taking into account the spectral relation

$$\lambda(A) = 1 - \lambda(B)$$

and the form of the matrix B, we obtain

$$\alpha(A) = 1 - \beta(B), \qquad \beta(A) = 1 + \beta(B). \tag{1.45}$$

Using (1.45) we have

$$\beta(B) = \frac{\beta(A) - \alpha(A)}{\beta(A) + \alpha(A)}$$

or

$$\beta(B) = 1 - \frac{2}{p+1} \tag{1.46}$$

where

$$p = \beta(A)/\alpha(A).$$

Restraining ourselves to the main factors, (1.44) can be expressed as

$$q = \frac{1}{\left(1 + \dfrac{2}{\sqrt{p}}\right)^2},$$

where $q = \max|\lambda(G)| = \beta(G)$. Since $s = -\ln q$, we have

$$s = \frac{4}{\sqrt{p}}. \tag{1.47}$$

A comparison of (1.47) and (1.30) shows that in the present case the over-relaxation method converges twice as fast as the Chebyshev acceleration method. This significant result tells us that there is a stationary iterative method for which only the greatest eigenvalue of the operator $B = S + T$ has to be computed. As in the case of Chebyshev acceleration the over-relaxation method is convenient for repeated computations, when the operator A stays the same and the input data change.

Having considered the formal algorithm above, we are now going to illustrate the over-relaxation method on some simple examples. To begin with, consider the difference equation

$$\phi = B\phi + f,$$

where the matrix B is such, that the equation can be written in terms of its components as

$$\phi_{k,l} = a_{k,l}\phi_{k-1,l} + b_{k,l}\phi_{k,l-1} + c_{k,l}\phi_{k+1,l} + d_{k,l}\phi_{k,l+1} + f_{k,l}. \tag{1.48}$$

Assume that all the coefficients $a_{k,l}, b_{k,l}, c_{k,l}, d_{k,l}$ are positive, and

$$a_{k,l} + b_{k,l} + c_{k,l} + d_{k,l} \leqslant 1,$$

Some Iterative Methods and their Optimization

the strict inequality holding for at least one pair (k, l). This condition is typical of difference analogs of elliptic equations. Note that the boundary conditions in (1.48) are automatically satisfied by the special choice of the coefficients in the equation. The over-relaxation method for (1.48) can be written as follows:

$$\phi_{k,l}^{j+1/2} = a_{k,l}\phi_{k-1,l}^{j+1} + b_{k,l}\phi_{k,l-1}^{j+1} + c_{k,l}\phi_{k+1,l}^{j} + d_{k,l}\phi_{k,l+1}^{j} + f_{k,l};$$
$$\phi_{k,l}^{j+1} = \tau\phi_{k,l}^{j+1/2} + (1-\tau)\phi_{k,l}^{j}. \tag{1.49}$$

The indices are taken in the following order: if $k_1 < k_2$, then $\phi_{k,l}^{j+1}$ from (1.49) is computed first at (k_1, l_1) and then at (k_2, l_2); if $k_1 = k_2$, then (k_1, l_1) comes first if $l_1 < l_2$. This procedure leads to a triangular block-matrix form for $E - B$ (see Wasov and Forsythe [3], Varga [3]).

The realization scheme of the algorithm is as follows: The first step is to compute the largest eigenvalue of B by solving a particular spectral problem:

$$\phi_{k,l}^{(n+1)} = a_{k,l}\phi_{k-1,l}^{(n)} + b_{k,l}\phi_{k,l-1}^{(n)} + c_{k,l}\phi_{k+1,l}^{(n)} + d_{k,l}\phi_{k,l+1}^{(n)},$$
$$\phi_{k,l}^{(0)} = g_{k,l}, \tag{1.50}$$

where $g_{k,l}$ is an arbitrary vector in general. In the case $\alpha(B) = -\beta(B)$, the Lyusternik method yields

$$\beta(B) = \lim_{n \to \infty} \frac{\|\phi^{(n+1)}\|}{\|\phi^{(n)}\|}. \tag{1.51}$$

Once $\beta(B)$ is found, τ is determined from the formula

$$\tau = \frac{2}{1 + \sqrt{1 - \beta^2(B)}}. \tag{1.52}$$

Problem (1.49) is now solved using the optimal τ above. The over-relaxation method turns out to be very simple in applications.

Let us make an important remark of an algorithmic character. In order to determine the largest eigenvalue of the operator B, needed for the over-relaxation method, one can use the successive approximations method of (1.50), (1.51). Indeed, this eigenvalue can be computed by the iterative process of (1.49) with $\tau = 1$. With the help of (1.41) we can relate $\beta(G)$ and $\beta(B)$ by the equation

$$\beta^2(B) = \beta(G).$$

Therefore, our problem now is to find $\beta(G)$ using the iterative process of (1.49) with $\tau = 1$. In contrast to the homogeneous Equation (1.50) we have to deal with a nonhomogeneous problem. A trivial generalization of the Lyusternik method gives us

$$\beta(G) = \lim_{n \to \infty} \frac{\|\phi^{(n+1)} - \phi^{(n)}\|}{\|\phi^{(n)} - \phi^{(n-1)}\|},$$

where ϕ^n solves (1.49) for $\tau = 1$. Finally, τ_0 is found in terms of $\beta(G)$ by the formula

$$\tau_0 = \frac{2}{1 + \sqrt{1 - \beta(G)}}. \tag{1.52a}$$

The following method, called a *block-relaxation method*, is very effective in many cases. It will be illustrated with a three-dimensional example below:

$$\phi_{k,l,m} = a_{k,l,m}\phi_{k-1,l,m} + b_{k,l,m}\phi_{k,l-1,m} + c_{k,l,m}\phi_{k,l,m-1}$$
$$+ d_{k,l,m}\phi_{k+1,l,m} + e_{k,l,m}\phi_{k,l+1,m} + f_{k,l,m}\phi_{k,l,m+1} + F_{k,l,m}, \tag{1.53}$$

where all the coefficients from this difference equation are positive and satisfy the condition

$$a_{k,l,m} + b_{k,l,m} + c_{k,l,m} + d_{k,l,m} + e_{k,l,m} + f_{k,l,m} \leq 1, \tag{1.54}$$

the strict inequality holding at least for one triple (k, l, m). The block-relaxation method is as follows:

$$\phi_{k,l,m}^{j+1/2} - c_{k,l,m}\phi_{k,l,m-1}^{j+1/2} - f_{k,l,m}\phi_{k,l,m+1}^{j+1/2} = a_{k,l,m}\phi_{k-1,l,m}^{j+1/2}$$
$$+ b_{k,l,m}\phi_{k,l-1,m}^{j+1} + d_{k,l,m}\phi_{k+1,l,m}^{j} + e_{k,l,m}\phi_{k,l+1,m}^{j} + F_{k,l,m} \tag{1.55}$$

$$(k = 1, 2, \ldots, l = 1, 2, \ldots, m = 1, 2, \ldots);$$

$$\phi_{k,l,m}^{j+1} = \tau \phi_{k,l,m}^{j+1/2} + (1 - \tau)\phi_{k,l,m}^{j}.$$

Before we can solve (1.55), we have to determine the parameter τ. Thus we have to solve first the following auxiliary problem:

$$\phi_{k,l,m}^{(n+1)} - c_{k,l,m}\phi_{k,l,m-1}^{(n+1)} - f_{k,l,m}\phi_{k,l,m+1}^{(n+1)} = a_{k,l,m}\phi_{k-1,l,m}^{(n)}$$
$$+ b_{k,l,m}\phi_{k,l-1,m}^{(n)} + d_{k,l,m}\phi_{k+1,l,m}^{(n)} + e_{k,l,m}\phi_{k,l+1,m}^{(n)} \tag{1.56}$$

$$\phi_{k,l,m}^{(0)} = g_{k,l,m}.$$

From here

$$\beta(B) = \lim_{n \to \infty} \frac{\|\phi^{(n+1)}\|}{\|\phi^{(n)}\|},$$

and τ is then found by Formula (1.52).

In order to be able to exploit Formula (1.52a), it is necessary to compute $\lambda(G)$ using (1.55) with $\tau = 1$. Note that for given k, l, m one always has to solve a one-dimensional difference equation. This can be done very effectively by the factorization method. If the parameter $\lambda(B)$ has not been precomputed, one usually takes $\tau_0 = 1.8$, which is quite close to the optimal choice in many cases.

The block-relaxation method can be used, of course, for two-dimensional difference equations of the form (1.48); at the same time the convergence will be faster than that of the point-relaxation method (1.49). The over-relaxation method is fairly completely explained in the monographs by Wasov and Forsythe [3], Varga [3], and others.

3.1.5. A Comparison of the Asymptotic Rate of Convergence for Various Iterative Methods

The asymptotic rate of convergence of iterative methods can be related to the conditioning of matrices, $p = \beta/\alpha$. Of great interest is the dependence of the rate estimate on the mesh size h of the net. While this latter kind of dependence can be judged upon only by looking at specific examples, the relation between the rate of convergence and the conditioning does not depend on any particular feature of the problem. The table below shows the relation between the rate of convergence s and the conditioning number p:

Iterative method	s
Simple iteration ($\tau_0 = 1/\beta$)	$1/p$
Displacement method ($\tau_0 = 2/(\alpha + \beta)$)	$2/p$
Chebychev acceleration	$2/\sqrt{p}$
Over-relaxation method	$4/\sqrt{p}$

There are still other iterative methods with even faster convergence than above. Their effectiveness, however, has been only established for the particular case of the Poisson equation and other simple problems. Generalizations of these results to broader class of problems encounter serious difficulties.

3.2. Gradient Iterative Methods

Present day computational facilities have further stimulated the development of certain optimal iterative methods which exploit the current *a posteriori* information about the solution in the course of computations. Such iterations may be called adaptive or selfadjusting to the optimal rate of convergence. In contrast to the spectral optimization methods, which only use *a priori* information about the spectrum of certain operators, the adaptive methods make an implicit use of both the information about the spectrum and that about the residual amplitudes of the corresponding components from the Fourier series expansion, while the iterative process evolves. This results in a significant improvement in the efficiency, since the residual components are being suppressed more purposefully, the emphasis being put on the components with the largest amplitudes.

Gradient methods have been initiated by Kantorovich [11], Hestenes and Stiefel [11], Krasnoselskii and Krein [11], and Lanczos [3] and continued by many others.

Among the adaptive algorithms we note first of all the gradient methods based on variational optimization.

3.2.1. The Residual Method

Let us focus our attention on one of the variants of gradient methods—the *residual* method. Essentially, it is as follows:

Consider a system of linear algebraic equations

$$A\phi = f. \tag{2.1}$$

Let $A > 0$. In order to solve this equation, let us use the successive approximations method:

$$\phi^{j+1} = \phi^j - \tau_j(A\phi^j - f), \quad \phi^0 = 0. \tag{2.2}$$

The residual

$$\xi^j = A\phi^j - f$$

satisfies in turn

$$\xi^{j+1} = (E - \tau_j A)\xi^j, \quad \xi^0 = -f. \tag{2.3}$$

Clearly, the solution of (2.2) can be expressed by means of the residuals as follows;

$$\phi^j = -\sum_{i=0}^{j-1} \tau_i \xi^i. \tag{2.4}$$

In order to choose the parameters τ_i, take the space R of n-dimensional real vectors, and define the scalar product

$$(\phi, \psi) = \sum_{i=1}^{n} \phi_i \psi_i, \tag{2.5}$$

the corresponding norm being given by

$$\|\xi\| = \sqrt{(\xi, \xi)}.$$

With the help of the recursive relations (2.3) we have

$$\|\xi^{j+1}\| = q_j \|\xi^j\|, \tag{2.6}$$

where

$$q_j = \sqrt{1 - 2\tau_j \frac{(A\xi^j, \xi^j)}{(\xi^j, \xi^j)} + \tau_j^2 \frac{(A\xi^j, A\xi^j)}{(\xi^j, \xi^j)}}. \tag{2.7}$$

The relaxation parameter τ_j is found by minimizing the function $q_j(\tau)$. To do that, differentiate Equation (2.7) with respect to τ_j and equate the result to zero. Thus

$$\tau_j = \frac{(A\xi^j, \xi^j)}{(A\xi^j, A\xi^j)}, \tag{2.8}$$

which is the optimal relaxation parameter corresponding to step j. Next, substituting Equation (2.8) into (2.7), we find that

$$q_j = \sqrt{1 - \rho_j^2}, \tag{2.9}$$

where

$$\rho_j^2 = \frac{(A\xi^j, \xi^j)^2}{(A\xi^j, A\xi^j)(\xi^j, \xi^j)}. \tag{2.10}$$

We will now show that for $A > 0$ we have $0 \leq q_j < 1$. Indeed, $q_j \geq 0$ by (2.6), and therefore

$$0 \leq \rho_j^2 \leq 1. \tag{2.11}$$

Note that ρ_j^2 will not converge to zero and is thus bounded from below. This follows from the fact that

$$(A\xi^j, \xi^j) = \left(\frac{A + A^*}{2} \xi^j, \xi^j\right),$$

and hence

$$\inf_{\xi^j \in R} \frac{(A\xi^j, \xi^j)}{(\xi^j, \xi^j)} = \min_n \left\{\lambda_n\left(\frac{A + A^*}{2}\right)\right\} > 0.$$

Thus

$$0 \leq q_j < q < 1, \tag{2.12}$$

as required. ($q < 0$ is a given number, independent of j.)

Krasnoselskii and Krein [11] have shown that the asymptotic rate of convergence of the residual method is given by

$$s = 2/p.$$

At first it seems that this method converges very slowly and is therefore inferior to the Chebyshev acceleration method and the over-relaxation method. This is true as far as the asymptotic behavior is concerned. During the beginning iterations, however, the approximations approach the exact solution much faster than when using the other methods. This fact was noted initially by Kantorovich [11] and later by Godunov and Prokopov [11]. The asymptotic performance is thus somewhat coarse for the purposes of estimating the number of the arithmetic operations required for obtaining the result with a given accuracy. Unfortunately, because of the nonlinear nature of the method, it would be very hard indeed to make the characteristics of the residual method more accurate.

In regard to the residual components from the Fourier expansion, the first components to be suppressed strongly are those with frequencies corresponding to the eigenvalues from inside the spectral interval. They are suppressed very quickly. The frequencies which correspond to the eigenvalues α and β (and their neighborhoods) are suppressed more slowly. Thus at some stage the iterative process reaches the asymptotic regime. In order to accelerate the method we need to combine the one-step residual process with a two-step process.

3.2.2. The Two-Step Residual Method

So far, the matrix A appearing in the linear algebraic system has been assumed positive. However, when solving Equation (2.1) we often obtain matrices A, the spectra of which, though real, may include negative parts. In such a case, generally speaking, the above-described methods of successive approximations may diverge. As suggested already by Gauss, this difficulty can be side stepped by transforming the original problem into an equivalent one with a positive matrix. This is known as *Gauss transformation*.

Let us consider this transformation. To this end multiply Equation (2.1) by the adjoint A^* of A from the left:

$$A^*A\phi = A^*f. \qquad (2.13)$$

Denoting $A^*f = g$, we have

$$A^*A\phi = g. \qquad (2.14)$$

Assume $\lambda = 0$ is not in the spectrum of A (and A^* for that matter). Hence there exists an inverse $(A^*A)^{-1} = A^{-1}A^{*-1}$, and the formal solution of Equation (2.14) can be written as

$$\phi = A^{-1}A^{*-1}g = A^{-1}A^{*-1}A^*f = A^{-1}f. \qquad (2.15)$$

At the same time it is not difficult to verify that the symmetric matrix A^*A is positive definite, since

$$(A^*A\xi, \xi) = (A\xi, A\xi) > 0, \qquad (2.16)$$

if $\lambda = 0$ is not in the spectrum of A. Thus, instead of Equation (2.1), which could not be solved by the iterative methods considered so far, we now have Equation (2.14) to which a variety of iterations may be successfully applied. It is well known, however, that the Gauss transformation leads as a rule to ill-conditioned matrices A^*A. To see this, let $A = A^* > 0$, and let the spectrum be given by

$$0 < \alpha \leqslant \lambda_n(A) \leqslant \beta.$$

The spectrum of A^*A is then as follows:

$$0 < \alpha^2 \leqslant \lambda_n(A^*A) \leqslant \beta^2.$$

If $\beta/\alpha = p \gg 1$, then $\beta^2/\alpha^2 \gg \beta/\alpha$, and hence the conditioning sharply declines. This means, therefore, that the iterative process will converge

Gradient Iterative Methods

slowly, even if the effective optimization methods are used. Nevertheless, it is advisable in many cases to solve Equation (2.14) via the residual-type iterative process

$$\phi^{j+1} = \phi^j - \tau_j(A^*A\phi^j - g), \tag{2.17}$$

where

$$\tau_j = \frac{(A^*A\xi^j, \xi^j)}{(A^*A\xi^j, A^*A\xi^j)} \tag{2.18}$$

and

$$\xi^j = A^*A\phi^j - g. \tag{2.19}$$

To summarize, an effective method for the case $A > 0$ is the residual method in the form of (2.2), whereas if A is arbitrary, one can use the residual method in the form of (2.17). Both iterations converge to the exact solution of Equation (2.1), provided zero does not belong to the spectrum of A.

It may be conjectured that a two-step residual process is superior to the two processes above. Consider the sequential approximations of the form

$$\phi^{j+1} = \phi^j - \tau_j(A\phi^j - f) - \gamma_j A^*(A\phi^j - f), \tag{2.20}$$

where τ_j and γ_j are yet to be determined. By (2.20) the residual ξ_j must satisfy

$$\xi^{j+1} = \xi^j - \tau_j A\xi^j - \gamma_j AA^*\xi^j. \tag{2.21}$$

The optimal choice of τ_j, γ_j will be based on the minimization of the functional

$$\begin{aligned}(\xi^{j+1}, \xi^{j+1}) = (\xi^j, \xi^j) &- 2\tau_j(A\xi^j, \xi^j) - 2\gamma_j(AA^*\xi^j, \xi^j) \\ &+ 2\tau_j\gamma_j(A\xi^j, AA^*\xi^j) + \tau_j^2(A\xi^j, A\xi^j) \\ &+ \gamma_j^2(AA^*\xi^j, AA^*\xi^j).\end{aligned} \tag{2.22}$$

with respect to τ_j and γ_j. There results a system of linear algebraic equations

$$\tau_j(y^j, y^j) + \gamma_j(y^j, z^j) = (x^j, y^j); \qquad \tau_j(y^j, z^j) + \gamma_j(z^j, z^j) = (z^j, x^j), \tag{2.23}$$

where

$$x^j = \xi^j; \qquad y^j = A\xi^j; \qquad z^j = AA^*\xi^j. \tag{2.24}$$

The solution is given by

$$\begin{aligned}\tau_j &= \frac{(x^j, y^j)(z^j, z^j) - (x^j, z^j)(y^j, z^j)}{(y^j, y^j)(z^j, z^j) - (y^j, z^j)^2}; \\ \gamma_j &= \frac{(x^j, z^j)(y^j, y^j) - (x^j, y^j)(y^j, z^j)}{(y^j, y^j)(z^j, z^j) - (y^j, z^j)^2}.\end{aligned} \tag{2.25}$$

Thus, in order to chose the optimal parameters, one has to consider the following functionals: (x^j, y^j), (x^j, z^j), (y^j, z^j), (y^j, y^j) and (z^j, z^j).

Let us now discuss some properties of the two-step iterative process of (2.20).

We note first that for a positive matrix A the process will converge faster than any one of the one-step Processes (2.2) or (2.17), since these are its special instances.

Secondly, the two-step Process (2.20) has a universal nature, since it is applicable to algebraic systems with arbitrary matrices.

System (2.23) is always solvable. It can be easily seen from the Schwartz inequality that

$$(A\xi^j, AA^*\xi^j)^2 \leq (A\xi^j, A\xi^j)(AA^*\xi^j, AA^*\xi^j), \tag{2.26}$$

with the equality sign holding if and only if

$$AA^*\xi^j = \lambda A\xi^j, \quad \text{that is} \quad A^*\xi^j = \lambda \xi^j. \tag{2.27}$$

Because of the fact that $\lambda = 0$ is not an eigenvalue of the operator A, the equality can thus take place if and only if a single frequency component corresponding to an eigenvalue of A^* has been left nontrivial during the course of the iterative process. It is easily seen however that in this case both denominator and numerator in (2.25) become zero. Thus in order to avoid fractions like 0/0, it is sometimes desirable to combine the two-step process with a one-step process which quickly damps that particular component.

Thus, here again arises the idea of combining the one- and two-step residual processes.

The n-step residual methods can be introduced similarly. They are discussed in a number of contributions by Kuznetsov [11]. Note in conclusion, that the optimization functionals may be based on more complicated residual norms, such as $\|\xi\|_C^2 = (C\xi, \xi)$, $C > 0$, rather than the simple norm used above. These norms have led to a number of new effective adaptive algorithms.

A distinguished case, when A is symmetric with an arbitrary spectrum, can be conveniently handled by conjugate gradients type methods (Hestenes and Stiefel [11], Godunov and Prokopov [11], and others).

3.2.3. The Method of Conjugate Gradients

One of the effective methods for solving the linear equations of the form

$$A\phi = f \tag{2.28}$$

with a symmetric and positive matrix A is provided by the *conjugate gradients* method. Let us briefly examine the origins of this procedure.

Let E_n be the space of n-dimensional vectors, with the inner product

$$(\phi, \psi)_D = (D\phi, \psi),$$

Gradient Iterative Methods

where $(u, v) = \sum_{i=1}^{n} u_i v_i$, $(u, v \in E_n)$, and D is a symmetric, positive definite matrix. As usual, the norm on E_n is defined by

$$\|u\|_D = (Du, u)^{1/2}.$$

Let us introduce an m-dimensional subspace $M \subseteq E_n$ with a basis $\{g_i\}_{i=1}^{m}$. Finding an approximate solution of Equation (2.28) can now be regarded as finding a vector $\hat{\phi} \in M$ for which

$$\|\phi^* - \hat{\phi}\|_D = \min_{\phi \in M}\|\phi^* - \phi\|_D = \min_{\alpha_1, \ldots, \alpha_m} \left\|\phi^* - \sum_{i=1}^{m} \alpha_i g_i\right\|_D, \quad (2.29)$$

where $\phi^* = A^{-1}f$ is the exact solution of Equation (2.28). The coefficients α_i^* in the expansion

$$\hat{\phi} = \sum_{i=1}^{m} \alpha_i^* g_i$$

are determined from a system of the form

$$B\alpha = S, \quad (2.30)$$

where $B = (b_{i,j})$ is an m-by-m matrix with the elements

$$b_{i,j} = (g_i, g_j)_D \quad (i, j = 1, \ldots, m),$$

and $S = (S_1, \ldots, S_m)$ is a vector with the components

$$S_i = (\phi^*, g_i)_D \quad (i = 1, \ldots, m).$$

It is readily seen from this, that the simplest case corresponds to $(g_i, g_j)_D = \delta_{ij}\|g_i\|_D^2$ (δ_{ij} is the Kronecker symbol); that is, the case when $\{g_i\}_{i=1}^{m}$ is a D-orthogonal basis of the subspace M. Then

$$\alpha_i^* = \frac{(\phi^*, g_i)_D}{\|g_i\|_D^2} \quad (i = 1, \ldots, m).$$

For the numerical realization of such a process we need to know the vector $D\phi^*$; for instance, $D = A$ insures this. Let us make Equation (2.29) somewhat more specific to suit our purposes.

Suppose we are given a nonzero vector $\phi^0 \in E_n$, $D = A$; the subspace M is the linear hull of the system of linearly independent vectors $\{A^i(\phi^* - \phi^0)\}_{i=1}^{m}$—the basis of M. Then Equation (2.29) can be reformulated as follows: find a vector $\hat{u} \in M$ such that

$$\|(\phi^* - \phi^0) - \hat{u}\|_A = \min_{u \in M}\|(\phi^* - \phi^0) - u\|_A. \quad (2.31)$$

As we have shown earlier, if $\{g_i\}_{i=1}^{m}$ is an A-orthogonal basis in M, then the approximation $\hat{\phi}$ of ϕ^* is given by the formulas

$$\hat{\phi} = \phi^0 + \sum_{i=1}^{m} \alpha_i g_i,$$

$$\alpha_i = \frac{(\phi^* - \phi^0, g_i)_A}{(g_i, g_i)_A} = -\frac{(A\phi^0 - f, g_i)}{(g_i, Ag_i)} \quad (i = 1, \ldots, m). \quad (2.32)$$

This process can also be written as

$$\phi^k = \phi^{k-1} - \alpha_k g_k, \quad \alpha_k = \frac{(\xi^{k-1}, g_k)}{(Ag_k, g_k)} \quad (k = 1, \ldots, m), \quad (2.33)$$

where $\xi^k = (A\phi^k - f)$ is the residual, and $\hat{\phi} = \phi^m$.

The best known method of constructing a basis in spaces like M is the Schmitt orthogonalization. For higher-order matrices, however, this procedure requires a large number of algebraic operations and a large computer memory for its realization. An effective way of constructing an A^2-orthogonal basis (when $D = A^2$) for M in the case of general, symmetric matrices is to use the Lanczos minimal iterations method. It is described in detail in the monograph by Faddeev and Faddeeva [8]. In the case of symmetric, positive definite matrices the most economical method for the A-orthogonalization of the vectors $\{A^i(\phi^0 - \phi^*)\}_{i=1}^m \equiv \{A^{i-1}(A\phi^0 - f)\}_{i=1}^m$ is provided by the conjugate gradients method. In the framework of (2.31) this latter method is formulated as follows:

$$g_k = \begin{cases} \xi^0, & \text{if } k = 1, \\ \xi^{k-1} - b_k g_{k-1}, & \text{if } k > 1, \end{cases}$$

$$b_k = \frac{(A\xi^{k-1}, g_{k-1})}{(Ag_{k-1}, g_{k-1})}, \quad \phi^k = \phi^{k-1} - \alpha_k g_k, \quad (2.34)$$

$$\alpha_k = \frac{(\xi^{k-1}, g_k)}{(Ag_k, g_k)} \quad (k = 1, \ldots, m),$$

where $\{\xi^k = A\phi^k - f\}_{k=1}^m$ are the residual vectors.

We will prove that the vectors $\{g_k\}_{k=1}^m$ so constructed form an A-orthogonal basis for M, if the vectors $\{A^{k-1}\xi^0\}_{k=1}^m$ are linearly independent. First of all, all the vectors $\{g_k\}_{k=1}^m$ are nontrivial. Indeed, after a succession of eliminations Expression (2.34) becomes

$$g_k = A^{k-1}\xi^0 + \sum_{j=1}^{k-1} \beta_{kj} A^{j-1}\xi^0 \quad (k = 1, \ldots, m) \quad (2.35)$$

with some coefficients $\{\beta_{kj}\}$. Hence $g_k = 0$ would contradict the linear independence of $\{A^{k-1}\xi^0\}_{k=1}^m$.

It remains to prove that $\{g_k\}_{k=1}^m$ are orthogonal. For this assume that for some $k \geq 2$ we have

$$\begin{aligned} (Ag_k, g_j) &= 0 & (j = 1, \ldots, k-1), \\ (\xi^k, g_j) &= 0 & (j = 1, \ldots, k), \\ \alpha_j &> 0 & (j = 1, \ldots, k), \end{aligned} \quad (2.36)$$

Gradient Iterative Methods

(these equations hold for $k = 1, 2$, as can be verified directly). We will show that (2.36) hold true for $(k + 1)$. The following relations will be of help:

$$Ag_j = \frac{1}{\alpha_j}[g_j + b_j g_{j-1} - g_{j+1} - b_{j+1} g_j] \qquad (2.37)$$
$$= \varepsilon_j g_{j+1} + \beta_j g_j + \gamma_j g_{j-1}$$
$$(j = 1, \ldots, k-1),$$

$g_0 = 0$. They can be derived using (2.34) and the relations

$$\xi^j = \xi^{j-1} - \alpha_j Ag_j,$$
$$g_{j+1} = \xi^j - b_{j+1} g_j,$$
$$g_j = \xi^{j-1} - b_j g_{j-1},$$
$$(j = 1, \ldots, k-1).$$

Since

$$(Ag_{k+1}, g_j) = (A\xi^k - b_{k+1} Ag_k, g_j)$$
$$= (A\xi^k, g_j) - b_{k+1} \times (Ag_k, g_j)$$
$$= (\xi^k, Ag_j) = (\xi^k, \varepsilon_j g_{j+1} + \beta_j g_j + \gamma_j g_{j-1}) = 0$$
$$(j = 1, \ldots, k-1)$$

by assumption, and since $(Ag_{k+1}, g_k) = 0$ by construction, we have

$$(Ag_{k+1}, g_j) = 0 \qquad (2.38)$$

for all $j \leq k$. Moreover, since $(\xi^{j+1}, g_{k+1}) = 0$, and since

$$(\xi^{k+1}, g_j) = (\xi^k, g_j) - \alpha_{k+1}(Ag_{k+1}, g_j) \qquad (j = 1, \ldots, k), \qquad (2.39)$$

by construction and using (2.36), it follows that

$$(\xi^{k+1}, g_j) = 0$$

for any $1 \leq j \leq k + 1$. Equations (2.38), (2.39) and the inequality

$$\alpha_{k+1} = \frac{(\xi^k, g_{k+1})}{(Ag_{k+1}, g_{k+1})} = \frac{(\xi^k, \xi^k)}{(Ag_{k+1}, g_{k+1})} > 0$$

imply the validity of (2.36) at the $(k + 1)$st step, completing the induction. So the vectors $\{g_k\}$ are orthogonal, and (2.31) is thus solved.

In conclusion consider the singular case, when for some $k \geq 1$ the system $\{A^{i-1}\xi^0\}_{i=1}^k$ is linearly independent, but $\{A^{i-1}\xi^0\}_{i=1}^{k+1}$ is dependent:

$$-\sum_{j=0}^{k} C_j A^j \xi^0 \equiv \sum_{j=0}^{k} C_j A^{j+1}(\phi^* - \phi^0) = 0, \qquad (2.40)$$

where not all the coefficients $\{C_j\}_{j=0}^k$ are zero. The coefficient C_0 is nonzero, since otherwise

$$\sum_{j=0}^k C_j A^j(\phi^* - \phi^0) = 0,$$

as can be seen by multiplying (2.40) by A^{-1}. This would contradict the independence of $\{A^j(\phi^* - \phi^0)\}_{j=1}^k$. From (2.40) we now have

$$\phi^* - \phi^0 = -\frac{1}{C_0} A^{-1} \sum_{j=1}^k C_j A^{j+1}(\phi^* - \phi^0) = \sum_{j=1}^k \frac{C_j}{C_0} A^{j-1} \xi^0,$$

which means that $\phi^* - \phi^0 \in M_A$. From here and (2.36) it follows that the approximation $\hat{u}$ necessarily equals $\phi^* - \phi^0$ ($\phi^* = \phi^0 + \hat{u} = \phi^0 + \phi^k$). In other words, the conjugate gradients method can be used in this case to find the exact solution of (2.28) already at the kth step.

The conjugate gradients method, as well as other orthogonalization procedures can be widely exploited for the convergence acceleration of stationary iterative methods. For instance, the iterations

$$\phi^{k+1} = \phi^k - B(A\phi^k - f) \qquad (k = 1, 2, \ldots) \tag{2.41}$$

with symmetric, positive matrices A and B can be accelerated as follows:

$$g_k = \begin{cases} B\xi^0, & \text{if } k = 1, \\ B\xi^{k-1} - b_k g_{k-1}, & \text{if } k > 1, \end{cases}$$

$$b_k = \frac{(AB\xi^{k-1}, g_{k-1})}{(Ag_{k-1}, g_{k-1})}, \quad \phi^k = \phi^{k-1} - \alpha_k g_k, \quad \alpha_k = \frac{(\xi^{k-1}, g_k)}{(Ag_k, g_k)} \tag{2.42}$$

$$(k = 1, 2, \ldots).$$

The convergence analysis of the method of conjugate gradients shows that the asymptotic rate of convergence is not lower than that corresponding to the Chebyshev acceleration

$$s = \frac{2}{\sqrt{p(A)}}. \tag{2.43}$$

Assume that the original matrix is positive but not symmetric. In order to be able to apply the conjugate gradients method, one can use the Gauss transformation to symmetrize the matrix, and instead of

$$A\phi = f$$

consider the transformed equation

$$A^*A\phi = F, \tag{2.44}$$

where $F = A^*f$.

Assume A has a real spectrum, and let $p = \beta(A)/\alpha(A)$ be its conditioning number. Then $A^*A = AA^*$ has the conditioning $p(A^*A) = p^2(A)$. Hence the iterative process corresponding to the symmetrized problem has the following rate of convergence:

$$s = \sqrt{2}/p(A), \qquad (2.45)$$

while the Chebyshev acceleration gives the rate shown in Equation (2.43). A comparison of (2.43) and (2.45) shows that in the above case the Chebyshev acceleration has a definite advantage as compared to the conjugate gradients method. However, if A is of arbitrary structure (and spectrum), then the two methods can be applied to the symmetrization (2.44) with relatively same success.

In the case of symmetric and positive matrices the conjugate gradients method is unquestionably more advantageous. The damping of the residual in the course of the initial iterations progresses much more effectively and faster than in the Chebyshev case. The difference between the rates of convergence of the two methods narrows down asymptotically when the number of iterations is large, although, as a rule, the residual in the conjugate gradients is asymptotically much smaller in magnitude, due to the fast initial convergence. This property is typical of all gradient methods.

3.3. The Splitting-Up Method

Douglas, Peaceman, and Rachford have suggested an iterative method which has been widely applied to stationary problems of mathematical physics under the name of *alternating direction* method. By now this method has acquired a large number of modifications and corresponding realization schemes. The alternating direction method is essentially based on certain special relaxation processes which allow one to reduce the complicated problem into a sequence of simple problems. All such methods will be referred to as *splitting-up* methods. It is this point of view which we take when looking at the alternating direction method—by now a classic.

Let us begin with a rather simple problem

$$A\phi = f, \qquad (3.1)$$

where $A = A_1 + A_2$, $A_1 > 0$, $A_2 > 0$.

Equation (3.1) is usually solved using the iterative process

$$\frac{\phi^{j+1} - \phi^j}{\tau} + A\phi^j = f, \qquad \phi^0 = 0. \qquad (3.2)$$

The parameter τ is chosen so as to maximize the asymptotic rate of convergence of the iterative process. This usually leads to a very slow convergence. The attempts to accelerate the procedure have resulted eventually in the following iterative method (Douglas and Gunn [15], Dyakonov [15], Samarskii [3, 15], Marchuk [15], Yanenko [3, 15], and others):

$$B \frac{\phi^{j+1} - \phi^j}{\tau} + A\phi^j = f, \qquad \phi^0 = 0, \tag{3.3}$$

where B is a new positive definite operator, the form of which is to be determined.

Note that the problem is solved in one iteration if we choose $B/\tau = A$. Indeed,

$$\phi^{j+1} = \phi^j - \tau B^{-1}(A\phi - f). \tag{3.4}$$

Taking into account that

$$\tau B^{-1} A = E$$

is the identity operator, we conclude that for arbitrary ϕ^j

$$\phi^{j+1} = A^{-1} f. \tag{3.5}$$

This is a formal expression for the exact solution of the problem. Even though the method looks impressive, its realization requires computation of the inverse A^{-1}, a task whose difficulty equals that of solving the original problem. For that reason the iterative process above is not constructive. It suggests, however, various other approaches for choosing the operator B/τ which, in a sense, would be close to the operator A.

Before we turn to studying such processes, let us consider a somewhat simpler scheme

$$\frac{\phi^{j+1} - \phi^j}{\tau} + A \frac{\phi^{j+1} + \phi^j}{2} = f, \qquad \phi^0 = 0. \tag{3.6}$$

Here A is a difference operator. Let us write this iterative process in the form

$$\left(E + \frac{\tau}{2} A\right) \phi^{j+1} = \left(E - \frac{\tau}{2} A\right) \phi^j + \tau f. \tag{3.7}$$

Solving for ϕ^{j+1} in Equation (3.7), we obtain

$$\phi^{j+1} = \left(E + \frac{\tau}{2} A\right)^{-1} \left[\left(E - \frac{\tau}{2} A\right) \phi^j + \tau f\right]. \tag{3.8}$$

Assuming that A is a matrix and that $A > 0$, Process (3.8) will converge for any positive τ. This follows by a consequence of Kellogg's lemma, which says that the norm of the transition operator is less than unity:

$$\left\| \left(E + \frac{\tau}{2} A\right)^{-1} \left(E - \frac{\tau}{2} A\right) \right\| < 1.$$

An examination of (3.8) shows that in order to implement the process under consideration, we have to invert a rather complicated operator

The Splitting-Up Method

$E + \tau A/2$ at each iteration. In fact, this is what makes the iteration Process (3.6) nonconstructive. Nevertheless, the process is very useful for theoretical purposes. Indeed, assume that

$$Au = \lambda u, \qquad A^*u^* = \lambda^* u^* \tag{3.9}$$

defines a complete set of eigenelements u_n and corresponding eigenvalues $\lambda_n > 0$. Let $\alpha(A) = \min_n \lambda_n(A) > 0$ and $\max_n \lambda_n(A) = \beta(A) > 0$. Let us represent the elements ϕ and f in the form

$$\phi = \sum_n \phi_n u_n; \qquad f = \sum_n f_n u_n, \tag{3.10}$$

where $\phi_n = (\phi, u_n^*)$ and $f_n = (f, u_n^*)$ are the Fourier coefficients in Expansion (3.10).

Substituting (3.10) into (3.8) and then taking the inner product of the result so obtained with the element u_n^*, we get the following recursive relation for the Fourier coefficients:

$$\phi_n^{j+1} = \frac{1 - \dfrac{\tau \lambda_n}{2}}{1 + \dfrac{\tau \lambda_n}{2}} \phi_n^j + \frac{\tau}{1 + \dfrac{\tau \lambda_n}{2}} f_n, \qquad \phi_n^0 = 0. \tag{3.11}$$

With the help of (3.11) we obtain

$$\phi_n^{j+1} = \tau \frac{1 - T_n^{j+1}}{1 - T_n} \cdot \frac{f_n}{1 + \dfrac{\tau \lambda_n}{2}}, \tag{3.12}$$

where

$$T_n = \frac{1 - \dfrac{\tau \lambda_n}{2}}{1 + \dfrac{\tau \lambda_n}{2}}. \tag{3.13}$$

Naturally, the convergence of (3.12) to its limit

$$\phi_n^\infty = \frac{f_n}{\lambda_n} \tag{3.14}$$

will be faster for a smaller T_n. Therefore we have to choose τ so as to attain

$$\min_\tau \left(\max_n \left\{ \left| \frac{1 - \dfrac{\tau \lambda_n}{2}}{1 + \dfrac{\tau \lambda_n}{2}} \right| \right\} \right). \tag{3.15}$$

It is easily shown that the function under the magnitude sign is monotonic with respect to the argument $x = \lambda_n/2$. For a fixed τ it thus achieves its maximum on the boundary of the interval, that is $\lambda_n = \alpha(A)$ or $\lambda_n = \beta(A)$. Hence it is enough to find τ from

$$\min_\tau \left(\max \left\{ \left| \frac{1 - \frac{\tau\alpha(A)}{2}}{1 + \frac{\tau\alpha(A)}{2}} \right|, \left| \frac{1 - \frac{\tau\beta(A)}{2}}{1 + \frac{\tau\beta(A)}{2}} \right| \right\} \right). \tag{3.16}$$

It is easy to verify that

$$\tau = \frac{2}{\sqrt{\alpha\beta}}. \tag{3.17}$$

in which case

$$q = \frac{1 - \frac{\tau\alpha}{2}}{1 + \frac{\tau\alpha}{2}} = \frac{1 - \sqrt{\frac{\alpha}{\beta}}}{1 + \sqrt{\frac{\alpha}{\beta}}}. \tag{3.18}$$

With this choice of τ we thus have the following uniform estimate:

$$|T_n| \leq \frac{1 - \sqrt{\frac{\alpha}{\beta}}}{1 + \sqrt{\frac{\alpha}{\beta}}}. \tag{3.19}$$

If $\alpha \ll \beta$, we obtain

$$q = 1 - 2/\sqrt{p} \tag{3.20}$$

up to the higher order terms of $1/\sqrt{p}$, and hence the asymptotic rate of convergence is given by

$$s = 2/\sqrt{p}. \tag{3.21}$$

This means that the asymptotic rate of convergence of (3.6) coincides with that corresponding to the Chebyshev acceleration, and it is one half the rate corresponding to the over-relaxation method.

In order to conclude the theoretical analysis of (3.6) let us consider the expression

$$T_n = \frac{1 - \frac{\lambda_n}{\sqrt{\alpha\beta}}}{1 + \frac{\lambda_n}{\sqrt{\alpha\beta}}}. \tag{3.22}$$

The Splitting-Up Method

In many problems of mathematical physics the parameter λ_n has the dimension $1/cm^k$, where k is the order of the highest derivatives with respect to the spatial variables; it follows from this that $\sqrt{\alpha\beta}$ is the inverse of the quantity characterizing the physical dimension of the "average" excitations expressed in units of the eigenvalues of A. Let us use the notation

$$\bar{\lambda} = \sqrt{\alpha\beta}. \tag{3.23}$$

Thus we have

$$T_n = \frac{1 - \frac{\lambda_n}{\bar{\lambda}}}{1 + \frac{\lambda_n}{\bar{\lambda}}},$$

and consequently the solution of (3.12) has the form

$$\phi^j = \sum_n \left[1 - \left(\frac{1 - \frac{\lambda_n}{\bar{\lambda}}}{1 + \frac{\lambda_n}{\bar{\lambda}}} \right)^{j+1} \right] \frac{f_n}{\lambda_n} u_n. \tag{3.24}$$

Since $\lambda_n > 0$,

$$\lim_{j \to \infty} \left(\frac{1 - \frac{\lambda_n}{\bar{\lambda}}}{1 + \frac{\lambda_n}{\bar{\lambda}}} \right)^j = 0,$$

and hence

$$\phi^\infty = \sum_n \frac{f_n}{\lambda_n} u_n \tag{3.25}$$

is the exact solution of Problem (3.1).

Let us now consider the splitting-up schemes. Assume that

$$A = A_1 + A_2$$

where $A_1 > 0$, $A_2 > 0$. Consider Equation (3.3) and choose the operator B therein as follows:

$$B = \left(E + \frac{\tau}{2} A_1 \right)\left(E + \frac{\tau}{2} A_2 \right).$$

We obtain the following scheme:

$$\left(E + \frac{\tau}{2} A_1 \right)\left(E + \frac{\tau}{2} A_2 \right)\frac{\phi^{j+1} - \phi^j}{\tau} + A\phi^j = f, \quad \phi^0 = 0. \tag{3.26}$$

Solving for ϕ^{j+1} we obtain

$$\phi^{j+1} = \left(E + \frac{\tau}{2} A_2\right)^{-1} \left(E + \frac{\tau}{2} A_1\right)^{-1} \left[\left(E - \frac{\tau}{2} A_1\right)\left(E - \frac{\tau}{2} A_2\right)\phi^j + \tau f\right]. \tag{3.27}$$

The realization scheme of the iterative process is very simple:

$$\xi^j = A\phi^j - f; \quad \left(E + \frac{\tau}{2} A_1\right)\xi^{j+1/3} = \xi^j;$$

$$\left(E + \frac{\tau}{2} A_2\right)\xi^{j+2/3} = \xi^{j+1/3}; \quad \phi^{j+1} = \phi^j - \tau\xi^{j+2/3}. \tag{3.28}$$

Regarding the iterative process of (3.26) we will prove the following assertion: if A_1 and A_2 are matrices, and if $A_1 > 0$, $A_2 > 0$, then the process converges for any positive τ. For proof let us introduce the new elements

$$\psi^j = \left(E + \frac{\tau}{2} A_2\right)\phi^j; \quad g = \left(E + \frac{\tau}{2} A_1\right)^{-1} f. \tag{3.29}$$

With their help the recursive Equation (3.27) can be written in the form

$$\psi^{j+1} = T\psi^j + \tau g, \quad \psi^0 = 0, \tag{3.30}$$

where

$$T = \left(E + \frac{\tau}{2} A_1\right)^{-1}\left(E - \frac{\tau}{2} A_1\right)\left(E + \frac{\tau}{2} A_2\right)^{-1}\left(E - \frac{\tau}{2} A_2\right) \tag{3.31}$$

is the transition operator.

Using (3.30) we obtain the estimate

$$\|\psi^{j+1}\| \leq \|T\|\|\psi^j\| + \tau\|g\|, \quad \|\psi^0\| = 0. \tag{3.32}$$

Hence it follows that

$$\|\psi^j\| \leq \tau \frac{1 - \|T\|^{j+1}}{1 - \|T\|} \|g\|. \tag{3.33}$$

Kellogg's lemma and the assumption $A_1 > 0$, $A_2 > 0$ imply

$$\|T\| < 1.$$

Consequently, as $j \to \infty$ we obtain

$$\|\psi^\infty\| \leq \frac{\tau}{1 - \|T\|} \|g\|. \tag{3.34}$$

It thus follows that the iterative process converges. Since the difference operator $(E + \tau A_2/2)$ has an inverse, and since Equation (3.1) has a unique solution (by the positivity of A), we have the following important result: the iterative method of (3.26) converges for any $\tau > 0$.

The Splitting-Up Method

We show next that the iterative Process (3.26) asymptotically coincides with Process (3.6) even for the most severe disturbances in the residuals ($\lambda_n \ll \beta$). To this end let us introduce the operators (without any physical dimension)

$$A_1 = \frac{1}{H}\bar{A}_1, \qquad A_2 = \frac{1}{H}\bar{A}_2,$$

where H has the dimension of $1/\lambda_n$, that is, a certain power of length as shown above.

The expressions of (3.8) and (3.27) will become respectively

$$\phi^{j+1} = (E + \varepsilon\bar{A})^{-1}[(E - \varepsilon\bar{A})\phi^j + \tau f] \qquad (3.35)$$

and

$$\phi^{j+1} = (E + \varepsilon\bar{A} + \varepsilon^2\bar{A}_1\bar{A}_2)^{-1}[(E - \varepsilon\bar{A} + \varepsilon^2\bar{A}_1\bar{A}_2)\phi^j + \tau f], \qquad (3.36)$$

where

$$\varepsilon = \frac{\tau}{2H}. \qquad (3.37)$$

has no dimension. With the parameter τ fixed we are now going to vary the character scale of the disturbance H. From (3.37) it follows that the parameter ε decreases as H increases, and approaches asymptotically zero as $H \to \infty$. This means that for small ε, (3.36) asymptotically coincides with (3.35), for which the optimal τ was established to be

$$\tau = 2/\sqrt{\alpha\beta}. \qquad (3.38)$$

If this choice of τ is also used in (3.36), then, at least for the residual disturbances of the highest character dimensionality, the asymptotic rate of convergence turns out to be

$$s = 2/\sqrt{p}. \qquad (3.39)$$

As far as the lower-scale disturbances are concerned, the corresponding residuals will go to zero, too, with progressing iterations, although at possibly slower rate than (3.39). Since the rate of convergence of an iterative process is usually limited by the rate at which the residuals with the highest character dimensionality (or, pictorially, with the "largest sphere of influence") diminish, it then turns out that (3.38) is a natural choice of τ for the splitting up schemes of the type of (3.26). Note that our qualitative considerations have not required so far any assumptions regarding commutativity of the operators A_1 and A_2, or their selfadjointness. Thus the relations obtained can be used for solving quite general problems. We have only required that the operators (matrices) satisfy the conditions $A_1 > 0$, $A_2 > 0$ and that A has a positive spectrum.

For a more accurate evaluation of the rate of convergence of the successive approximations (3.28) with τ chosen according to (3.38), let us look at a particular case where the operators A_1 and A_2 commute and generate a common basis. Then the solution may be constructed using the Fourier expansion of (3.10) in terms of the eigenvectors u_n. In terms of coefficients, Problem (3.26) becomes

$$\phi_n^{j+1} = \frac{1 - \frac{\lambda_{1n}}{\bar{\lambda}}}{1 + \frac{\lambda_{1n}}{\bar{\lambda}}} \cdot \frac{1 - \frac{\lambda_{2n}}{\bar{\lambda}}}{1 + \frac{\lambda_{2n}}{\bar{\lambda}}} \phi_n^j + \frac{\tau f_n}{\left(1 + \frac{\lambda_{1n}}{\bar{\lambda}}\right)\left(1 + \frac{\lambda_{2n}}{\bar{\lambda}}\right)}, \quad (3.40)$$

where

$$\bar{\lambda} = \sqrt{\alpha\beta}, \quad (3.41)$$

and λ_{1n}, and λ_{2n} are defined by the equations $A_1 u_n = \lambda_{1n} u_n$, $A_2 u_n = \lambda_{2n} u_n$, and related to λ_n by $\lambda_n = \lambda_{1n} + \lambda_{2n}$.

Assume further that

$$0 < \alpha \leqslant \lambda_n \leqslant \beta < \infty.$$

Consider now the transition operator from (3.40) which corresponds to the sinusoidal excitation with the index n:

$$T_n = \frac{1-x}{1+x} \cdot \frac{1-y}{1+y}, \quad (3.42)$$

where

$$x = \frac{\lambda_{1n}}{\bar{\lambda}}; \quad y = \frac{\lambda_{2n}}{\bar{\lambda}}.$$

It is not difficult to verify that $T_n(x, y)$ is a monotonic function for $x > 0$ and $y > 0$, and it therefore achieves its extremal values on the boundary of its domain. Let us compute T_n for the smallest values of the parameters $\lambda_{1n} = \alpha_1$, $\lambda_{2n} = \alpha_2$, and for the largest ones, $\lambda_{1n} = \beta_1$, $\lambda_{2n} = \beta_2$; denote them correspondingly q_α and q_β. We have

$$q_\alpha = \frac{1 - \sqrt{\frac{\alpha}{\beta}} + \frac{\alpha_1 \alpha_2}{\alpha\beta}}{1 + \sqrt{\frac{\alpha}{\beta}} + \frac{\alpha_1 \alpha_2}{\alpha\beta}} \quad (3.43)$$

and

$$q_\beta = \frac{1 - \sqrt{\frac{\beta}{\alpha}} + \frac{\beta_1 \beta_2}{\alpha\beta}}{1 + \sqrt{\frac{\beta}{\alpha}} + \frac{\beta_1 \beta_2}{\alpha\beta}}. \quad (3.44)$$

The Splitting-Up Method

Taking $\beta_i \gg \alpha_i$, we obtain the following relations for the estimate: for the largest disturbances we have

$$q_\alpha = 1 - 2\sqrt{\frac{\alpha}{\beta}}, \qquad (3.45)$$

and for the smallest

$$q_\beta = 1 - 2\sqrt{\frac{\alpha}{\beta}\frac{\beta^2}{\beta_1\beta_2}}. \qquad (3.46)$$

In (3.45) and (3.46) we have disregarded the terms with higher order smallness relative to $\sqrt{\alpha/\beta}$.

As shown earlier, (3.45) and (3.46) directly imply that the asymptotic rate of convergence of the residual component with the lowest frequency is given by

$$s_\alpha = \frac{2}{\sqrt{p}}, \qquad (3.47)$$

and for the highest

$$s_\beta = \frac{2}{\sqrt{p}} \cdot \frac{\beta^2}{\beta_1\beta_2}. \qquad (3.48)$$

Since $\beta = \beta_1 + \beta_2$ and $\beta^2 \leq 4\beta_1\beta_2$, it follows that

$$s_\beta \geq \frac{8}{\sqrt{p}}. \qquad (3.49)$$

Thus, for a given $\tau = 2/\sqrt{\alpha\beta}$, the asymptotic rate of convergence does not fall below

$$s = 2/\sqrt{p} \qquad (3.50)$$

for all components.

It may be useful to note that in the particular case of commuting operators and under the assumption that $\alpha_i = \alpha/2$ and $\beta_i = \beta/2, i = 1, 2$, a better choice for τ is as follows:

$$\tau = \frac{2}{\sqrt{\alpha_1\beta_1}} = \frac{2}{\sqrt{\alpha_2\beta_2}} = \frac{4}{\sqrt{\alpha\beta}}. \qquad (3.51)$$

In this case it can be established, as was done previously, that for all components

$$s = 4/\sqrt{p} \qquad (3.52)$$

(see Section 6.1.3).

The optimization problem can also be approached from a somewhat different point of view. Rather than (3.26), consider a more general process

$$(E + \sigma A_1)(E + \sigma A_2)\frac{\phi^{j+1} - \phi^j}{\tau} + A\phi^j = f, \tag{3.53}$$

with two free parameters σ and τ. Solving for ϕ^{j+1}, we get

$$\phi^{j+1} = \phi^j - \tau B^{-1}(A\phi^j - f), \tag{3.54}$$

where

$$B = (E + \sigma A_1)(E + \sigma A_2). \tag{3.55}$$

With the notation $B^{-1}A = \Lambda$, $B^{-1}f = F$ we obtain the usual form

$$\phi^{j+1} = \phi^j - \tau(\Lambda\phi^j - F). \tag{3.56}$$

The optimization problem for (3.56) is taken up in the following chapter. We only note that our problem is to determine σ. Numerical experiments suggest to choose σ from the interval

$$\frac{1}{\sqrt{\alpha\beta}} \leqslant \sigma \leqslant \frac{2}{\sqrt{\alpha\beta}}. \tag{3.57}$$

The parameter τ is chosen independently in connection with the features of the optimization process.

In some cases with a positive and commutative A_i one can achieve the asymptotic rate of convergence $s = c/\ln p$, $c \simeq 1$, by a special choice of σ_j and τ_j (Peaceman and Rachford [15] and Wachspress [15]). For

$$\sigma = \frac{1}{\sqrt{\alpha\beta}}$$

and τ selected according to the Chebyshev acceleration one obtains

$$\left(E + \frac{1}{\sqrt{\alpha\beta}}A_1\right)\left(E + \frac{1}{\sqrt{\alpha\beta}}A_2\right)\frac{\phi^{j+1} - \phi^j}{\tau_j} + A\phi^j = f. \tag{3.58}$$

Here τ_j is given by Equation (1.28) in which α and β are the spectral boundaries of $B^{-1}A$.

In this case, assuming in addition that A_1 and A_2 commute, it is possible to achieve the rate

$$s = O(1/\sqrt[4]{p}). \tag{3.59}$$

The convergence estimates for the splitting-up iterations obtained under the assumption of commutativity make one believe that the method can also be successfully applied to the case where the operators A_1 and A_2 do not commute. Although there are no strict and sufficiently precise estimates of the rate of convergence in the absence of commutativity, the mathematical experiments show that in many cases the splitting-up methods yield very efficient results.

In conclusion let us note that if

$$A = \sum_{\gamma=1}^{n} A_\gamma, \quad A_\gamma > 0,$$

the iterative process can be taken in the form

$$B \frac{\phi^{i+1} - \phi^i}{\tau} + A\phi^i = f, \quad (3.60)$$

where

$$B = \prod_{\gamma=1}^{n} \left(E + \frac{1}{\sqrt{\alpha\beta}} A_\gamma \right).$$

Here, as usual, α and β are positive spectral boundaries of the operator A. The iterations (3.60) converge fairly quickly if τ is chosen according to (1.28) (Chebyshev acceleration), where now α and β are the spectral boundaries of $B^{-1}A$. In order to use this combination, however, $B^{-1}A$ must have a real and positive spectrum. Thus the implementation of Algorithm (3.60) coupled with the Chebyshev optimization involves computation of the spectral boundaries $\alpha(A)$, $\beta(A)$, $\alpha(B^{-1}A)$, $\beta(B^{-1}A)$ and the parameters τ_j. The realization scheme is as follows:

$$\xi^j = (A\phi^j - f),$$

$$\left(E + \frac{1}{\sqrt{\alpha\beta}} A_1 \right) \xi^{j+(1/n+1)} = \xi^j,$$

$$\cdots \cdots \cdots \cdots \cdots \cdots \cdots \cdots \cdots \cdots \cdots \quad (3.61)$$

$$\left(E + \frac{1}{\sqrt{\alpha\beta}} A_n \right) \xi^{j+(n/n+1)} = \xi^{j+(n-1/n+1)},$$

$$\phi^{j+1} = \phi^j - \tau_j \xi^{j+(n/n+1)}.$$

Let us point out that the *a priori* constants $\alpha(A)$, $\beta(A)$, $\alpha(B^{-1}A)$, and $\beta(B^{-1}A)$ can be determined using the methods from Section 1.1.

A number of interesting splitting-up optimization algorithms for stationary problems have been published by Douglas [15], Wachspress [15], Varga [15], Samarskii [3, 15], Dyakonov [15], and others.

3.4. The Splitting-Up Method with Variational Optimization

Consider the equation

$$A\phi = f \quad (4.1)$$

where $A > 0$, assuming again that (4.1) is actually a system of algebraic equations already obtained from the original problem of mathematical physics.

Consider the successive approximations

$$B\frac{\phi^{j+1} - \phi^j}{\tau_j} + A\phi^j = f, \qquad \phi^0 = 0, \qquad (4.2)$$

where B is to be determined later, and τ_j is a parameter of the nonstationary iterative process.

Assuming

$$A = \sum_{\alpha=1}^{n} A_\alpha, \qquad A_\alpha > 0, \qquad (4.3)$$

let us take the matrix B in the following form:

$$B = \prod_{\alpha=1}^{n} (E + \sigma A_\alpha)^m, \qquad (4.4)$$

where $m \geqq 1$. The prime reason for this choice has been computational economy. Note that in Equation (4.4) there is one more parameter σ which can be chosen on the basis of *a posteriori* information about the solution inclusive the optimization of the algorithm.

No prior knowledge of the spectral properties of A and B will be assumed. The problem is to optimize the parameters τ_j and σ from the *a posteriori* information about the approximate solution, using the residual method. To do that, let us multiply the left-hand side of (4.2) by B^{-1}. With the help of the residual relation

$$\xi^j = A\phi^j - f,$$

we obtain

$$\frac{\phi^{j+1} - \phi^j}{\tau_j} + B^{-1}\xi^j = 0, \qquad \xi^0 = -f. \qquad (4.5)$$

Applying the operator A to both sides of this equation, followed by a subsequent addition and subtraction of the term f/τ_j, we obtain the following relation for the residual of the iterative process:

$$\frac{\xi^{j+1} - \xi^j}{\tau_j} + AB^{-1}\xi^j = 0, \qquad \xi^0 = -f, \qquad (4.6)$$

or,

$$\xi^{j+1} = (E - \tau_j AB^{-1})\xi^j, \qquad \xi^0 = -f. \qquad (4.7)$$

Consider the inner product (ξ^{j+1}, ξ^{j+1}). Equations (4.7) imply

$$(\xi^{j+1}, \xi^{j+1}) = q_j(\xi^j, \xi^j), \qquad (4.8)$$

where

$$q_j = 1 - 2\tau_j \frac{(AB^{-1}\xi^j, \xi^j)}{(\xi^j, \xi^j)} + \tau_j^2 \frac{(AB^{-1}\xi^j, AB^{-1}\xi^j)}{(\xi^j, \xi^j)}. \qquad (4.9)$$

Assume that B is fixed, that is, σ is known. The parameter τ_j will then be chosen by minimizing the functional q_j. Consider the equations

$$\frac{dq_j}{d\tau_j} = 0, \qquad \frac{d^2 q_j}{d\tau_j^2} > 0. \tag{4.10}$$

The second relation above holds true if we assume $AB^{-1} > 0$. The first one will then yield

$$\tau_j = \frac{(AB^{-1}\xi^j, \xi^j)}{(AB^{-1}\xi^j, AB^{-1}\xi^j)}, \tag{4.11}$$

Let us introduce the notation

$$y^{j+1} = B^{-1}\xi^j, \qquad z^{j+1} = Ay^{j+1} \tag{4.12}$$

and consider the realization scheme of the splitting-up method. We take ξ^j as given. The functions y^{j+1} are then found with the help of the relation

$$By^{j+1} = \xi^j. \tag{4.13}$$

This equation is reduced to the system

$$\begin{aligned}
(E + \sigma A_1)^m y^{j+(1/n)} &= \xi^j, \\
(E + \sigma A_2)^m y^{j+(2/n)} &= y^{j+(1/n)}, \\
&\cdots\cdots\cdots\cdots\cdots \\
(E + \sigma A_n)^m y^{j+1} &= y^{j+(n-1/n)}.
\end{aligned} \tag{4.14}$$

Once y^{j+1} has been found, we compute

$$z^{j+1} = Ay^{j+1}, \tag{4.15}$$

and

$$\tau_j = \frac{(z^{j+1}, \xi^j)}{(z^{j+1}, z^{j+1})}. \tag{4.16}$$

The new residual is now obtained by the formula

$$\xi^{j+1} = \xi^j - \tau_j z^{j+1}. \tag{4.17}$$

and the solution ϕ^{j+1} by

$$\phi^{j+1} = \phi^j - \tau_j y^{j+1}. \tag{4.18}$$

It remains to choose the parameter σ.

In general, the optimal choice of σ is still an open problem. We will make this choice on the basis of certain heuristic methods drawn from a systematic analysis of numerical experiments. In the absence of *a priori* information about the spectrum of the positive operator A, the parameter σ will be considered as changing from iteration to iteration and will be chosen in the form

$$\sigma_j = \tau_{j-1}$$

or

$$\sigma_j = \frac{1}{j-1} \sum_{i=1}^{j-1} \tau_i,$$

where $j = 1, 2, \ldots$.

A special arrangement of the algorithm makes the computation of σ_1 superfluous for numerical purposes. The computational scheme is determined by the choice of the operator B which we will assume to depend on the index j. Let

$$B_1 = \prod_{\alpha=1}^{n} A_\alpha, \quad B_j = \prod_{\alpha=1}^{n} \left(E + \frac{\sigma_j}{2} A_\alpha \right),$$

where $j = 2, 3, \ldots$.

Note that for the first iteration one can also take $\sigma_1 = 0$. At this step we thus obtain the usual residual method.

In a well-known case the operator B is taken in the form

$$B = \prod_{\gamma=1}^{n} \left(E + \frac{1}{\sqrt{\alpha\beta}} A_\gamma \right),$$

where $\alpha = \alpha(A)$, $\beta = \beta(A)$. The splitting-up method then turns out to be close to (3.61) with the Chebyshev τ_j and with the distinction that the preliminary computation of the *a priori* constants $\alpha(B^{-1}A)$, $\beta(B^{-1}A)$ is no longer needed.

3.5. Equations with Singular Operators

In Sections 3.1–3.4 we have considered various iterative methods under the assumption that the difference operator A in the equation

$$A\phi = f \tag{5.1}$$

was positive and had a real spectrum.

In applications, however, we often encounter difference equations or linear algebraic systems with nonnegative, singular operators, that is, such that

$$(A\phi, \phi) \geq 0, \tag{5.2}$$

the equality possibly holding for a nontrivial admissible element $\phi = \phi_0 \in \Phi$:

$$(A\phi_0, \phi_0) = 0, \quad \phi_0 \neq 0. \tag{5.3}$$

Clearly, if ϕ_0 is an eigenfunction of A, then the spectrum of the operator A is contained in the interval

$$0 \leq \lambda_n \leq \beta. \tag{5.4}$$

Equations with Singular Operators

An instance of this is the difference analog of the Neumann problem, or the Poincare problem for the Laplace equation. As is well known, in these cases Equality (5.3) is achieved on the elements $\phi_0 \equiv \text{const}$.

Let us see what happens if we solve Equation (5.1) using one of the iterations considered in this chapter. Since $\alpha(A) = 0$, successive approximations do not converge as a rule. In order to understand the reason let us somewhat analyze the matter. Consider the simple iterative method (1.11)

$$\phi^{j+1} = \phi^j - \tau(A\phi^j - f), \qquad \phi^0 = 0, \tag{5.5}$$

assuming that the spectral problem of (1.10) admits a normed eigenfunction basis $\{u_n\}$, $\{u_n^*\}$. Let

$$\phi = \sum_n \phi_n u_n, \qquad f = \sum_n f_n u_n, \tag{5.6}$$

where

$$\phi_n = (\phi, u_n^*), \qquad f_n = (f, u_n^*). \tag{5.7}$$

Substituting (5.6) into (5.5) and then taking the inner product with u_n^*, we obtain a system of equations for the Fourier coefficients:

$$\begin{aligned}\phi_n^{j+1} &= \phi_n^j - \tau(\lambda_n \phi_n^j - f_n), \\ \phi_n^0 &= 0 \qquad (n = 0, 1, 2, \ldots, m).\end{aligned} \tag{5.8}$$

Assume now that A has a real, nonnegative spectrum with $\lambda_0 = 0$. Then, as is well known, for a solution of (5.1) to exist it is necessary that

$$f_0 = 0. \tag{5.9}$$

If (5.9) is true, then (5.8) implies

$$\phi_0^j = 0; \qquad \phi_n^j = \tau \frac{1 - q_n^j}{1 - q_n} f_n \qquad (n = 1, 2, \ldots, m), \tag{5.10}$$

where

$$q_n = 1 - \tau \lambda_n. \tag{5.11}$$

There results

$$\phi^j = \sum_{n=1}^m [1 - (1 - \tau\lambda_n)^j] \frac{f_n}{\lambda_n} u_n. \tag{5.12}$$

If τ is chosen from the condition

$$|1 - \tau\beta(A)| < 1, \tag{5.13}$$

then (5.12) converges to the solution of (5.1) as $j \to \infty$, and we obtain the relation

$$\phi^\infty = \sum_{n=1}^m \frac{f_n}{\lambda_n} u_n. \tag{5.14}$$

Note that for realization of the iterations (5.5) we do not represent the approximate solution in the form (5.12). Otherwise it would be necessary to solve the auxiliary eigenvalue problems

$$Au = \lambda u, \qquad A^*u^* = \lambda u^*. \tag{5.15}$$

Only such a realization method guarantees that in the course of the numerical process based on the formula (5.12) will there be no members of the sum involving the eigenfunction u_0, since it is excluded from the consideration. Indeed, for the realization of (5.5) we do not decompose the solution into separate harmonic components as corresponding to different eigenelements. All these elements (including u_0) are implicitly superposed in the approximate solution.

Taking into account that (5.9) is satisfied up to the round-off errors, the actual numerical algorithm involves the formula

$$\phi^j = \sum_{n=0}^{m} [1 - (1 - \tau\lambda_n)^j] \frac{f_n}{\lambda_n} u_n, \tag{5.16}$$

rather than (5.12). The term in (5.16), corresponding to u_0, has the amplitude f_0/λ_0 generated by the round-off errors, and becomes thus random for all practical considerations. The approximate solution is corrupted by this error to the extent that it becomes useless. But most importantly, the random character of the zero-indexed component prevents us from achieving the desired accuracy of the solution of (5.1).

In order for the iterative process of (5.5) to converge, it is necessary to modify the algorithm so as to exclude possible presence of the component corresponding to u_0. Such a modification will be performed with the help of an orthogonalization method, the essence of which is as follows.

Denote by Φ the subspace of all admissible elements, and by Φ_0 a proper subspace of Φ such that Φ_0 and u_0 generate Φ. It is not difficult to see that

$$(A\phi, \phi) > 0, \quad \text{if} \quad \phi \in \Phi_0. \tag{5.17}$$

Thus, the restriction of A to Φ_0 has now a positive spectrum:

$$0 < \alpha^*(A) \leq \lambda_n \leq \beta(A), \tag{5.18}$$

where $\alpha^*(A)$ coincides with the first eigenvalue $\lambda_1(A)$ of A. This means that any of the iterative procedures considered in the present chapter will converge on the elements of Φ_0. Except for the new spectral interval $[\alpha^*, \beta]$ the optimization procedure is as before.

The orthogonalization procedure is constructed as follows. Assume that u_0, u_0^* are known in advance and

$$(u_0, u_0^*) = 1. \tag{5.19}$$

Equations with Singular Operators

Instead of the one-step iterative process (5.5) let us construct the following one:

$$\phi^{j+1/2} = \phi^j - \tau(A\phi^j - f);$$
$$\phi^{j+1} = \phi^{j+1/2} - (\phi^{j+1/2}, u_0^*)u_0. \tag{5.20}$$

At this point we will assume that only the integer-indexed iterations make sense. We will show that at each step the process (5.20) restricted to the integer indices makes the u_0-component of the approximate solution zero (up to the round-off errors). In order to see that, it is enough to take the inner product of the second equation of (5.20) with u_0^* and use (5.19). Then

$$(\phi^{j+1}, u_0^*) = 0. \tag{5.21}$$

But the expression on the left in the above equation is the Fourier coefficient ϕ_0^{j+1} corresponding to u_0. Hence, the claim is proved.

Let us next describe the realization scheme for the iterative process with orthogonalization. The first thing to be done is the preliminary computation of the eigenelements u_0 and u_0^* of the operators A and A^*, and the evaluation of the spectral boundaries $\alpha^*(A)$ and $\beta(A)$ (they will be needed later for optimization purposes). These computations can be made using the Lyusternik iterative process (see Section 1.1). Let

$$\omega^{(n+1)} = A \frac{\omega^{(n)}}{\|\omega^{(n)}\|}. \tag{5.22}$$

The largest eigenvalue β can be found by putting

$$\beta = \lim_{n \to \infty} \|\omega^{(n)}\|. \tag{5.23}$$

Once β has been computed, we construct new operators

$$B = \beta E - A, \qquad B^* = \beta E - A^* \tag{5.24}$$

in correspondence with the algorithm from Section 1.1 and define the iterative processes

$$u^{(n)} = \frac{1}{c_n} Bu^{(n-1)}; \qquad u^{*(n)} = \frac{1}{c_n} B^* u^{*(n-1)}, \tag{5.24a}$$

where c_n is a normalizing sequence obtained from the conditions

$$(u^{(n)}, u^{*(n)}) = 1. \tag{5.24b}$$

Hence

$$c_n = \sqrt{|(Bu^{(n-1)}, B^* u^{*(n-1)})|}. \tag{5.25}$$

Clearly, both processes must be computed simultaneously. We thus obtain

$$u_0 = \lim_{n \to \infty} u^{(n)}; \qquad u_0^* = \lim_{n \to \infty} u^{*(n)}. \tag{5.26}$$

There is no need here to compute $\alpha(A) = \alpha(A^*)$, since the lower spectral boundary is *a priori* known to be zero.

At the next stage of the realization scheme it is necessary to compute $\alpha^*(A) = \alpha^*(A^*) = \alpha^*$—the eigenvalue of A and A^* which comes first after the zero eigenvalue $\alpha(A) = \alpha(A^*) = 0$. For this purpose we use successive approximations (5.24a), taking into account the orthogonalization with respect to the zero eigenfunction. We obtain

$$v^{(n-1/2)} = B \frac{v^{(n-1)}}{\|v^{(n-1)}\|}; \qquad v^{(n)} = v^{(n-1/2)} - (v^{(n-1/2)}, u_0^*)u_0. \qquad (5.27)$$

As can be easily seen, this iterative process makes it possible to evaluate α^* as a limit

$$\alpha^* = \lim_{n \to \infty} \frac{\|Av^{(n)}\|}{\|v^{(n)}\|}. \qquad (5.28)$$

Thus, we have at our disposal *a priori* information about u_0, u_0^*, α^*, and β. Supplemented by the additional orthogonalization algorithm (5.20), this information is enough to implement any of the iterative schemes of this chapter. Note, that everywhere in the optimization procedures one has to take α^* in place of α. The additional computation of α^* is not needed in the case of ordinary iterations or the two-step residual method, since the orthogonalization needs only be applied with reference to the eigenvector corresponding to $\lambda = 0$.

3.6. Iterative Methods for Inaccurate Input Data

Consider a problem of linear algebra

$$A\phi = f, \qquad (6.1)$$

where $A > 0$ and f are given.

Up to now we have assumed that the matrix A and the vector f are known without error, so that our method of solving equations like (6.1) tacitly presupposed an exact input data. In practice, however, we often have to deal with input data which are known only approximately: rather than (6.1) we have the equation

$$A^h \phi^h = f^h, \qquad (6.2)$$

where the index h indicates that the given information depends on the approximation error or on various random errors. We will assume that the operator- and vector-related errors are known; more precisely, we will assume *a priori* knowledge of the following kind of estimates:

$$\|(A - A^h)\phi\| \leqslant \xi(h), \qquad \|f - f^h\| \leqslant \eta(h). \qquad (6.3)$$

We now attempt to solve the problem of Equation (6.1) via Equation (6.2) and the *a priori* information (6.3). Since our results will trivially carry over

Iterative Methods for Inaccurate Input Data

for most of the iterative processes considered earlier, our description will thus be restricted to the algorithm based on the simple scheme. With this in mind let us consider an iterative process of the form

$$[\phi^h]^{j+1} = [\phi^h]^j - \tau(A^h[\phi^h]^j - f^h), \qquad [\phi^h]^0 = \tau f^h, \qquad (6.4)$$

where the parameter τ is assumed to satisfy

$$q = \|E - \tau A^h\| < 1. \qquad (6.5)$$

This brings up the following question: In view of the estimates (6.3), for how long is the iterative process to be continued? It is natural to assume that for given errors the successive approximations should continue until the error of the iterative process becomes near to the error originating from the approximation. If the number of iterations is restricted so that these errors of different nature become approximately equal at the last iteration, then the error of the approximate solution will turn out unimprovable. Moreover, if the matrix A is ill conditioned in addition (so that the inverse $[A^h]^{-1}$ may differ considerably from A^{-1}), the continuing iterations may in fact lead to significantly worse results rather than improvements. Thus, we have the following problem: with the input errors given, find the optimal number of iterations for which all the errors of various nature match each other. This problem has been considered by Marchuk and Vasilyev [16].

We argue as follows: Consider the formal solutions of Equations (6.1) and (6.2):

$$\phi = A^{-1}f; \qquad \phi^h = [A^h]^{-1}f^h. \qquad (6.6)$$

and write the identity

$$\phi^h - \phi = [A^h]^{-1}[f^h - f + (A - A^h)\phi]. \qquad (6.7)$$

Hence

$$\|\phi^h - \phi\| \leq \|[A^h]^{-1}\|[\|f - f^h\| + \|(A - A^h)\phi\|],$$

or, accounting for the *a priori* facts of (6.3),

$$\|\phi^h - \phi\| \leq \|[A^h]^{-1}\|[\xi(h) + \eta(h)]. \qquad (6.8)$$

Considering the iterative process (6.4), the equation below is not difficult to obtain:

$$\phi^h - [\phi^h]^{j+1} = (E - \tau A^h)^{j+2}[A^h]^{-1}f^h.$$

Hence

$$\|\phi^h - [\phi^h]^{j+1}\| \leq q^{j+2}\|[A^h]^{-1}\|\|f^h\|. \qquad (6.9)$$

But

$$\|\phi - [\phi^h]^{j+1}\| \leq \|[\phi^h]^{j+1} - \phi^h\| + \|\phi^h - \phi\|,$$

by the triangle inequality, and thus we find

$$\|\phi - [\phi^h]^{j+1}\| \leq q^{j+2}\|[A^h]^{-1}\|\|f^h\| + \|[A^h]^{-1}\|[\xi(h) + \eta(h)]. \qquad (6.10)$$

The first summand on the right of (6.10) represents the error estimate of the iterative process, while the second term estimates the error due to the inaccurate input data. We want the two errors to be the same:

$$q^{j+2}\|[A^h]^{-1}\|\|f^h\| = \|[A^h]^{-1}\|[\xi(h) + \eta(h)], \tag{6.11}$$

Consequently, the index $j = j_0$ of the terminal iteration should satisfy

$$j_0 = \frac{1}{\ln q} \ln \frac{\xi(h) + \eta(h)}{\|f^h\|} - 2. \tag{6.12}$$

Since j_0 must be an integer, it is natural to take for j_0 the integer part of the expression above.

Let us note that the norm of the inverse operator does not enter Equation (6.12). This simplifies considerably the computation of the optimal number of iterations.

We see that besides the *a priori* information $\xi(h)$, $\eta(h)$, $\|f^h\|$ Equation (6.12) also contains $q = \|E - \tau A^h\|$. This quantity can be found with the help of the largest eigenvalue of the operator T^*T, where $T = E - \tau A^h$; namely

$$q = \sqrt{\beta(T^*T)}.$$

The upper spectral boundary of T can be evaluated by the Lyusternik method described in Section 1.1.

Various approaches to the problems of linear algebra with ill-conditioned matrices and inaccurate input data have been considered by Tichonov [16], Faddeyeva [8], Lavrentiev [16], Voevodin [8], and others.

3.7. The Fast Fourier Transform

Lately there have been numerous attempts to apply the discrete Fourier method for solving equations in finite differences. This technique was also used earlier, but only very rarely. The reason was that, because of the number of the necessary arithmetic operations involved, the method could not compete with the other methods—both direct and iterative. The discrete Fourier method necessitates large amounts of computations for the system of eigenfunctions, the Fourier coefficients, and summing the series.

The ideas of the fast Fourier transform have been brought up many times However, the great interest in the method came only after Cooley and Tukey [13] generalized Gold's algorithm, thereby reducing significantly the number of necessary algebraic operations.

Thus, let $f(k)$ be a function of a discrete argument, $k = 0, 1, \ldots, N - 1$. Let us represent this function by means of a finite Fourier series

$$f(k) = \sum_{n=0}^{N-1} A(n)W^{kn}, \quad A(n) = \frac{1}{N}\sum_{k=0}^{N-1} f(k)W^{-kn}. \tag{7.1}$$

The Fast Fourier Transform

Here, following Cooley and Tukey, we have denoted by

$$W = \exp\frac{2\pi i}{N}$$

the principal Nth order root of unity.

By an *operation* we shall mean an addition followed by a multiplication in the complex arithmetic. Thus it follows from Equations (7.1) that for a given $A(n)$ and W^{kn}, we need a total of N^2 operations for computing $f(k)$.

The essence of Cooley's and Tukey's idea consists of the following: If N is not a prime number, then the number of operations can be significantly reduced by representing Equations (7.1) as a multiple series.

Consider the case $N = N_1 N_2$, where N_1, N_2 are integers. In order to write (7.1) as a double series, let us represent k and n in the form

$$\begin{array}{llll} k = k_1 N_1 + k_0; & k_1 = 0, \ldots, N_2 - 1; & k_0 = 0, \ldots, N_1 - 1; \\ n = n_1 N_2 + n_0; & n_1 = 0, \ldots, N_1 - 1; & n_0 = 0, \ldots, N_2 - 1. \end{array} \quad (7.2)$$

Since

$$W^{k_1 N_1 n_1 N_2} = (W^N)^{k_1 n_1} = 1,$$

we have

$$W^{kn_1 N_2} = W^{k_0 n_1 N_2}$$

and

$$f(k) = f(k_1, k_0) = \sum_{n_0=0}^{N_2-1}\left[\left(\sum_{n_1=0}^{N_1-1} A(n_1, n_0) W^{k_0 n_1 N_2}\right) W^{kn_0}\right]. \quad (7.3)$$

Hence finding the sum of the series in (7.1) is equivalent to summing the double series (7.3), or, what is the same, summing in succession the series

$$A_1(n_0, k_0) = \sum_{n_1=0}^{N_1-1} A(n_1, n_0) W^{k_0 n_1 N_2}; \quad (7.4)$$

$$f(k_1, k_0) = \sum_{n_0=0}^{N_2-1} A_1(n_0, k_0) W^{(k_1 N_1 + k_0) n_0}. \quad (7.5)$$

But from (7.2) and (7.4) it follows that the vector A_1 has the dimension N and hence we need $N \times N_1$ operations for its computation. With A_1 known we can compute $f(k)$ using $N \times N_2$ operations. Consequently, we need $N \times (N_1 + N_2)$ operations altogether. The larger the N we take the more reduction there is in the number of necessary operations.

It is easily seen that if one of the two numbers, say N_2, is not prime, then we can further reduce the number of operations by applying the Cooley-Tukey transformation to Series (7.5). In general, if $N = N_1 \times N_2 \times \cdots \times N_m$, then rather than N^2 operations it is enough to perform $N \times (N_1 + \cdots + N_m)$ operations, the extreme reduction being achieved when $N_i = 2, 3,$ or 4.

Thus, for instance, if $N = 256 = 2^8$, the number of operations is reduced $256/(8 \times 2) = 16$ times; for $N = 243 = 3^5$, the reduction is by the factor of $243/(3 \times 5) = 16.2$. From the programming point of view the best case is $N_i = 2$ ($i = 1, \ldots, m$), although there are economic variants for other N_i ($N_i = 4, 8$).

Let us consider the case $N = 2^m$. In order to obtain the corresponding formulas, we use repeatedly the following process: take $N_1 = 2, N_2 = 2^{m-1}$ and obtain a series of the type (7.4) and (7.5). Writing

$$k = \overline{k_{m-1}k_{m-2}\cdots k_1 k_0} \equiv k_{m-1}2^{m-1} + k_{m-2}2^{m-2} + \cdots + k_1 2 + k_0,$$
$$n = \overline{n_{m-1}n_{m-2}\cdots n_1 n_0} \equiv n_{m-1}2^{m-1} + n_{m-2}2^{m-2} + \cdots + n_1 2 + n_0,$$

where k_i, n_i are zero or one, we have

$$f(k_{m-1},\ldots,k_0) = \sum_{n_0=0}^{1} \left\{ \sum_{n_1=0}^{1} \left[\cdots \sum_{n_{m-1}=0}^{1} (A(n_{m-1},\ldots,n_0) \right. \right.$$
$$\left. \left. \times W^{kn_{m-1}2^{m-1}}) \cdots W^{kn_1 2} \right] W^{kn_0} \right\}. \quad (7.6)$$

Since

$$W^{kn_{m-1}2^{m-1}} = W^{k_0 n_{m-1}2^{m-1}}, \qquad W^{kn_{m-2}2^{m-2}} = W^{\overline{k_1 k_0} n_{m-2}2^{m-2}},$$

etc., finding the sum of (7.6) is equivalent to summing the m series below:

$$A_1(k_0, n_{m-2},\ldots,n_0) = \sum_{n_{m-1}=0}^{1} A(n_{m-1},\ldots,n_0) W^{k_0 n_{m-1}2^{m-1}},$$

$$A_2(k_1, k_0, n_{m-3},\ldots,n_0) = \sum_{n_{m-2}=0}^{1} A_1(k_0, n_{m-2},\ldots,n_0) \times W^{\overline{k_1 k_0} n_{m-2}2^{m-2}},$$

$$\cdots\cdots\cdots\cdots\cdots\cdots\cdots\cdots\cdots\cdots\cdots\cdots\cdots\cdots\cdots\cdots\cdots\cdots$$

$$A_m(k_{m-1},\ldots,k_0) = \sum_{n_0=0}^{1} A_{m-1}(k_{m-2},\ldots,k_0, n_0) W^{\overline{k_{m-1}\cdots k_0} n_0},$$

$$f(k) = A_m(k_{m-1},\ldots,k_0). \quad (7.7)$$

In conclusion let us note that the fast Fourier transform is an effective tool in the correlation analysis of random samples $f(k)$ ($k = 0, 1, \ldots, N-1$).

Consider now the Dirichlet problem for the equation

$$\begin{aligned} -\Delta\phi + \mu\phi &= f \quad \text{in} \quad D, \\ \phi &= 0 \quad \text{on} \quad \partial D, \end{aligned} \quad (7.8)$$

Here μ is a given constant, and f a given sufficiently smooth function with the domain $D = \{0 \leq x \leq 1, 0 \leq y \leq 1\}$.

The Fast Fourier Transform

The corresponding difference problem is taken as follows:

$$\frac{-\phi_{k-1,l} - \phi_{k+1,l} - \phi_{k,l-1} - \phi_{k,l+1} + 4\phi_{k,l}}{h^2} + \mu\phi_{k,l} = f_{k,l} \quad \text{in} \quad D_h,$$

$$\phi_{k,l} = 0 \quad \text{on} \quad \partial D_h, \tag{7.9}$$

$$0 \leq k \leq \frac{1}{h} = N, \quad 0 \leq l \leq \frac{1}{h} = N.$$

If $\mu \geqq 0$, then (7.8), (7.9) has a unique solution. In case $\mu < 0$ one has to place additional requirements on μ and f in order to have the existence.

Let us suppose that (7.8) and (7.9) can be solved uniquely. Write (7.9) in the matrix form:

$$(\Lambda_1 + \Lambda_2 + \mu E)\phi = f, \tag{7.10}$$

where

$$\phi = \begin{Vmatrix} \phi_1 \\ \vdots \\ \phi_{N-1} \end{Vmatrix}, \quad \phi_l = \begin{Vmatrix} \phi_{1,l} \\ \vdots \\ \phi_{N-1,l} \end{Vmatrix}, \quad \Lambda_1 = \frac{1}{h^2} \begin{Vmatrix} A & 0 & \cdots & 0 \\ 0 & A & \cdots & 0 \\ \cdots & \cdots & \cdots & \cdots \\ 0 & 0 & \cdots & A \end{Vmatrix},$$

$$\Lambda_2 = \frac{1}{h^2} \begin{Vmatrix} 2E & -E & 0 & \cdots & 0 & 0 \\ -E & 2E & -E & \cdots & 0 & 0 \\ \cdots & \cdots & \cdots & \cdots & \cdots & \cdots \\ 0 & 0 & 0 & \cdots & -E & 2E \end{Vmatrix},$$

$$f = \begin{Vmatrix} f_1 \\ \vdots \\ f_{N-1} \end{Vmatrix}, \quad f_l = \begin{Vmatrix} f_{1,l} \\ \vdots \\ f_{N-1,l} \end{Vmatrix},$$

$$A = \begin{Vmatrix} 2 & -1 & 0 & \cdots & 0 & 0 \\ -1 & 2 & -1 & \cdots & 0 & 0 \\ \cdots & \cdots & \cdots & \cdots & \cdots & \cdots \\ 0 & 0 & 0 & \cdots & -1 & 2 \end{Vmatrix},$$

$$E - \begin{Vmatrix} 1 & 0 & 0 & \cdots & 0 & 0 \\ 0 & 1 & 0 & \cdots & 0 & 0 \\ \cdots & \cdots & \cdots & \cdots & \cdots & \cdots \\ 0 & 0 & 0 & \cdots & 0 & 1 \end{Vmatrix}.$$

Equation (7.10) is derived in detail in Section 6.1.3. The dimension of the matrices A and E as well as that of the block matrices Λ_1, Λ_2 is equal to $N - 1$. The vector ϕ_l has for its components the elements $\phi_{k,l}$ in one row, i.e. for fixed $y = lh$.

The basic idea of Hockney [13] is as follows: Assuming N is even, we find first the solutions ϕ_l involving the even l only. This is done using the system of equations for ϕ_l, l even, obtained from (7.10) by simple manipulations. The solutions ϕ_l with l odd are found by solving the same system, by expanding into a Fourier series (using the fast Fourier transform). Indeed, having computed ϕ_l with l even we form the following system

$$B\phi_l = h^2 f_l + \phi_{l-1} + \phi_{l+1}, \qquad l = 1, 3, \ldots, N-1, \qquad (7.11)$$

where

$$B = A + (2 + \mu h^2)E.$$

For the notational simplicity we did not exclude the cases $l = 1$ and $l = N - 1$, assuming that $\phi_0 = \phi_N = 0$.

In order to complete the description of Hockney's method, it remains to obtain the system of equations with even-indexed ϕ_l. To this end let us write three consecutive matrix equations (7.11) as follows:

$$-\phi_{l-2} + B\phi_{l-1} - \phi_l = h^2 f_{l-1},$$
$$-\phi_{l-1} + B\phi_l - \phi_{l+1} = h^2 f_l, \qquad l = 4, 6, \ldots, N-4,$$
$$-\phi_l + B\phi_{l+1} - \phi_{l+2} = h^2 f_{l+1}.$$

Multiplying the second equation above by the three-diagonal matrix B and adding the result to the remaining two, we obtain

$$-\phi_{l-2} + (B^2 - 2E)\phi_l - \phi_{l+2} = g_l,$$

where

$$g_l = h^2(f_{l-1} + Bf_l + f_{l+1}).$$

Similarly one can obtain equations relating ϕ_2 with ϕ_4, and ϕ_{N-4} with ϕ_{N-2}. Hence we arrive at the system

$$(B^2 - 2E)\phi_2 - \phi_4 = g_2$$
$$-\phi_{l-1} + (B^2 - 2E)\phi_l - \phi_{l+2} = g_l, \qquad l = 4, 6, \ldots, N-4, \qquad (7.12)$$
$$-\phi_{N-4} + (B^2 - 2E)\phi_{N-2} = g_{N-2}.$$

Let us note that $(B^2 - 2E)$ is a five-diagonal matrix, sharing the eigenvector basis with B and A.

A complete solution of the eigenvalue problem

$$Au^{(m)} = \lambda_m(A)u^{(m)}$$

has the form

$$\lambda_m(A) = 2\left(1 - \cos\frac{m\pi}{N}\right), \qquad u_k^{(m)} = \sqrt{\frac{2}{N}} \sin\frac{m\pi k}{N},$$

where $u_k^{(m)}$ is the kth component of the eigenvector $u^{(m)}$, $m = 1, \ldots, N-1$. $k = 1, \ldots, N-1$. Note that the multiplicative factor $\sqrt{2/N}$ has been introduced in order to have

$$\|u^{(m)}\| = \sqrt{\sum_{k=1}^{N-1} (u_k^{(m)})^2} = 1.$$

Thus

$$\lambda_m(B) = \lambda_m(A) + (2 + \mu h^2) = 2\left(2 - \cos\frac{m\pi}{N}\right) + \frac{\mu}{N^2},$$

$$\lambda_m = \lambda_m(B^2 - 2E) = [\lambda_m(B)]^2 - 2 = \left[2\left(2 - \cos\frac{m\pi}{N}\right) + \frac{\mu}{N^2}\right]^2 - 2. \quad (7.13)$$

Since the vectors $u^{(m)}$ form an orthonormal basis in the $(N - 1)$-dimensional space, the vectors ϕ_l and g_l ($l = 2, 4, \ldots, N - 2$) may be represented as

$$\phi_l = \sum_{m=1}^{N-1} \Phi_{m,l} u^{(m)},$$

$$g_l = \sum_{m=1}^{N-1} G_{m,l} u^{(m)}. \quad (7.14)$$

Substituting these expressions into (7.12) and multiplying both sides of the system by u^m, we obtain the following system with a three-dimensional matrix for each fixed m:

$$\lambda_m \Phi_{m,2} - \Phi_{m,4} = G_{m,2},$$
$$-\Phi_{m,l-2} + \lambda_m \Phi_{m,l} - \Phi_{m,l+2} = G_{m,l}, \quad l = 2, 4, \ldots, N - 4, \quad (7.15)$$
$$-\Phi_{m,N-4} + \lambda_m \Phi_{m,N-2} = G_{m,N-2}.$$

where

$$\lambda_m = \lambda_m(B^2 - 2E).$$

The algorithm for solving directly the Helmholtz problem, as we have described above, is applicable not only to the Dirichlet conditions but also to Neumann boundary conditions and the periodicity conditions of $\phi(x, y)$ on the boundary of the square for equations of the form

$$a(y)\frac{\partial^2 \phi}{\partial x^2} + \frac{\partial}{\partial y}\left(b(y)\frac{\partial \phi}{\partial y}\right) - \mu(y)\phi = f(x, y).$$

The method of the fast Fourier transform can be conveniently combined with the successive approximation method whenever a problem of mathematical physics is to be solved via difference methods. To see this, consider

$$A\phi = f \quad \text{in} \quad D,$$
$$\phi = 0 \quad \text{on} \quad \partial D, \quad (7.16)$$

where $A > 0$ is a differential operator with a real spectrum such that $\alpha(A) < \lambda(A)$. With the help of difference approximations we transform the problem of (7.16) into a problem of linear algebra:

$$A^h \phi^h = f^h \quad \text{in} \quad D_h,$$
$$\phi^h = 0 \quad \text{on} \quad \partial D_h, \quad (7.17)$$

with $A^h > 0$, the spectrum $\lambda(A^h)$ being real and

$$\alpha(A^h) \leq \lambda(A^h) \leq \beta(A^h).$$

Let us now imbed the region D_h along with its boundary ∂D_h into a rectangle E_h in such a way that the ratio of volumes Vol D_h/Vol E_h is as large as possible. With the rectangle E_h so defined let us consider the spectral problem

$$\begin{aligned} -\Delta^h u &= \lambda u \quad \text{in} \quad E_h, \\ u &= 0 \quad \text{on} \quad \partial E_h, \end{aligned} \tag{7.18}$$

where Δ^h is the difference analog of the Laplace operator.

Assume now that $\alpha(A^h)$, $\beta(A^h)$ have been found by methods considered in Section 1.1 and that $\alpha(-\Delta^h)$, $\beta(-\Delta^h)$ have been obtained by Fourier analysis. With those at hand we construct the iterative process

$$\phi^{j+1} = \phi^j - \tau[B^h]^{-1}(A^h \phi^j - f), \tag{7.19}$$

where

$$B^h = aE - b\Delta^h. \tag{7.20}$$

The requirement that the spectral boundaries of A^h and B^h coincide (or, following the terminology of Dyakonov, that the two operators are *spectral equivalent*) implies

$$a = \frac{\alpha(A^h)\beta(-\Delta^h) - \beta(A^h)\alpha(-\Delta^h)}{\beta(-\Delta^h) - \alpha(-\Delta^h)}; \qquad b = \frac{\beta(A^h) - \alpha(A^h)}{\beta(-\Delta^h) - \alpha(-\Delta^h)}. \tag{7.21}$$

Let us extend the domain of definition of the solution of (7.17) from D_h onto E_h in such a manner that the solution becomes zero outside D_h. The iterative process can then be put into the following two stages. First, with a ϕ^j fixed and zero outside D_h along with f, we find the residual

$$\xi^j = \begin{cases} A^h \phi^j - f & \text{in} \quad D_h, \\ 0 & \text{in} \quad E_h \setminus D_h. \end{cases} \tag{7.22}$$

At the second stage we solve the equation

$$(aE - b\Delta^h) r^{j+1} = -\xi^j, \tag{7.23}$$

where

$$r^{j+1} = \phi^{j+1} - \phi^j. \tag{7.24}$$

Equation (7.23) can be solved by the fast Fourier transform. Since the spectral boundaries of A^h and B^h coincide, the spectral components in the residual will be eliminated rapidly.

We will not venture into the foundations of this algorithm. Interesting related convergence results as applied to the Helmholtz equation can be found in the contribution by Kuznetsov and Matzokin [13].

3.8. Factorization of Difference Equations

There is much mathematical literature devoted to methods for solving equations in finite differences. One of the methods, the so-called *factorization* method has turned out to be very efficient in the particular case of one-dimensional difference equations. This method has been described with sufficient completeness by Vladimirov [12], Gelfand and Lokutzievskii [12], Godunov [12], Abramov and Andreyev [12], Aintz [12], Richtmyer [3], and others. The factorization method as applied to net equations is essentially a kind of realization of the Gauss elimination method.

A further development of the method, related to systems of ordinary differential equations of the first order with arbitrary linear constraints (including point conditions and boundary conditions), can be found in Fage [12].

The essence of the factorization method is as follows: Let

$$A\phi = f, \tag{8.1}$$

where A is a linear operator. For the sake of simplicity let us assume that the operator in (8.1) is in fact a matrix, so that we actually deal with a system of linear algebraic equations. Write $A = A_1 A_2$. In such a case the problem of (8.1) is split into two successive problems, namely

$$A_1 z = f, \quad A_2 \phi = z. \tag{8.2}$$

The factorization method becomes especially efficient if A_1 and A_2 turn out to be a lower and upper triangular matrices respectively. The decomposition of A into simpler matrices thus facilitates the transition from (8.1) into the two simpler problems of (8.2). A formal solution of (8.2) can be written as

$$\phi = A_2^{-1} A_1^{-1} f. \tag{8.3}$$

Of course, for solvability of (8.2) we need that A_1^{-1} and A_2^{-1} be bounded operators.

The idea of factorization supports many direct methods of linear algebra. Leaving the general problem aside, we focus our attention on a simple problem which is frequently encountered in applications, namely when A is a three-diagonal matrix. In this case the equations of (8.1) take the form

$$a_k \phi_{k-1} - b_k \phi_k + c_k \phi_{k+1} = f_k \quad (k = 1, 2, \ldots, n) \tag{8.4}$$

where

$$a_1 = 0, \quad c_n = 0. \tag{8.5}$$

Assume that the coefficients a_k, b_k, and c_k satisfy the conditions

$$a_k > 0, \quad b_k > 0, \quad c_k > 0, \quad a_k + c_k \leqslant b_k, \tag{8.6}$$

where the last inequality holds at least for one index k.

It is not difficult to verify that the factorization method of (8.2) as applied to (8.4) yields the formulas

$$\beta_1 = 0, \qquad \beta_{k+1} = \frac{c_k}{b_k - a_k \beta_k} \qquad (k = 1, \ldots, n-1), \tag{8.7}$$

$$z_1 = 0, \qquad z_{k+1} = \frac{a_k z_k - f_k}{b_k - a_k \beta_k} \qquad (k = 1, \ldots, n), \tag{8.8}$$

followed by

$$\phi_n = z_{n+1}, \tag{8.9}$$

and

$$\phi_k = \beta_{k+1}\phi_{k+1} + z_{k+1} \qquad (k = n-1, \ldots, 1). \tag{8.10}$$

Godunov and Ryabenkii [3] have shown that the conditions of (8.6) are sufficient for the countable stability of (8.7)–(8.10). Thus there will be no increase in round-off errors in the course of the computations.

In conclusion let us note that many problems of mathematical physics lead eventually to a difference equation of the following type:

$$a_k \phi_{k-1} - b_k \phi_k + c_k \phi_{k+1} = f_k,$$
$$a_1 = 0, \qquad c_n = 0, \tag{8.11}$$

where a_k, b_k, and c_k are square matrices of the order r, and ϕ_k and f_k are vectors. In this case we again encounter Problem (8.1) with a three-diagonal block-matrix A. A similar development as above results in the following triples of difference vector-matrix equations with a factorized operator:

$$\beta_{k+1} = (b_k - a_k \beta_k)^{-1} c_k,$$
$$z_{k+1} = (b_k - a_k \beta_k)^{-1}(a_k z_k - f_k), \tag{8.12}$$
$$\phi_k = \beta_{k+1}\phi_{k+1} + z_{k+1}$$

where

$$z_1 = \beta_1 = 0, \qquad a_1 = 0, \qquad c_n = 0, \qquad \phi_n = z_{n+1}. \tag{8.13}$$

The factorization method for the matrix differential equation (8.11) has become known as a *matrix factorization* method. It is efficient in solving boundary value problems for ordinary differential equations and certain partial differential equations. In particular using the matrix approach described above, the method can be easily formulated for more complicated difference equations. As an example, consider the system with a five-diagonal matrix:

$$a_k \phi_{k-2} + b_k \phi_{k-1} + c_k \phi_k + d_k \phi_{k+1} + e_k \phi_{k+2} = f_k$$
$$(k = 1, 2, \ldots, n),$$
$$a_1 = b_1 = a_2 = e_{n-1} = e_{n-2} = d_{n-2} = 0.$$

Factorization of Difference Equations

Write the solution in the form
$$\phi_k = \alpha_k \phi_{k+1} + \beta_k \phi_{k+2} + z_k.$$
The coefficients α_k, β_k, and z_k are then computed according to the formulas
$$\alpha_k = -\frac{d_k + a_k \alpha_{k-2} \beta_{k-1} + b_k \beta_{k-1}}{c_k + a_k \alpha_{k-2} \alpha_{k-1} + a_k \beta_{k-2} + b_k \alpha_{k-1}},$$
$$\beta_k = -\frac{e_k}{c_k + a_k \alpha_{k-2} \alpha_{k-1} + a_k \beta_{k-2} + b_k \alpha_{k-1}},$$
$$z_k = \frac{f_k - a_k \alpha_{k-2} z_{k-1} - a_k z_{k-2} - b_k z_{k-1}}{c_k + a_k \alpha_{k-2} \alpha_{k-1} + a_k \beta_{k-2} + b_k \alpha_{k-1}}.$$

The matrix factorization method has found broad applications in gas dynamics and the theory of wave propagation (Babenko and Rusanov [12], Chentzov and Marchuk [17], and others).

Chapter 4

Methods for Solving Nonstationary Problems

The main object to be considered is the following evolution problem of mathematical physics:

$$\frac{\partial \phi}{\partial t} + A\phi = f \quad \text{in} \quad D \times D_t, \qquad \phi = g \quad \text{in} \quad D \quad \text{at} \quad t = 0,$$

where $A \geq 0$, and the solution ϕ as well as the functions f and g are assumed sufficiently smooth. In addition, the solution of the problem is further required to satisfy certain boundary conditions on ∂D (the boundary of the region D).

4.1. Second-Order-Approximation Difference Schemes with Time-Varying Operators

Consider the evolution equation

$$\frac{\partial \phi}{\partial t} + A\phi = 0 \quad \text{in} \quad D \times D_t, \qquad \phi = g \quad \text{in} \quad D \quad \text{at} \quad t = 0. \quad (1.1)$$

The corresponding difference equation is taken in the form

$$\frac{\phi^{j+1} - \phi^j}{\tau} + A \frac{\phi^{j+1} + \phi^j}{2} = 0, \qquad \phi^0 = g. \quad (1.2)$$

Provided the solution is sufficiently smooth it is not difficult to verify that (1.2) approximates Problem (1.1) with second-order accuracy with respect to τ. The difference Scheme (1.2) is usually referred to as a *central differences* scheme (in time), or the *Crank–Nicholson* scheme. It is interesting to note that (1.2) is obtained by applying alternately the first-order accuracy schemes (explicit or implicit) which are considered on the intervals $t_j \leq t \leq t_{j+1/2}$ and $t_{j+1/2} \leq t \leq t_{j+1}$ correspondingly, the linear operator A being assumed independent of t:

$$\begin{aligned}\frac{\phi^{j+1/2} - \phi^j}{\tau/2} + A\phi^j &= 0, \\ \frac{\phi^{j+1} - \phi^{j+1/2}}{\tau/2} + A\phi^{j+1} &= 0.\end{aligned} \quad (1.3)$$

The Crank–Nicholson scheme is obtained from this by eliminating the unknowns $\phi^{j+1/2}$.

Second-Order-Approximation Difference Schemes

Assume now that the operator A depends on time and consider Equation (1.1) with A replaced by the approximating difference operator denoted by Λ. Solutions ϕ of the problem will be interpreted in this case as vector functions whose components are the values of the approximating solutions at the net points of the space. Thus we have the problem of linear algebra

$$\frac{\phi^{j+1} - \phi^j}{\tau} + \Lambda^j \frac{\phi^{j+1} + \phi^j}{2} = 0, \qquad \phi^0 = g, \tag{1.4}$$

$$(\Lambda^j \phi, \phi) \geq 0 \tag{1.5}$$

for any function from the subspace Φ.

Solving for ϕ^{j+1} in Equation (1.4) we obtain

$$\phi^{j+1} = \left(E + \frac{\tau}{2}\Lambda^j\right)^{-1}\left(E - \frac{\tau}{2}\Lambda^j\right)\phi^j, \tag{1.6}$$

or

$$\phi^{j+1} = T^j \phi^j, \tag{1.7}$$

where T^j is the transition operator:

$$T^j = \left(E + \frac{\tau \Lambda^j}{2}\right)^{-1}\left(E - \frac{\tau \Lambda^j}{2}\right). \tag{1.8}$$

For proof of countable stability we do not necessarily need to evaluate the norms of the transition operators T^j. Indeed, taking the inner product of Equation (1.4) with $(\phi^{j+1} + \phi^j)/2$, we obtain

$$\frac{(\phi^{j+1}, \phi^{j+1}) - (\phi^j, \phi^j)}{2\tau} + \left(\Lambda^j \frac{\phi^{j+1} + \phi^j}{2\tau}, \frac{\phi^{j+1} + \phi^j}{2}\right) = 0. \tag{1.9}$$

Since Λ^j is positive semidefinite by assumption [see Equation (1.5)], we have

$$\|\phi^{j+1}\| \leq \|\phi^j\|, \tag{1.10}$$

i.e., stability.

By no means does this lessen the importance of norm estimates of transitions operators in the analysis of difference schemes. In the present case such an estimate can be obtained with the help of Relation (1.10):

$$\|T^j\| \leq 1. \tag{1.11}$$

Using (1.8) and the relations $\Lambda^j \geq 0$, $\tau > 0$, we note that the above inequality is also an immediate consequence of the Kellogg lemma [see Equation (1.1)]. If Λ^j is antisymmetric, that is if

$$(\Lambda^j \phi, \phi) = 0,$$

then

$$\|\phi^{j+1}\| = \|\phi^j\|, \tag{1.12}$$

or we have a strict equality in (1.10).

One can show, using a procedure similar to the one above, that in this case

$$\|T^j\| = 1. \tag{1.13}$$

Next let us discuss the Crank–Nicholson approximation scheme for the case where A changes with time. Define the operators H and H^j by

$$H\phi \equiv \frac{\partial \phi}{\partial t} + A\phi, \tag{1.14}$$

$$H^j_\tau(\phi)^j = \frac{(\phi)^{j+1} - (\phi)^j}{\tau} + \Lambda^j \frac{(\phi)^{j+1} + (\phi)^j}{2}, \tag{1.15}$$

where $(\phi)^j$ denotes the projection of the exact solution of (1.1) on the net D_τ. The following norm is convenient for evaluating the approximation of the operator H:

$$\|(H\phi)^j - H^j_\tau(\phi)^j\|_{c_\tau} = \|H^j_\tau(\phi)^j\|_{c_\tau} = \max_{t_j} \|H^j_\tau(\phi)^j\|, \tag{1.16}$$

where $\|\cdot\|$ is the norm on the net elements (for $t = t_j$). In order to estimate the norm (1.16) we first expand the solution of the original Equation (1.1) into the Taylor series:

$$(\phi)^{j+1} = (\phi)^j + \tau(\phi_t)^j + \frac{\tau^2}{2}(\phi_{tt})^j + \cdots. \tag{1.17}$$

Note the immediate relations

$$\phi_t = -A\phi, \quad \phi_{tt} = A^2\phi - A_t\phi, \tag{1.18}$$

where we use the convention $A_t \equiv \partial A/\partial t$. With the above in mind, the Taylor series (1.17) can be rewritten as

$$(\phi)^{j+1} = (\phi)^j - \tau A^j(\phi)^j + \frac{\tau^2}{2}[(A^j)^2(\phi)^j - A^j_t(\phi)^j] - \cdots. \tag{1.19}$$

Substituting (1.19) into (1.16), and noting (1.15) we obtain

$$\|(H\phi)^j - H^j_\tau(\phi)^j\|_{c_\tau} = \max_{t_j} \left\| \Lambda^j(\phi)^j - A^j(\phi)^j \right.$$
$$\left. + \frac{\tau}{2}\{(A^j)^2 - A^j_t - \Lambda^j A^j\}(\phi)^j + O(\tau^2) \right\|. \tag{1.20}$$

If we take

$$\Lambda^j = A^j = A(t_j) \tag{1.21}$$

for the approximating operator Λ^j, then (1.20) implies

$$\|(H\phi)^j - H^j_\tau(\phi)^j\|_{c_\tau} = \frac{\tau}{2} \max_{t_j \in D_\tau} \|A^j_t(\phi)^j\| + O(\tau^2),$$

and the approximation is first order. Note that in the particular case of a time-invariant A the approximation (1.21) guarantees second order with respect to τ on the class of sufficiently smooth functions.

Assume now that the approximating operator Λ^j is chosen in the form

$$\Lambda^j = A^j + \frac{\tau}{2} A_t^j. \tag{1.22}$$

In this case

$$\|(H\phi)^j - H_\tau^j(\phi)^j\|_{c_\tau} = O(\tau^2).$$

Let us note that the second-order approximation of the Crank–Nicholson scheme is also obtained by choosing the operator Λ^j in the form

$$\Lambda^j = A^{j+1/2}, \tag{1.23}$$

or

$$\Lambda^j = \tfrac{1}{2}(A^{j+1} + A^j). \tag{1.24}$$

The forms (1.22), (1.23), and (1.24) of the second-order approximation of the operator A are used in many applications, in particular for numerical integration of quasilinear equations.

4.2. Nonhomogeneous Equations of the Evolution Type

The foregoing paragraph dealt with homogeneous equations. Now we will consider the nonhomogeneous equation

$$\frac{\partial \phi}{\partial t} + A\phi = f \quad \text{in} \quad D \times D_t, \tag{2.1}$$

$$\phi = g \quad \text{in} \quad D \quad \text{for} \quad t = 0.$$

Assuming the hypothesis discussed in Section 4.1, the difference approximation for (2.1) based on the Crank–Nicholson scheme has the form

$$\frac{\phi^{j+1} - \phi^j}{\tau} + \Lambda^j \frac{\phi^{j+1} + \phi^j}{2} = f^j, \qquad \phi^0 = g, \tag{2.2}$$

where

$$f^j = f(t_{j+1/2}).$$

It is easy to see that the difference problem of (2.2) is the second-order approximation in τ of (2.1). On each interval let us write the formal solution of (2.2) as

$$\phi^{j+1} = T^j \phi^j + \tau \left(E + \frac{\tau}{2} \Lambda^j\right)^{-1} f^j. \tag{2.3}$$

In the case of the homogeneous equation it was shown in Section 4.1 that for $\Lambda^j \geq 0$ the following estimate holds:

$$\|T^j\| \leq 1. \tag{2.4}$$

Naturally, this norm estimate is independent of the right-hand side f. From (2.3) it follows that

$$\|\phi^{j+1}\| \leq \|T^j\| \|\phi^j\| + \tau \left\|\left(E + \frac{\tau}{2}\Lambda^j\right)^{-1}\right\| \|f^j\|. \tag{2.5}$$

In order to establish stability we exploit the estimate of Equation (1.25) from Chapter 1. Since $\tau > 0$ and

$$(\Lambda^j \phi^j, \phi^j) \geq 0, \tag{2.6}$$

we obtain

$$\left\|\left(E + \frac{\tau}{2}\Lambda^j\right)^{-1}\right\| \leq 1, \tag{2.7}$$

and hence inequality (2.5) becomes

$$\|\phi^{j+1}\| \leq \|\phi^j\| + \tau \|f^j\|, \tag{2.8}$$

using (2.4) and (2.7). Setting $\|\phi^0\| = \|g\|$ and $\|f\| = \max_j \|f_j\|$, the recursive Relation (2.8) yields

$$\|\phi^j\| \leq \|g\| + j\tau \|f\|, \quad j\tau \leq \text{const.} \tag{2.9}$$

The stability of the difference scheme is thus established. Furthermore, Relation (2.9) is an *a priori* norm estimate of the solution.

4.3. Splitting-Up Methods for Nonstationary Problems

Complicated problems of mathematical physics can often be reduced to those consisting of a chain of simpler problems which can be effectively solved with a computer. This kind of reduction is possible in the cases where the original positive semidefinite operator characterizing the problem can be decomposed into a sum of positive semidefinite operators with a simple structure. Such methods will be referred to as *splitting-up* methods. Splitting-up methods were initiated by Douglas, Peaceman and Rachford [15], and then further developed by Soviet mathematicians Bagrinovskii and Godunov [15], Yanenko [3, 15], Samarskii [3, 15], Dyakonov [3, 15], Saulyev [3], and Marchuk [15].

Splitting-up methods were originally formulated and theoretically justified for simple problems with commuting positive-definite operators. It has become clear that the various splitting-up methods as introduced by various authors for these kind of problems are either essentially equivalent (the only difference being the actual realization of the scheme), or otherwise quite similar. For this reason we will not personalize any of these by now classical methods.

The circle of nontrivial applications of splitting-up methods has been considerably enlarged. At present the splitting-up methods have become a powerful tool for solving highly complicated problems of mathematical physics. The most complete theory has been given for the case when the operator corresponding to the original problem has a representation as a sum of two simpler operators. We will begin our exposition with this latter case. In our opinion the most important method for applications is the *component-by-component splitting-up* method. We hope that the reader takes notice of this claim while reading the present chapter.

Thus consider the evolution equation

$$\frac{\partial \phi}{\partial t} + A\phi = f \quad \text{in} \quad D \times D_t, \quad (3.1)$$

$$\phi = g \quad \text{in} \quad D \quad \text{for} \quad t = 0,$$

where the operator $A \geq 0$ does not depend on time and has the representation

$$A = A_1 + A_2, \quad (3.2)$$

$$A_1 \geq 0, A_2 \geq 0. \quad (3.3)$$

Assume further that the solution of (3.1) is sufficiently smooth. Wherever it is necessary for a proof we will assume that (3.1) is already reduced to a difference form and therefore A, A_1, and A_2 are matrices.

4.3.1. The Stabilization Method

Let us consider first the approximate solution of Equations (3.1)–(3.3) under the condition that $f = 0$:

$$\left(E + \frac{\tau}{2} A_1\right)\left(E + \frac{\tau}{2} A_2\right)\frac{\phi^{j+1} - \phi^j}{\tau} + A\phi^j = 0, \quad \phi^0 = g. \quad (3.4)$$

It is not difficult to see that under sufficient smoothness assumptions Equations (3.4) approximate the original problem expressed in Equations (3.1)–(3.3) up to the second order of accuracy in τ. Indeed, with the help of some algebra, Equations (3.4) can be rewritten as

$$\left(E + \frac{\tau^2}{4} A_1 A_2\right)\frac{\phi^{j+1} - \phi^j}{\tau} + A\frac{\phi^{j+1} + \phi^j}{2} = 0, \quad \phi^0 = g. \quad (3.5)$$

From this we see that the difference Equation (3.5) coincides in approximation order with the Crank–Nicholson scheme

$$\frac{\phi^{j+1} - \phi^j}{\tau} + A\frac{\phi^{j+1} + \phi^j}{2} = 0, \quad \phi^0 = g, \quad (3.6)$$

provided that the solution is smooth enough. This conclusion follows from the fact that (3.6) itself is a second-order approximation in τ, and that (3.5) and (3.6) are equivalent.

Let us turn next to the stability analysis of (3.4). To this end rewrite this equation in the form

$$\left(E + \frac{\tau}{2}A_1\right)\left(E + \frac{\tau}{2}A_2\right)\phi^{j+1} = \left(E - \frac{\tau}{2}A_1\right)\left(E - \frac{\tau}{2}A_2\right)\phi^j, \quad \phi^0 = g. \tag{3.7}$$

Solving for ϕ^{j+1}, we obtain

$$\phi^{j+1} = \left(E + \frac{\tau}{2}A_2\right)^{-1}\left(E + \frac{\tau}{2}A_1\right)^{-1}\left(E - \frac{\tau}{2}A_1\right)\left(E - \frac{\tau}{2}A_2\right)\phi^j. \tag{3.8}$$

For each unknown ϕ^j define a new unknown ψ^j by the relation

$$\psi^j = \left(E + \frac{\tau}{2}A_2\right)\phi^j. \tag{3.9}$$

Thus the new variable ψ^j satisfies

$$\psi^{j+1} = T\psi^j \tag{3.10}$$

where the transition operator T is given by

$$T = \left(E + \frac{\tau}{2}A_1\right)^{-1}\left(E - \frac{\tau}{2}A_1\right)\left(E - \frac{\tau}{2}A_2\right)\left(E + \frac{\tau}{2}A_2\right)^{-1}. \tag{3.11}$$

With the help of (3.10) we obtain the following estimate in the energy norm:

$$\|\psi^{j+1}\| \leq \|T\|\|\psi^j\|. \tag{3.12}$$

Let us estimate the norm of T:

$$\|T\| \leq \|T_1\|\|T_2\|,$$

where

$$T_\alpha = \left(E - \frac{\tau}{2}A_\alpha\right)\left(E + \frac{\tau}{2}A_\alpha\right)^{-1}, \quad (\alpha = 1, 2). \tag{3.13}$$

Here use was made of the relationship

$$\left(E + \frac{\tau}{2}A_\alpha\right)^{-1}\left(E - \frac{\tau}{2}A_\alpha\right) = \left(E - \frac{\tau}{2}A_\alpha\right)\left(E + \frac{\tau}{2}A_\alpha\right)^{-1},$$

which follows from the immediate identity

$$\left(E - \frac{\tau}{2}A_\alpha\right)^{-1}\left(E - \frac{\tau}{2}A_\alpha\right) = \left(E + \frac{\tau}{2}A_\alpha\right)\left(E + \frac{\tau}{2}A_\alpha\right)^{-1}. \tag{3.14}$$

Splitting-Up Methods for Nonstationary Problems

Indeed, multiplying (3.14) by $[E + (\tau/2)A_\alpha]^{-1} \times [E - (\tau/2)A_\alpha]$ from the left and using the fact that the operators $[E - (\tau/2)A_\alpha]$ and $[E + (\tau/2)A_\alpha]$ commute (direct verification) gives the claimed property.

Thus the problem of stability has been reduced to estimating the norms of T_α.

Application of Kellogg's lemma for estimating $\|T_1\|$ and $\|T_2\|$ in (3.13) yields

$$\|T\| \leq 1 \qquad (3.15)$$

and consequently

$$\|\psi^{j+1}\| \leq \|\psi^j\|. \qquad (3.16)$$

Our final goal however is to establish the stability of the original difference problem (3.4). For that we rewrite relation (3.16) in the form

$$\left\|\left(E + \frac{\tau}{2} A_2\right)\phi^{j+1}\right\| \leq \left\|\left(E + \frac{\tau}{2} A_2\right)\phi^j\right\|. \qquad (3.17)$$

Introducing the notation

$$\left\|\left(E + \frac{\tau}{2} A_2\right)\phi\right\| = (C_2 \phi, \phi)^{1/2} = \|\phi\|_{C_2}, \qquad (3.18)$$

where

$$C_\alpha = \left(E + \frac{\tau}{2} A_\alpha^*\right)\left(E + \frac{\tau}{2} A_\alpha\right), \qquad \alpha = 1, 2,$$

it is easy to see that $C_\alpha > 0$ and that $\|\cdot\|_{C_2}$ is indeed a norm.

Hence it follows that in the given norm we have absolute stability:

$$\|\phi^{j+1}\|_{C_2} \leq \|\phi^j\|_{C_2}. \qquad (3.19)$$

We thus conclude that if $A_1 \geq 0$ and $A_2 \geq 0$, and if the entries of these matrices are independent of time, then, assuming the solution of Problem (3.1) is smooth enough, the difference scheme is absolutely stable and has a solution of second-order accuracy in τ.

In conclusion let us point out that the difference scheme of the stabilization method can be conveniently implemented on a computer, For this purpose the difference equation is put in the form

$$F^j = A\phi^j$$

$$\left(E + \frac{\tau}{2} A_1\right)\xi^{j+1/2} = -F^j,$$

$$\left(E + \frac{\tau}{2} A_2\right)\xi^{j+1} = \xi^{j+1/2}, \qquad (3.20)$$

$$\phi^{j+1} = \phi^j - \tau\xi^{j+1}.$$

Here $\xi^{j+1/2}$ and ξ^{j+1} are certain auxiliary variables which facilitate the reduction of (3.4) to the sequence of simpler Problems (3.20). Note that the first and last relations in (3.20) are explicit. It means that the operator inversion is only needed in the second and third equations which involve the simple operators A_1 and A_2.

Consider next the nonhomogeneous problem

$$\frac{\partial \phi}{\partial t} + A\phi = f, \qquad \phi = g \quad \text{for} \quad t = 0, \tag{3.21}$$

where $A = A_1 + A_2$, $A_1 \geq 0$, $A_2 \geq 0$. In this case the stabilization method scheme is written as follows:

$$\left(E + \frac{\tau}{2} A_1\right)\left(E + \frac{\tau}{2} A_2\right)\frac{\phi^{j+1} - \phi^j}{\tau} + A\phi^j = f^j, \qquad \phi^0 = g, \tag{3.22}$$

where

$$f^j = f(t_{j+1/2}). \tag{3.23}$$

It can be shown that under the assumption of (3.23) the difference Problem (3.22) approximates the original Problem (3.21) up to the second order in τ.

Let us investigate the stability of the difference scheme. To this end we write Equation (3.22) in the form

$$\psi^{j+1} = T\psi^j + \tau\left(E + \frac{\tau}{2} A_1\right)^{-1} f^j, \tag{3.24}$$

where

$$\psi^j = \left(E + \frac{\tau}{2} A_2\right)\phi^j. \tag{3.25}$$

From (3.24) it follows that

$$\|\psi^{j+1}\| \leq \|T\|\|\psi^j\| + \tau\left\|\left(E + \frac{\tau}{2} A_1\right)^{-1}\right\|\|f^j\|. \tag{3.26}$$

Since it has already been established that for a homogeneous equation

$$\|T\| \leq 1,$$

it follows that

$$\|\psi^{j+1}\| \leq \|\psi^j\| + \tau\left\|\left(E + \frac{\tau}{2} A_1\right)^{-1}\right\|\|f^j\|. \tag{3.27}$$

A simple manipulation yields

$$\|f^j\| = \left\|\left(E + \frac{\tau}{2} A_2\right)^{-1}\left(E + \frac{\tau}{2} A_2\right)f^j\right\| \leq \left\|\left(E + \frac{\tau}{2} A_2\right)^{-1}\right\|\left\|\left(E + \frac{\tau}{2} A_2\right)f^j\right\|. \tag{3.28}$$

Taking (3.18), (3.25), and (3.28) into account, (3.27) implies

$$\|\phi^{j+1}\|_{C_2} \leq \|\phi^j\|_{C_2} + \tau \left\|\left(E + \frac{\tau}{2} A_1\right)^{-1}\right\| \left\|\left(E + \frac{\tau}{2} A_2\right)^{-1}\right\| \|f^j\|_{C_2}. \quad (3.29)$$

Furthermore, making use of the norm estimate

$$\left\|\left(E + \frac{\tau}{2} A_\alpha\right)^{-1}\right\| \leq 1, \quad (3.30)$$

and the inequality $A_\alpha \geq 0$ [established in (1.25); (see Chapter 1)], we arrive at

$$\|\phi^{j+1}\|_{C_2} \leq \|\phi^j\|_{C_2} + \tau \|f^j\|_{C_2}. \quad (3.31)$$

With the help of recursive relations we obtain the estimate

$$\|\phi^j\|_{C_2} \leq \|g\|_{C_2} + j\tau \|f\|_{C_2}, \quad (3.32)$$

where

$$\|f\|_{C_2} = \max_j \|f^j\|_{C_2}. \quad (3.33)$$

Therefore, if the matrices A_1, A_2 are nonnegative and their entries are independent of time, then, under the assumption of sufficient smoothness of the solution ϕ of (3.1) and of the function f, the difference Scheme (3.22) is absolutely stable and its solution is of second-order accuracy in τ.

In conclusion let us point out once more that the proof given above holds if and only if the original operator A is time invariant.

4.3.2. The Predictor-Corrector Method

Let us discuss a splitting method known as the *predictor-corrector* method. The essence of the method is to decompose the whole interval $0 \leq t \leq T$ into subintervals $t_j \leq t \leq t_{j+1}$ and then to solve on each of these Problem (3.1) in two ways as follows: First, using the first-order approximation scheme with a comparatively large "degree" of stability, we find an approximate solution at the time instant $t_{j+1/2} = t_j + \tau/2$. After that we write the second-order scheme (corrector) on the whole interval (t_j, t_{j+1}), the main feature being that for its construction we make use of the "coarse" solution at $t_{j+1/2}$ obtained by means of the predictor (the first step).

Consequently, the predictor-corrector scheme can be written in the form

$$\frac{\phi^{j+1/4} - \phi^j}{\tau/2} + A_1 \phi^{j+1/4} = 0,$$

$$\frac{\phi^{j+1/2} - \phi^{j+1/4}}{\tau/2} + A_2 \phi^{j+1/2} = 0, \quad (3.34)$$

$$\frac{\phi^{j+1} - \phi^j}{\tau} + A \phi^{j+1/2} = 0,$$

where we assume $\phi^0 = g$.

Let us look at the scheme in some detail. First of all the elimination of the auxiliary variable $\phi^{j+1/4}$ from the first two equations in (3.34) reduces this system to

$$\left(E + \frac{\tau}{2} A_1\right)\left(E + \frac{\tau}{2} A_2\right)\phi^{j+1/2} = \phi^j,$$

$$\frac{\phi^{j+1} - \phi^j}{\tau} + A\phi^{j+1/2} = 0. \tag{3.35}$$

Furthermore, eliminating $\phi^{j+1/2}$, we obtain

$$\frac{\phi^{j+1} - \phi^j}{\tau} + A\left(E + \frac{\tau}{2} A_2\right)^{-1}\left(E + \frac{\tau}{2} A_1\right)^{-1}\phi^j = 0, \qquad \phi^0 = g. \tag{3.36}$$

Let us now analyse the approximation problem. To this end let us rewrite (3.36) in the form

$$\left(E + \frac{\tau}{2} A_1\right)\left(E + \frac{\tau}{2} A_2\right)\frac{\phi^{j+1} - \phi^j}{\tau} + \Lambda\phi^j = 0,$$

where

$$\Lambda = \left(E + \frac{\tau}{2} A_1\right)\left(E + \frac{\tau}{2} A_2\right)A\left(E + \frac{\tau}{2} A_2\right)^{-1}\left(E + \frac{\tau}{2} A_1\right)^{-1}.$$

Expanding into a power series in τ and assuming that

$$\frac{\tau}{2}\|A_\alpha\| < 1, \qquad \alpha = 1, 2,$$

one easily obtains

$$\Lambda = A + O(\tau^2).$$

Using the estimate used for the stabilization method we conclude that the predictor-corrector method yields a second-order approximation in τ.

Consider next the stability of this method. First let us write (3.36) in the form

$$\left(E + \frac{\tau}{2} A_1\right)\left(E + \frac{\tau}{2} A_2\right)\frac{\Phi^{j+1} - \Phi^j}{\tau} + A\Phi^j = 0, \tag{3.37}$$

where

$$\Phi^j = \left(E + \frac{\tau}{2} A_2\right)^{-1}\left(E + \frac{\tau}{2} A_1\right)^{-1}\phi^j. \tag{3.38}$$

Since

$$\|\Phi^{j+1}\|_{C_2} \le \|\Phi^j\|_{C_2}, \tag{3.39}$$

Splitting-Up Methods for Nonstationary Problems

the difference Equation (3.37) is stable. Substituting (3.38) into (3.39) and making use of (3.18) we obtain

$$\left\|\left(E + \frac{\tau}{2}A_1\right)^{-1}\phi^{j+1}\right\| \leq \left\|\left(E + \frac{\tau}{2}A_1\right)^{-1}\phi^j\right\| \qquad (3.40)$$

or

$$\|\phi^{j+1}\|_{C_1^{-1}} \leq \|\phi^j\|_{C_1^{-1}}, \qquad (3.41)$$

where

$$C_1^{-1} = \left(E + \frac{\tau}{2}A_1^*\right)^{-1}\left(E + \frac{\tau}{2}A_1\right)^{-1}.$$

We have thus established stability in the metric of (3.41). Consequently, if $A_1 \geq 0$, $A_2 \geq 0$, and if the entries of these matrices are time invariant, then the difference Scheme (3.34) is absolutely stable and yields an approximate solution of second order accuracy in τ, provided (3.1) has a sufficiently smooth solution.

For the nonhomogeneous problem the predictor-corrector method is formulated as follows:

$$\frac{\phi^{j+1/4} - \phi^j}{\tau/2} + A_1\phi^{j+1/4} = f^j,$$

$$\frac{\phi^{j+1/2} - \phi^{j+1/4}}{\tau/2} + A_2\phi^{j+1/2} = f^j, \qquad (3.42)$$

$$\frac{\phi^{j+1} - \phi^j}{\tau} + A\phi^{j+1/2} = f^j,$$

where

$$f^j = f(t_{j+1/2}).$$

Taking f^j as indicated it can be shown that (3.42) approximates the original problem with second-order accuracy in τ. The stability of (3.42) is established as follows. First, eliminate $\phi^{j+1/2}$ and $\phi^{j+1/4}$; the result is

$$\frac{\phi^{j+1} - \phi^j}{\tau} + A\left(E + \frac{\tau}{2}A_2\right)^{-1}\left[\left(E + \frac{\tau}{2}A_1\right)^{-1}\left(\phi^j + \frac{\tau}{2}f^j\right) + \frac{\tau}{2}f^j\right] = f^j.$$

(3.43)

Introduce the notation

$$\psi = \left(E + \frac{\tau}{2}A_1\right)^{-1}\phi.$$

Using this, the relation (3.43) can be written as

$$\frac{\psi^{j+1} - \psi^j}{\tau} + \left(E + \frac{\tau}{2}A_1\right)^{-1}A\left(E + \frac{\tau}{2}A_2\right)^{-1}$$

$$\times \left[\psi^j + \left(E + \frac{\tau}{2}A_1\right)^{-1}\frac{\tau}{2}f^j + \frac{\tau}{2}f^j\right] = \left(E + \frac{\tau}{2}A_1\right)^{-1}f^j$$

Hence

$$\psi^{j+1} = \left[E - \left(E + \frac{\tau}{2}A_1\right)^{-1} \tau A \left(E + \frac{\tau}{2}A_2\right)^{-1}\right]\psi^j$$

$$- \tau\left(E + \frac{\tau}{2}A_1\right)^{-1} A\left(E + \frac{\tau}{2}A_2\right)^{-1}\left[\left(E + \frac{\tau}{2}A_1\right)^{-1}\frac{\tau}{2}f^j + \frac{\tau}{2}f^j\right]$$

$$+ \tau\left(E + \frac{\tau}{2}A_1\right)^{-1} f^j \qquad (3.44)$$

The following relation holds:

$$E - \left(E + \frac{\tau}{2}A_1\right)^{-1} \tau A\left(E + \frac{\tau}{2}A_2\right)^{-1} = \left(E + \frac{\tau}{2}A_1\right)^{-1}$$

$$\times \left[\left(E + \frac{\tau}{2}A_1\right)\left(E + \frac{\tau}{2}A_2\right) - \tau A\right]\left(E + \frac{\tau}{2}A_2\right)^{-1}$$

$$= \left(E + \frac{\tau}{2}A_1\right)^{-1}\left[\left(E - \frac{\tau}{2}A_1\right)\left(E - \frac{\tau}{2}A_2\right)\right]\left(E + \frac{\tau}{2}A_2\right)^{-1}$$

$$= \left[\left(E + \frac{\tau}{2}A_1\right)^{-1}\left(E - \frac{\tau}{2}A_1\right)\right]\left[\left(E - \frac{\tau}{2}A_2\right)\left(E + \frac{\tau}{2}A_2\right)^{-1}\right]. \qquad (3.44a)$$

Thus, according to Kellogg's lemma, and using Equation (1.25) from Chapter 1, we obtain

$$\|\phi^j\|_{C_1^{-1}} \leqslant \|g\|_{C_1^{-1}} + 3\tau j\|f\|_{C_1^{-1}}, \qquad (3.45)$$

where $\|f\|_{C_1^{-1}} = \max_j \|f^j\|_{C_1^{-1}}$; that is, stability follows provided $0 \geqq t_j \geqq T$.

Hence if the matrices $A_1 \geq 0$, $A_2 \geq 0$ are time invariant, then the difference Scheme (3.42) is absolutely stable and provides a second-order approximation (in τ) of the solution, if only the right-hand side f in (3.1) along with the solutions are smooth enough.

4.3.3. The Component-by-Component Splitting-Up Method

The stabilization method and the predictor-corrector method are equivalent in accuracy and are absolutely stable provided $A_\alpha \geq 0$.

It is to be kept in mind, however, that we have assumed A_α to be time-invariant. This constraint has made it possible to analyze completely the stability, provided only the constructive assumption of positive semi-definiteness of the operators A_α. Unfortunately, in the time-dependent case the stability analysis of this kind can not be done in general. An exception is the component-by-component splitting-up method which we will formulate presently.

Consider Equations (3.1)–(3.3) and let $A_1 \geq 0$, $A_2 \geq 0$. Approximate these matrices on $t_j \leqq t \leqq t_{j+1}$ in the form

$$\Lambda_\alpha^j = A_\alpha(t_{j+1/2}),$$

Splitting-Up Methods for Nonstationary Problems   155

assuming that their entries are sufficiently smooth. The difference scheme below (suggested by Yanenko [15]) consists of a sequence of simple Crank–Nicholson schemes†:

$$\frac{\phi^{j+1/2} - \phi^j}{\tau} + \Lambda_1^j \frac{\phi^{j+1/2} + \phi^j}{2} = 0,$$

$$\frac{\phi^{j+1} - \phi^{j+1/2}}{\tau} + \Lambda_2^j \frac{\phi^{j+1} + \phi^{j+1/2}}{2} = 0. \quad (3.46)$$

By eliminating the auxiliary functions $\phi^{j+1/2}$, the system of difference equations of (3.46) can be reduced to one single equation

$$\phi^{j+1} = T^j \phi^j, \quad (3.47)$$

where

$$T^j = \left(E + \frac{\tau}{2}\Lambda_2^j\right)^{-1}\left(E - \frac{\tau}{2}\Lambda_2^j\right)\left(E + \frac{\tau}{2}\Lambda_1^j\right)^{-1}\left(E - \frac{\tau}{2}\Lambda_1^j\right). \quad (3.48)$$

At first, let us consider the approximation problem. To this end let us expand the operator T^j into a power series in τ and assume that

$$\frac{\tau}{2}\|\Lambda_\alpha^j\| < 1.$$

Performing some simple manipulations we obtain

$$T^j = E - \tau\Lambda^j + \frac{\tau^2}{2}[(\Lambda_1^j)^2 + 2\Lambda_2^j\Lambda_1^j + (\Lambda_2^j)^2] - \cdots. \quad (3.49)$$

If the operators Λ_α^j commute, that is, if $\Lambda_1^j\Lambda_2^j = \Lambda_2^j\Lambda_1^j$, then (3.49) can be written as

$$T^j = E - \tau\Lambda^j + \frac{\tau^2}{2}(\Lambda^j)^2 - \cdots. \quad (3.50)$$

Hence if the matrices $A_1(t) \geq 0$, $A_2(t) \geq 0$, and the solution ϕ of Equations (3.1)–(3.3) are smooth enough, then the difference Scheme (3.46) is absolutely stable, as implied by the inequality $\|T^j\| \leq 1$ (from Kellogg's lemma). Moreover, the scheme approximates the original equation [Equation (3.1)] with the second- or first-order accuracy in τ, according to whether Λ_1^j and Λ_2^j commute or not.

The next step is to approximate the operators $A_1(t)$ and $A_2(t)$ on the interval $t_j \leq t \leq t_{j+1}$ the same way as in (3.46), and on the interval $t_{j-1} \leq t \leq t_{j+1}$ by taking

$$\Lambda_\alpha^j = A_\alpha(t_j).$$

---

† The theoretical foundations and some modifications of the scheme were given by the author in a paper at the Symposium on Numerical Solutions of Partial Differential Equations (SYNSPADE, 1970, USA).

Consider the following two systems of difference equations:

$$\frac{\phi^{j-1/2} - \phi^{j-1}}{\tau} + \Lambda_1^j \frac{\phi^{j-1/2} + \phi^{j-1}}{2} = 0,$$

$$\frac{\phi^j - \phi^{j-1/2}}{\tau} + \Lambda_2^j \frac{\phi^j + \phi^{j-1/2}}{2} = 0,$$

(3.51)

$$\frac{\phi^{j+1/2} - \phi^j}{\tau} + \Lambda_2^j \frac{\phi^{j+1/2} + \phi^j}{2} = 0,$$

$$\frac{\phi^{j+1} - \phi^{j+1/2}}{\tau} + \Lambda_1^j \frac{\phi^{j+1} + \phi^{j+1/2}}{2} = 0.$$

(3.52)

The computational cycle consists of a sequential application of Schemes (3.51) and (3.52). In analogy with the above it can be shown that the full computational cycle, using (3.51) and (3.52), gives us

$$\phi^{j+1} = T^j \phi^{j-1},$$ (3.53)

where

$$T^j = \left(E + \frac{\tau}{2}\Lambda_1^j\right)^{-1} \left(E - \frac{\tau}{2}\Lambda_1^j\right) \left(E + \frac{\tau}{2}\Lambda_2^j\right)^{-1}$$

$$\times \left(E - \frac{\tau}{2}\Lambda_2^j\right) \left(E + \frac{\tau}{2}\Lambda_2^j\right)^{-1}$$

$$\times \left(E - \frac{\tau}{2}\Lambda_2^j\right) \left(E + \frac{\tau}{2}\Lambda_1^j\right)^{-1}$$

$$\times \left(E - \frac{\tau}{2}\Lambda_1^j\right) = E - 2\tau\Lambda^j + \frac{(2\tau)^2}{2}(\Lambda^j)^2 - \cdots.$$

Let us now compare the transition operator T^j above with that corresponding to the following Crank–Nicholson scheme:

$$\frac{\phi^{j+1} - \phi^{j-1}}{2\tau} + \Lambda^j \frac{\phi^{j+1} + \phi^{j-1}}{2} = 0.$$

It then follows that with an accuracy up to τ^2 the two operators T^j (corresponding to the two-cycle splitting scheme and to the Crank–Nicholson scheme on the doubled interval) coincide, no matter whether the operators A_α commute or not. Thus this approach avoids the considerable constraint of commutativity.

Next let us turn to the problem of countable stability. Consider Relation (3.47); from this

$$\|\phi^{j+1}\| \leq \|T^j\| \|\phi^j\|.$$

Since for $A_\alpha \geq 0$

$$\|T^j\| \leq 1$$

Splitting-Up Methods for Nonstationary Problems

as we have seen above, we obtain the estimate

$$\|\phi^{j+1}\| \leq \|\phi^j\|. \tag{3.54}$$

From this one can immediately show that

$$\|\phi^j\| \leq \|g\|. \tag{3.55}$$

In the case of the two-cycle method one has the estimate of (3.54) at each step of the cycle. This means that the two-cycle method is absolutely stable. (Note that a similar approach of symmetrization has been suggested independently by Strang [7] for the alternating-directions method.)

Consequently, if $A_1(t) \geq 0$ and $A_2(t) \geq 0$, then, assuming the entries of these matrices as well as the solution ϕ of Equations (3.1)–(3.3) are smooth enough, the difference system of (3.51), (3.52) is absolutely stable, and Scheme (3.53) represents a second-order approximation (in τ) of Equation (3.1).

Let us now seek the solution of the nonhomogeneous problem by means of the two-cycle, full splitting. With this in mind, consider the difference system of the form (3.51), (3.52), which can be written more conveniently as follows:

$$\left(E + \frac{\tau}{2} \Lambda_1^j\right) \phi^{j-1/2} = \left(E - \frac{\tau}{2} \Lambda_1^j\right) \phi^{j-1},$$

$$\left(E + \frac{\tau}{2} \Lambda_2^j\right)(\phi^j - \tau f^j) = \left(E - \frac{\tau}{2} \Lambda_2^j\right) \phi^{j-1/2},$$

$$\left(E + \frac{\tau}{2} \Lambda_2^j\right) \phi^{j+1/2} = \left(E - \frac{\tau}{2} \Lambda_2^j\right)(\phi^j + \tau f^j), \tag{3.56}$$

$$\left(E + \frac{\tau}{2} \Lambda_1^j\right) \phi^{j+1} = \left(E - \frac{\tau}{2} \Lambda_1^j\right) \phi^{j+1/2},$$

where $f^j = f(t_j)$. Solving for ϕ^{j+1}, we obtain

$$\phi^{j+1} = T^j \phi^{j-1} + 2\tau T_1^j T_2^j f^j, \tag{3.57}$$

where

$$T^j = T_1^j T_2^j T_2^j T_1^j \tag{3.58}$$

$$T_\alpha^j = \left(E + \frac{\tau}{2} \Lambda_\alpha^j\right)^{-1} \left(E - \frac{\tau}{2} \Lambda_\alpha^j\right). \tag{3.59}$$

Using a power-series expansion in the small parameter τ, expression (3.57) becomes

$$\phi^{j+1} = \left[E - 2\tau \Lambda^j + \frac{(2\tau)^2}{2}(\Lambda^j)^2\right] \phi^{j-1} + 2\tau(E - \tau \Lambda^j) f^j + O(\tau^3), \tag{3.60}$$

which can further be written in the form

$$\frac{\phi^{j+1} - \phi^{j-1}}{2\tau} + \Lambda^j(E - \tau \Lambda^j) \phi^{j-1} = (E - \tau \Lambda^j) f^j + O(\tau^2). \tag{3.61}$$

Expanding into a Taylor series in the vicinity of t_{j-1}, we may eliminate ϕ^{j-1} from the last equation. Thus

$$\phi^j = \phi^{j-1} + \left(\frac{\partial \phi}{\partial t}\right)^{j-1} \tau + O(\tau^2). \tag{3.62}$$

With the help of the relation

$$\left(\frac{\partial \phi}{\partial t}\right)^{j-1} = -\Lambda^j \phi^{j-1} + f^j + O(\tau)$$

we further eliminate the derivative $\partial \phi / \partial t$. Hence substituting (3.63) into (3.62) we obtain

$$\phi^j = (E - \tau \Lambda^j)\phi^{j-1} + \tau f^j + O(\tau^2).$$

Thus

$$(E - \tau \Lambda^j)\phi^{j-1} = \phi^j - \tau f^j + O(\tau^2). \tag{3.64}$$

Using (3.64) in (3.61), we have finally

$$\frac{\phi^{j+1} - \phi^{j-1}}{2\tau} + \Lambda^j \phi^j = f^j + O(\tau^2). \tag{3.65}$$

Clearly, (3.65) is a second-order approximation (in τ) of the original Equation (3.1) on the interval $t_{j-1} \leq t \leq t_{j+1}$. We have thus found a difference approximation of the nonhomogeneous evolution equation of second order by means of the two-cycle method.

The stability is established by elementary manipulations, if we use the energy norms. Indeed, from the norm estimate of (3.57)

$$\|\phi^{j+1}\| \leq \|T^j\| \|\phi^{j-1}\| + 2\tau \|T_1^j\| \|T_2^j\| \|f^j\|. \tag{3.66}$$

It has been shown above that $\|T_\alpha^j\| \leq 1$, and hence

$$\|T^j\| \leq \|T_1^j\| \|T_2^j\| \|T_2^j\| \|T_1^j\| \leq 1.$$

Therefore

$$\|\phi^{j+1}\| \leq \|\phi^{j-1}\| + 2\tau \|f^j\|. \tag{3.67}$$

Using the recursive Relation (3.63) we have

$$\|\phi^j\| \leq \|g\| + \tau j \|f\|, \tag{3.68}$$

where

$$\|f\| = \max_j \|f^j\|.$$

The countable stability of the scheme on any finite time interval now follows from Relation (3.68).

The Equations (3.56) can also be equivalently written as

$$\left(E + \frac{\tau}{2}\Lambda_1^j\right)\phi^{j-2/3} = \left(E - \frac{\tau}{2}\Lambda_1^j\right)\phi^{j-1},$$

$$\left(E + \frac{\tau}{2}\Lambda_2^j\right)\phi^{j-1/3} = \left(E - \frac{\tau}{2}\Lambda_2^j\right)\phi^{j-2/3}.$$

$$\phi^{j+1/3} = \phi^{j-1/3} + 2\tau f^j, \qquad (3.69)$$

$$\left(E + \frac{\tau}{2}\Lambda_2^j\right)\phi^{j+2/3} = \left(E - \frac{\tau}{2}\Lambda_2^j\right)\phi^{j+1/3},$$

$$\left(E + \frac{\tau}{2}\Lambda_1^j\right)\phi^{j+1} = \left(E - \frac{\tau}{2}\Lambda_1^j\right)\phi^{j+2/3}.$$

Eliminating the unknowns indexed by fractions we arrive at the explicit formula

$$\phi^{j+1} = T_1^j T_2^j T_2^j T_1^j \phi^{j-1} + 2\tau T_1^j T_2^j f^j, \qquad (3.70)$$

which coincides with (3.57). Of the two forms (3.57) and (3.69), the latter is preferable in some cases.

In summary, if the matrices $A_1(t) \geq 0$, $A_2(t) \geq 0$, along with the function $f(t)$ and the solution ϕ are smooth enough, then the difference system (3.56) is absolutely stable on $0 \leq t \leq T$ and approximates the original equation with second-order accuracy in τ.

4.3.4. Some General Remarks

To begin with, let us compare the splitting-up methods of this section under the assumption that A, A_1, and A_2 are independent of time, $A_1 \geq 0$, $A_2 \geq 0$, and $A_1 A_2 = A_2 A_1$. It is namely this simple case which is fairly completely discussed in the literature. Let us assume that the splitting-up differential schemes are already formally resolved with respect to the solution sought at each step. It is not difficult to see that all the splitting-up schemes for the homogeneous ($f = 0$) evolution Problem (3.1) are mutually equivalent and hence represent only various realization schemes. In particular, they can be written in the form

$$\phi^{j+1} = T\phi^j,$$

where

$$T = \left(E + \frac{\tau}{2}A_1\right)^{-1}\left(E - \frac{\tau}{2}A_1\right)\left(E + \frac{\tau}{2}A_2\right)^{-1}\left(E - \frac{\tau}{2}A_2\right).$$

The splitting-up schemes for nonhomogeneous evolution problems are equivalent only as far as the order of approximation accuracy is concerned. This means that the various splitting-up schemes which approximate the nonhomogeneous problem with the second-order of accuracy in τ lead to different results, the difference being within the limits of order $O(\tau^2)$.

The second remark concerns the case where $A_1 \geq 0$, $A_2 \geq 0$ are time invariant, but $A_1 A_2 \neq A_2 A_1$. As shown above, any one of the three splitting-up schemes (that is, the stabilization method, the predictor-corrector method, and the component-by-component method) can be used to obtain an approximate solution of the evolution problem. Although equivalent in order of accuracy, these methods differ considerably. Even in the homogeneous case the transition operators are quite different. It is still much too early for any recommendation as to the areas of the most effective applications of one scheme or another, since this problem has not yet been adequately analyzed. However, the fact that there are three independent techniques available enhances the confidence when attempting to solve complicated problems.

The third remark deals with the most general case where the operators $A_1 \geq 0, A_2 \geq 0$ may depend on time and may not commute. We recommend in this case the component-by-component splitting. In its two-cycle form it gives a second-order approximation of the solution. It should be noted at the same time that if at least one of the operators A_1, A_2 is time invariant, second-order accuracy can be achieved by the ordinary one-cycle realization method.

4.4. Multicomponent Splitting

So far we have assumed that the original operator A is represented as a sum of two operators of simpler structure. As it is often the case, more complex physical problems require that A be decomposed into a large number of components. In general we have

$$A = \sum_{\alpha=1}^{n} A_\alpha, \tag{4.1}$$

where $A_\alpha \geq 0$. We will consider only the case $n > 2$, since the case $n = 2$ has been dealt with at length in Section 4.3.

First of all it can be seen that the straightforward generalizations of the splitting-up methods considered above are not possible in general. Therefore we will impose additional assumptions, as necessary, which will allow the generalizations of the splitting-up algorithms.

4.4.1. The Stabilization Method

Assuming the validity of Equation (4.1), the stabilization method can be represented in the form

$$\prod_{\alpha=1}^{n}\left(E + \frac{\tau}{2} A_\alpha\right)\frac{\phi^{j+1} - \phi^j}{\tau} + A\phi^j = f^j, \quad \phi^0 = g, \tag{4.2}$$

where

$$f^j = f(t_{j+1/2}).$$

Multicomponent Splitting

The realization scheme of the algorithm is as follows:

$$F^j = -A\phi^j + f^j,$$

$$\left(E + \frac{\tau}{2}A_1\right)\xi^{j+(1/n)} = F^j,$$

$$\left(E + \frac{\tau}{2}A_2\right)\xi^{j+(2/n)} = \xi^{j+(1/n)}, \quad (4.3)$$

$$\dots\dots\dots\dots\dots\dots\dots\dots\dots\dots\dots$$

$$\left(E + \frac{\tau}{2}A_n\right)\xi^{j+1} = \xi^{j+(n-1/n)}$$

$$\phi^{j+1} = \phi^j + \tau\xi^{j+1}.$$

It is easy to check that the stabilization method is of second-order accuracy in τ, provided the solution is sufficiently smooth. The countable stability is assured if

$$\|T\| < 1, \quad (4.4)$$

where the transition operator T is given by

$$T = E - \tau \prod_{\alpha=n}^{1}\left(E + \frac{\tau}{2}A_\alpha\right)^{-1} A. \quad (4.5)$$

Unfortunately, the condition $A_\alpha \geq 0$ does not imply the stability in any norm, contrary to what we have in the case $n = 2$.

To establish stability, one usually uses the following simple algorithmic approach. Put f^j equal to zero in (4.2) and solve for ϕ^{j+1} to obtain

$$\phi^{j+1} = T\phi^j. \quad (4.6)$$

Since T is assumed independent of time (or the index j), we can solve (4.6) with the initial condition

$$\phi^0 = g \quad (4.7)$$

and for a fixed parameter τ, which ensures the necessary approximation. If the norm $\|\phi^j\|$ does not increase, then $\|T\| < 1$, and therefore the condition of countable stability can be taken as satisfied. Having this, we may turn to the nonhomogeneous problem. Equation (4.2) can be rewritten as follows:

$$\phi^{j+1} = T\phi^j + \tau \prod_{\alpha=n}^{1}\left(E + \frac{\tau}{2}A_\alpha\right)^{-1} f^j. \quad (4.8)$$

Hence

$$\|\phi^{j+1}\| \leq \|T\|\|\phi^j\| + \tau \prod_{\alpha=n}^{1}\left\|\left(E + \frac{\tau}{2}A_\alpha\right)^{-1}\right\|\|f^j\|,$$

or, as a consequence of the inequalities (4.4) and (1.25) from Chapter 1, we have

$$\|\phi^{j+1}\| \leqslant \|\phi^j\| + \tau\|f^j\|. \tag{4.9}$$

Using the recursive relation we arrive at the condition of stability in the energy metric, namely

$$\|\phi\| \leqslant \|g\| + \tau j\|f\|, \tag{4.10}$$

where

$$\|f\| = \max_j \|f^j\|. \tag{4.11}$$

Note that when solving the homogeneous Equation (4.6) we have made use of the initial Condition (4.7). This was not necessary at all. Any function could have been taken in place of g.

4.4.2. The Predictor-Corrector Method

In this case the splitting-up scheme becomes

$$\left(E + \frac{\tau}{2} A_1\right) \phi^{j+(1/2n)} = \phi^j + \frac{\tau}{2} f^j$$

$$\left(E + \frac{\tau}{2} A_2\right) \phi^{j+(2/2n)} = \phi^{j+(1/2n)},$$

$$\ldots\ldots\ldots\ldots\ldots\ldots\ldots\ldots\ldots\ldots \tag{4.12}$$

$$\left(E + \frac{\tau}{2} A_n\right) \phi^{j+1/2} = \phi^{j+[(n-1)/2n]},$$

$$\frac{\phi^{j+1} - \phi^j}{\tau} + A\phi^{j+1/2} = f^j,$$

where again it is assumed that $A_\alpha \geq 0$ and $f^j = f(t_{j+1/2})$. The above system can be reduced to a single equation of the form

$$\frac{\phi^{j+1} - \phi^j}{\tau} + A \prod_{\alpha=n}^{1} \left(E + \frac{\tau}{2} A_\alpha\right)^{-1} \left(\phi^j + \frac{\tau}{2} f^j\right) = f^j \tag{4.13}$$

with

$$\phi^0 = g.$$

Provided there is sufficient smoothness, the predictor-corrector method is of second-order accuracy in τ. Let us rewrite (4.13) as follows:

$$\phi^{j+1} = T\phi^j + \frac{\tau}{2}(E + T)f^j, \tag{4.14}$$

where the transition operator T is given by

$$T = E - \tau A \prod_{\alpha=n}^{1} \left(E + \frac{\tau}{2} A_\alpha\right)^{-1}. \tag{4.15}$$

Multicomponent Splitting

The requirement of countable stability eventually comes down to estimating the norm of T. Unfortunately, also in this case the constructive condition $A_\alpha \geq 0$ does not yield the proof of stability for the scheme. This remains an open problem.

In order to complete the analysis of the two schemes considered above, let us take a simple case when the operators A_α are mutually commuting and have common bases. It turns out that this additional hypothesis along with $A_\alpha \geq 0$ are enough to prove stability of the considered schemes. Indeed, commutativity implies that the transition operators for the two schemes coincide. For the sake of simplicity consider the homogeneous Problem (4.6), (4.7), and let us seek the solution in the spectral form

$$\phi^j = \sum_k \phi^j_k u_k, \qquad (4.16)$$

where u_k are the eigenfunctions of Problem (1.7) in Chapter 1. $\phi^j_k = (\phi^j, u^*_k)$, where u^*_k are the eigenfunctions of the adjoint Problem (1.7) in Chapter 1. Since $\{u_k\}$ is the common basis, we have

$$A u_k = \lambda_k u_k, \qquad A_\alpha u_k = \lambda^\alpha_k u_k, \qquad \lambda_k = \sum_{\alpha=1}^n \lambda^\alpha_k. \qquad (4.17)$$

Substituting the expansions of (4.16) and the corresponding representations for g into (4.6), (4.7), we obtain the following expressions for the Fourier coefficients:

$$\phi^{j+1}_k = T_k \phi^j_k, \qquad \phi^0_k = g_k, \qquad (4.18)$$

where

$$T_k = 1 - \frac{\tau \lambda_k}{\prod_{\alpha=1}^n \left(1 + \frac{\tau}{2} \lambda^\alpha_k\right)}. \qquad (4.19)$$

The formula for T_k above can be written as

$$T_k = \frac{\mu_k - \frac{\tau}{2} \lambda_k}{\mu_k + \frac{\tau}{2} \lambda_k}, \qquad (4.20)$$

where μ_k are positive constants, provided $\lambda^\alpha_k \geq 0$. From (4.20) we have

$$|T_k| \leq 1, \qquad (4.21)$$

which proves the claim, in agreement with Section 1.3.

The stabilization method and the predictor-corrector method for the n-component splitting can also be applied to the case where A depends on time. The stability analysis however becomes much more complex. Therefore it is difficult to say to what extent we can justify the application of these schemes in general situations. In particular, this stimulated the present author to formulate a more or less universal approach for solving various complicated and sufficiently general problems, the basic idea being that of splitting. The two-cycle sequential splitting method is of this type.

4.4.3. The Component-by-Component Splitting-Up Method Based on the Elementary Schemes

Let us try to construct a second-order difference approximation of the problem so that it is absolutely stable. In accord with the assumptions regarding the component-by-component splitting let us suppose that

$$\Lambda^j = \sum_{\alpha=1}^{n} \Lambda_\alpha^j, \qquad (4.22)$$

where all Λ_α^j are nonnegative definite operators, that is $\Lambda_\alpha^j \geq 0$. Consider the following system:

$$\left(E + \frac{\tau}{2}\Lambda_\alpha^j\right)\Phi^{j+(\alpha/n)} = \left(E - \frac{\tau}{2}\Lambda_\alpha^j\right)\Phi^{j+[(\alpha-1)/n]} \qquad (\alpha = 1, 2, \ldots, n). \qquad (4.23)$$

Assume that $\Lambda_\alpha^j \geq 0$ commute and $\Lambda_\alpha^j = A_\alpha^{j+1/2}$ or $\Lambda_\alpha^j = \frac{1}{2}(A^{j+1} + A^j)$. Then (4.23) is an unconditionally stable, second-order approximation scheme. This can be seen quite easily using the Fourier method. In the noncommutative case, however, the scheme will only be of first-order accuracy in τ. Of more interest in applications is the following second-order approximation scheme suggested by Dyakonov [15]:

$$\Phi^{j+(\alpha/2n)} = \left(E - \frac{\tau}{2}\Lambda_\alpha^j\right)\Phi^{j+[(\alpha-1)/2n]} \qquad (\alpha = 1, 2, \ldots, n),$$

$$\left(E + \frac{\tau}{2}\Lambda_{2n-\alpha+1}^j\right)\Phi^{j+(\alpha/2n)} = \Phi^{j+[(\alpha-1)/2n]} \qquad (\alpha = n+1, n+2, \ldots, 2n). \qquad (4.24)$$

Let us try to find a special construction of the full splitting-up method based on (4.23) which would solve the Cauchy problem corresponding to nonnegative definite and noncommuting operators Λ_α^j, and which would be of second-order accuracy. In a sense, this would solve completely the splitting-up problem.

Note that (4.23) reduces to a single equation of the form

$$\Phi^{j+1} = \prod_{\alpha=1}^{n} \left(E + \frac{\tau}{2}\Lambda_\alpha^j\right)^{-1}\left(E - \frac{\tau}{2}\Lambda_\alpha^j\right)\Phi^j. \qquad (4.25)$$

Using (4.25), we find the norm estimates

$$\|\Phi^{j+1}\| \leq \prod_{\alpha=1}^{n} \left\|\left(E + \frac{\tau}{2}\Lambda_\alpha^j\right)^{-1}\left(E - \frac{\tau}{2}\Lambda_\alpha^j\right)\right\| \|\Phi^j\|. \quad (4.26)$$

From Kellogg's lemma

$$\|\Phi^{j+1}\| \leq \|\Phi^j\| \leq \cdots \leq \|g\|. \quad (4.27)$$

If the operators are skew symmetric, that is, $(\Lambda_\alpha^j \phi, \phi) = 0$, then

$$\|\Phi^{j+1}\| = \|\Phi^j\| = \cdots = \|g\|. \quad (4.28)$$

Thus we have absolute stability for the scheme.

To determine the order of approximation, let us expand the expression below in terms of powers of τ (assuming $(\tau/2)\|\Lambda_\alpha\| < 1$):

$$T^j = \prod_{\alpha=1}^{n} \left(E + \frac{\tau}{2}\Lambda_\alpha^j\right)^{-1}\left(E - \frac{\tau}{2}\Lambda_\alpha^j\right).$$

Since

$$T^j = \prod_{\alpha=1}^{n} T_\alpha^j,$$

and since T_α^j can be expanded into the series

$$T_\alpha^j = E - \tau\Lambda_\alpha^j + \frac{\tau^2}{2}(\Lambda_\alpha^j)^2 \cdots, \quad (4.29)$$

we have that

$$T^j = E - \tau\Lambda^j + \frac{\tau^2}{2}\left[(\Lambda^j)^2 + \sum_{\alpha=1}^{n}\sum_{\beta=\alpha+1}^{n}(\Lambda_\alpha^j\Lambda_\beta^j - \Lambda_\beta^j\Lambda_\alpha^j)\right] + O(\tau^3). \quad (4.30)$$

In the case where the operators Λ_α^j commute, the expression under the double sum is zero and (4.30) becomes

$$T^j = E - \tau\Lambda^j + \frac{\tau^2}{2}(\Lambda^j)^2 + O(\tau^3). \quad (4.31)$$

Comparing (4.31) with (1.13) and (1.22)–(1.24), we conclude that in this particular case Scheme (4.23) approximates with the second order in τ. If the operators Λ_α^j do not commute, then the splitting-up scheme turns out to be only of first-order accuracy in τ. To obtain the second order in this latter case it is necessary to use the scheme

$$\Phi^j = \prod_{\alpha=1}^{n} T_\alpha^j \Phi^{j-1}, \quad \Phi^{j+1} = \prod_{\alpha=n}^{1} T_\alpha^j \Phi^j, \quad (4.32)$$

rather than (4.23). In the language of algorithms, this says that the system of equations (4.23) is solved first on the interval $t_{j-1} \leq t \leq t_j$ for $\alpha = 1, 2, \ldots, n$, and then again on the interval $t_j \leq t \leq t_{j+1}$, but with the indices α

taken in the reversed order ($\alpha = n, n-1, \ldots, 1$):

$$\left(E + \frac{\tau}{2}\Lambda_\alpha^j\right)\Phi^{j+(\alpha/n)-1} = \left(E - \frac{\tau}{2}\Lambda_\alpha^j\right)\Phi^{j+[(\alpha-1)/n]-1}$$

$$(\alpha = 1, 2, \ldots, n), \qquad (4.33)$$

$$\left(E + \frac{\tau}{2}\Lambda_\alpha^j\right)\Phi^{j+1-[(\alpha-1)/n]} = \left(E - \frac{\tau}{2}\Lambda_\alpha\right)\Phi^{j+1-(\alpha/n)}$$

$$(\alpha = n, n-1, \ldots, 1).$$

For the full cycle (4.33) we have clearly

$$\Phi^{j+1} = T^j \Phi^{j-1},$$

where

$$T^j = \prod_{\alpha=1}^{n} T_\alpha^j \prod_{\alpha=n}^{1} T_\alpha^j = E - 2\tau\Lambda^j + \frac{(2\tau)^2}{2}(\Lambda^j)^2 + O(\tau^3).$$

Thus, if we take for Λ_α^j one of the analogs introduced in (1.22)–(1.24), Scheme (4.33) becomes a second-order approximation on the interval $t_{j-1} \leq t \leq t_{j+1}$.

Finally let us note that the difference system (4.33) is absolutely stable, provided $\Lambda_\alpha^j \geq 0$. In a sense, we have thus obtained an optimal multicomponent splitting-up algorithm.

Consider next the nonhomogeneous equation

$$\frac{\partial \phi}{\partial t} + A\phi = f,$$
$$\phi = g \quad \text{for} \quad t = 0, \qquad (4.34)$$

where $A(t) \geq 0$ and $A = \sum_{\alpha=1}^{n} A_\alpha$, $A_\alpha(t) \geq 0$. The splitting-up scheme on the interval $t_{j-1} \leq t \leq t_{j+1}$ is as follows:

$$\left(E + \frac{\tau}{2}\Lambda_1^j\right)\phi^{j-[(n-1)/n]} = \left(E - \frac{\tau}{2}\Lambda_1^j\right)\phi^{j-1},$$

$$\cdots\cdots\cdots\cdots\cdots\cdots\cdots\cdots\cdots\cdots\cdots\cdots$$

$$\left(E + \frac{\tau}{2}\Lambda_n^j\right)(\phi^j - \tau f^j) = \left(E - \frac{\tau}{2}\Lambda_n^j\right)\phi^{j-(1/n)},$$

$$\left(E + \frac{\tau}{2}\Lambda_n^j\right)\phi^{j+(1/n)} = \left(E - \frac{\tau}{2}\Lambda_n^j\right)(\phi^j + \tau f^j), \qquad (4.35)$$

$$\cdots\cdots\cdots\cdots\cdots\cdots\cdots\cdots\cdots\cdots\cdots\cdots$$

$$\left(E + \frac{\tau}{2}\Lambda_1^j\right)\phi^{j+1} = \left(E - \frac{\tau}{2}\Lambda_1^j\right)\phi^{j+[(n-1)/n]},$$

where
$$\Lambda_\alpha^j = A_\alpha(t_j).$$

It is not difficult to see that this scheme approximates with second order in τ and is absolutely stable, provided ϕ is sufficiently smooth.

The n-component system of Equations (4.35) can be rewritten (as in the case $\alpha = 2$) in the following equivalent form:

$$\left(E + \frac{\tau}{2}\Lambda_\alpha\right)\Phi^{j-[(n+1-\alpha)/(n+1)]} = \left(E - \frac{\tau}{2}\Lambda_\alpha\right)\Phi^{j-[(n+1-\alpha+1)/(n+1)]}$$

$$(\alpha = 1, 2, \ldots, n),$$

$$\Phi^{j+[1/(n+1)]} = \Phi^{j-[1/(n+1)]} + 2\tau f^j, \tag{4.36}$$

$$\left(E + \frac{\tau}{2}\Lambda_{n-\alpha+2}\right)\Phi^{j+[\alpha/(n+1)]} = \left(E - \frac{\tau}{2}\Lambda_{n-\alpha+2}\right)\Phi^{j+[(\alpha-1)/(n+1)]}$$

$$(\alpha = 2, 3, \ldots, n+1).$$

Let us now discuss the splitting-up method for the backward implicit difference approximations. To this end, consider the problem

$$\frac{\partial \phi}{\partial t} + A\phi = 0 \quad \text{in} \quad D \times D_t, \tag{4.37}$$

$$\phi = g \quad \text{in} \quad D \quad \text{for} \quad t = 0.$$

Suppose that $A = \sum_{\alpha=1}^{n} A_\alpha$, and the $A_\alpha \geq 0$ are time invariant. Take the splitting algorithm in the form

$$\frac{\phi^{j+(1/n)} - \phi^j}{\tau} + A_1 \phi^{j+(1/n)} = 0,$$

$$\cdots\cdots\cdots\cdots\cdots\cdots\cdots\cdots\cdots\cdots \tag{4.38}$$

$$\frac{\phi^{j+1} - \phi^{j+[(n-1)/n]}}{\tau} + A_n \phi^{j+1} = 0.$$

We will show that such an algorithm is absolutely stable. Indeed, consider the equation

$$\frac{\phi^{j+(\alpha/n)} - \phi^{j+[(\alpha-1)/n]}}{\tau} + A_\alpha \phi^{j+(\alpha/n)} = 0. \tag{4.39}$$

Take the scalar product of this equation with $\phi^{j+(\alpha/n)}$. Then

$$(\phi^{j+(\alpha/n)} - \phi^{j+[(\alpha-1)/n]}, \phi^{j+(\alpha/n)}) + \tau(A_\alpha \phi^{j+(\alpha/n)}, \phi^{j+(\alpha/n)}) = 0.$$

A_α being positive semidefinite, we have further

$$(\phi^{j+(\alpha/n)} - \phi^{j+[(\alpha-1)/n]}, \phi^{j+(\alpha/n)}) \leq 0$$

or

$$(\phi^{j+(\alpha/n)}, \phi^{j+(\alpha/n)}) \leq (\phi^{j+(\alpha/n)}, \phi^{j+[(\alpha-1)/n]}).$$

But since
$$(\phi^{j+(\alpha/n)}, \phi^{j+[(\alpha-1)/n]}) \leq \tfrac{1}{2}[(\phi^{j+(\alpha/n)}, \phi^{j+(\alpha/n)}) + (\phi^{j+[(\alpha-1)/n]}, \phi^{j+[(\alpha-1)/n]})],$$
the following inequality holds:
$$\|\phi^{j+(\alpha/n)}\|^2 \leq \|\phi^{j+[(\alpha-1)/n]}\|^2 \qquad (\alpha = 1, 2, \ldots, n).$$
Hence
$$\|\phi^{j+1}\| \leq \|\phi^j\|. \tag{4.40}$$
In other words, under the assumptions postulated, the computations based on the splitting-up Scheme (4.38) will be absolutely stable.

It is a simple matter to verify that System (4.38) represents a first-order approximation in τ of the original problem.

Next, consider the nonhomogeneous problem:
$$\begin{aligned}\frac{\partial \phi}{\partial t} + A\phi &= f \quad \text{in} \quad D \times D_t, \\ \phi &= g \quad \text{in} \quad D \quad \text{for} \quad t = 0.\end{aligned} \tag{4.41}$$

The splitting-up scheme can be taken in the form
$$\begin{aligned}\frac{\phi^{j+(1/n)} - \phi^j}{\tau} + A_1 \phi^{j+(1/n)} &= 0, \\ \cdots\cdots\cdots\cdots\cdots\cdots\cdots\cdots\cdots\cdots\cdots& \\ \frac{\phi^{j+1} - \phi^{j+[(n-1)/n]}}{\tau} + A_n \phi^{j+1} &= f^j.\end{aligned} \tag{4.42}$$

This type of splitting-up scheme approximates the original nonhomogeneous equation up to the first order of accuracy in τ.

Stability is shown as follows: Take the scalar products of each of Equations (4.42) with $\phi^{j+(1/n)}, \ldots, \phi^{j+1}$ correspondingly. Similarly as before, we obtain
$$\|\phi^{j+(\alpha/n)}\| \leq \|\phi^{j+(\alpha-1/n)}\| \qquad (\alpha = 1, 2, \ldots, n-1). \tag{4.43}$$

Let us take a closer look at the last equation in (4.42). We have
$$(\phi^{j+1}, \phi^{j+1}) = (\phi^{j+[(n-1)/n]}, \phi^{j+1}) - \tau(A_n \phi^{j+1}, \phi^{j+1}) + \tau(f^j, \phi^{j+1}).$$

Accounting for $A_n \geq 0$, we have further
$$(\phi^{j+1}, \phi^{j+1}) \leq (\phi^{j+[(n-1)/n]}, \phi^{j+1}) + \tau(f^j, \phi^{j+1}).$$

From the Cauchy–Schwartz inequality
$$|(\phi^{j+[(n-1)/n]}, \phi^{j+1})| \leq \|\phi^{j+[(n-1)/n]}\| \|\phi^{j+1}\|,$$
$$|(f^j, \phi^{j+1})| \leq \|f^j\| \|\phi^{j+1}\|.$$

Hence
$$\|\phi^{j+1}\|^2 \leq \|\phi^{j+[(n-1)/n]}\| \|\phi^{j+1}\| + \tau \|f^j\| \|\phi^{j+1}\|,$$
or, dividing by $\|\phi^{j+1}\|$,
$$\|\phi^{j+1}\| \leq \|\phi^{j+[(n-1)/n]}\| + \tau \|f^j\|.$$

Multicomponent Splitting

The elimination of fractional indices yields

$$\|\phi^{j+1}\| \leq \|\phi^j\| + \tau\|f^j\|. \tag{4.44}$$

Since

$$\|\phi^0\| = \|g\|,$$

the elimination of intermediate values of the solution gives

$$\|\phi^{j+1}\| \leq \|g\| + \tau j\|f\|,$$

where

$$\|f\| = \max_j \|f^j\|.$$

The absolute stability of the difference scheme for any instant of the time interval $0 \leq t_j \leq T$ follows from this.

The algorithm above can be generalized to cover the case where the operator A depends on time. This is done by taking a suitable difference approximation of this operator on every subinterval $t_j \leq t \leq t_{j+1}$.

4.4.4. Splitting-Up of Quasilinear Problems

Consider the evolution problem

$$\frac{\partial \phi}{\partial t} + A(t, \phi)\phi = 0 \quad \text{in} \quad D \times D_t, \tag{4.46}$$

$$\phi = g \quad \text{in} \quad D \quad \text{for} \quad t = 0,$$

where the operator A depends both on time and the solution of the problem. Suppose that this operator is a sum of nonnegative operators, that is,

$$A(t, \phi) = \sum_{\alpha=1}^n A_\alpha(t, \phi), \tag{4.47}$$

$A_\alpha(t, \phi) \geq 0$ and is sufficiently smooth. Suppose further that the solution ϕ is also a sufficiently smooth function of time. Take the interval $t_{j-1} \leq t \leq t_{j+1}$ and consider the following splitting-up scheme:

$$\frac{\phi^{j+(1/n)-1} - \phi^{j-1}}{\tau} + A_1^j \frac{\phi^{j+(1/n)-1} + \phi^{j-1}}{2} = 0,$$

$$\cdots\cdots\cdots\cdots\cdots\cdots\cdots\cdots\cdots\cdots\cdots$$

$$\frac{\phi^j - \phi^{j-(1/n)}}{\tau} + A_n^j \frac{\phi^j + \phi^{j-(1/n)}}{2} = 0,$$

$$\frac{\phi^{j+(1/n)} - \phi^j}{\tau} + A_n^j \frac{\phi^{j+(1/n)} + \phi^j}{2} = 0, \tag{4.48}$$

$$\cdots\cdots\cdots\cdots\cdots\cdots\cdots\cdots\cdots\cdots\cdots$$

$$\frac{\phi^{j+1} - \phi^{j+[(n-1)/n]}}{\tau} + A_1^j \frac{\phi^{j+1} + \phi^{j+[(n-1)/n]}}{2} = 0,$$

where

$$A^j = A(t_j, \tilde{\phi}^j),$$
$$\tilde{\phi}^j = \phi^{j-1} + \tau A^{j-1}(t_{j-1}, \phi^{j-1})\phi^{j-1}, \qquad (4.49)$$
$$\tau = t_j - t_{j-1}.$$

By methods similar to those used above for the linear operators which depend on the time variable only, one can easily prove that the splitting-up Scheme (4.48) and (4.49) is a second-order approximation in τ and is absolutely stable. The splitting-up method for the nonhomogeneous quasilinear equations is derived the same way. This opens up a large variety of possibilities for applying component-by-component splitting to the nonstationary quasilinear problems of hydrodynamics, meteorology, oceanology, and other significant fields.

4.5. General Approach to Component-by-Component Splitting

Many problems of mathematical physics can often be solved by splitting the original differential (integral, integrodifferential) equations into simpler ones and then reducing them further to difference forms by means of the algorithms given in this chapter. This technique is closely related to the *weak approximation* of the original equations by equations with a simpler structure; it has been described by Samarskii [3, 15], Yanenko [3], Demidov [15], Dyakonov [15], and subsequently expanded upon by many authors. This problem also will be the subject of our present discussion.

Assume that the following equations describe a certain problem of mathematical physics:

$$\frac{\partial \phi}{\partial t} + A\phi = 0 \quad \text{in} \quad D \times D_t,$$
$$\phi = g \quad \text{in} \quad D \quad \text{for} \quad t = 0. \qquad (5.1)$$

Assume further that

$$A = \sum_{\alpha=1}^{n} A_\alpha, \qquad (5.2)$$

with $A_\alpha \geq 0$. The solution ϕ and the function f are assumed to be sufficiently smooth. On each interval $\Theta_j = \{t_j \leq t \leq t_{j+1}\}$ let us now represent (5.1) in the form

$$\frac{\partial \phi_\alpha}{\partial t} + A_\alpha \phi_\alpha = 0 \quad \text{in} \quad D \times \Theta_j,$$
$$\phi_\alpha^j = \phi_{\alpha-1}^{j+1} \quad \text{in} \quad D \quad \text{for} \quad t = t_j \qquad (5.3)$$
$$(\alpha = 1, 2, \ldots, n).$$

We have used the notation

$$\phi_0^{j+1} = \phi^j, \qquad \phi_n^{j+1} = \phi^{j+1}. \tag{5.4}$$

As has been shown earlier, by applying the Crank–Nicholson scheme to each of the equations, we arrive at the system of difference equations

$$\frac{\phi^{j+(\alpha/n)} - \phi^{j+[(\alpha-1)/n]}}{\tau} + A_\alpha \frac{\phi^{j+(\alpha/n)} + \phi^{j-[(\alpha-1)/n]}}{2} = 0 \tag{5.5}$$

$$(\alpha = 1, 2, \ldots, n),$$

where

$$\phi^{j+(\alpha/n)} = \phi_\alpha^{j+1}; \qquad \phi^{j+1} = \phi_n^{j+1}. \tag{5.6}$$

Suppose that each of the operators A_α is itself represented as

$$A_\alpha = \sum_{\beta=1}^{m_\alpha} A_{\alpha\beta}, \tag{5.7}$$

where $A_{\alpha\beta} \geq 0$. At this point one may question the usefulness of splitting the operator A twice in a row. Is it not more straightforward to split the operator into $A_{\alpha\beta}$ right away at the outset? Let us note that disregarding the formal equivalence, it is useful in many cases to decompose first the complicated problem of mathematical physics into simpler problems and then reduce them later in an independent fashion into even simpler ones.

Consider System (5.3), and let us split it further into more elementary problems using (5.7):

$$\frac{\phi_\alpha^{j+(\beta/m_\alpha)} - \phi_\alpha^{j+[(\beta-1)/m_\alpha]}}{\tau} + A_{\alpha\beta} \frac{\phi_\alpha^{j+(\beta/m_\alpha)} + \phi_\alpha^{j+[(\beta-1)/m_\alpha]}}{2} = 0 \tag{5.8}$$

$$(\alpha = 1, 2, \ldots, n; \quad \beta = 1, 2, \ldots, m_\alpha),$$

where

$$\phi_1^j = \phi^j; \qquad \phi_\alpha^j = \phi_{\alpha-1}^{j+1} \quad (\alpha > 1); \qquad \phi_n^{j+1} = \phi^{j+1}.$$

It is not difficult to see that System (5.8) approximates the original Problem (5.1) with second-order accuracy in τ, provided the operators $A_{\alpha\beta}$ commute. For proof we rearrange first the components of A and write

$$A = \sum_{\alpha=1}^{n} \sum_{\beta=1}^{m_\alpha} A_{\alpha\beta} = \sum_{\gamma=1}^{p} A_\gamma.$$

Then we obtain

$$\frac{\phi^{j+(\gamma/p)} - \phi^{j+[(\gamma-1)/p]}}{\tau} + A_\gamma \frac{\phi^{j+(\gamma/p)} + \phi^{j+[(\gamma-1)/p]}}{2} = 0 \quad (\gamma = 1, 2, \ldots, p), \tag{5.9}$$

which represent a second-order approximation in τ of Problem (5.1), as we know from Section 4.4. This result remains true even if $A_{\alpha\beta}$ depend on time, in which case it is necessary to construct a second-order approximation of the operators $A_{\alpha\beta} = \Lambda_{\alpha\beta}^j$ on each of the intervals $t_j \le t \le t_{j+1}$. If $\Lambda_{\alpha\beta}^j$ do not commute, then using the two-cycle procedure described in Section 4.4, we obtain a second-order scheme on each $t_{j-1} \le t \le t_{j+1}$.

To summarize, the evolution problem of the type of (5.1) with $A_\alpha \ge 0$ can be considered [upon reduction to the particular evolution Problems (5.3)] as a set of new independent evolution problems; if at least one of the elementary evolution problems is represented by the first-order approximation, then the original Problem (5.1) is also approximated with first-order accuracy. If every elementary problem is approximated with a second-order accuracy, then in the framework of the two-cycle procedure in α and β we arrive at a second-order approximation of (5.1). Note that if the operators $A_{\alpha\beta}$ do not commute, we may still obtain a first-order approximation of (5.1) without any reference to the two-cycle procedure. Indeed, in the non-commutative case the original problem becomes

$$\frac{\partial \phi}{\partial t} + \sum_{\alpha=1}^n A_\alpha \phi = 0 \quad \text{in} \quad D \times \Theta_j, \tag{5.10}$$

$$\phi = \phi^i \quad \text{in} \quad D \quad \text{for} \quad t = t_j.$$

The above is reduced to the system

$$\frac{\partial \phi_\alpha}{\partial t} + A_\alpha \phi_\alpha = 0, \qquad \phi_\alpha^j = \phi_{\alpha-1}^{j+1} \qquad (\alpha = 1, 2, \ldots, n). \tag{5.11}$$

Let $A_\alpha = \sum_{\beta=1}^{m_\alpha} A_{\alpha\beta}$. For each of Problems (5.11) we use the two-cycle method:

$$\frac{\phi_\alpha^{j+(\beta/2m_\alpha)} - \phi_\alpha^{j+[(\beta-1)/2m_\alpha]}}{\tau/2} + A_{\alpha\beta} \frac{\phi^{j+(\beta/2m_\alpha)} + \phi^{j+[(\beta-1)/2m_\alpha]}}{2} = 0$$

$$(\beta = 1, 2, \ldots, m_\alpha), \tag{5.12}$$

$$\frac{\phi_\alpha^{j+(\beta/2m_\alpha)} - \phi_\alpha^{j+[(\beta-1)/2m_\alpha]}}{\tau/2} + A_{\alpha, 2m_\alpha+1-\beta} \frac{\phi^{j+(\beta/2m_\alpha)} + \phi^{j+[(\beta-1)/2m_\alpha]}}{2} = 0.$$

$$(\beta = m_\alpha + 1, m_\alpha + 2, \ldots, 2m_\alpha).$$

The initial conditions for (5.12) are correspondingly

$$\phi_1^j = \phi^j, \qquad \phi_\alpha^j = \phi_{\alpha-1}^{j+1} \qquad (\alpha = 2, \ldots, n). \tag{5.13}$$

It is easy to verify that Problem (5.12) approximates any of Problems (5.11) with an accuracy up to τ^2 on the interval $t_j \le t \le t_{j+1}$.

In order that the overall algorithm yields a solution with an accuracy of τ^2 it is necessary to rearrange the basic cycles in addition. Thus instead of (5.11) we have to take

$$\frac{\partial \phi_\alpha}{\partial t} + A_\alpha \phi_\alpha = 0 \quad (\alpha = 1, 2, \ldots, n),$$

$$\phi_1^{j-1} = \phi^{j-1}, \qquad \phi_\alpha^{j-1} = \phi_{\alpha-1}^j \quad (\alpha > 1), \qquad \phi^j = \phi_n^j \tag{5.14}$$

on $t_{j-1} \leq t \leq t_j$, and

$$\frac{\partial \phi_\alpha}{\partial t} + A_{n-\alpha+1}\phi_\alpha = 0 \quad (\alpha = 1, 2, \ldots, n), \tag{5.15}$$

$$\phi_1^j = \phi^j, \quad \phi_\alpha^j = \phi_{\alpha-1}^{j+1} \quad (\alpha > 1), \quad \phi^{j+1} = \phi_n^{j+1}$$

on $t_j \leq t \leq t_{j+1}$. We also assume that each of Problems (5.14) and (5.15) is solved by means of the two-cycle method of the form of (5.12). Note that under the condition $A_{\alpha\beta} \geq 0$ the component-by-component splitting-up method is absolutely stable.

In conclusion let us describe the general splitting-up scheme for the nonhomogeneous equation

$$\frac{\partial \phi}{\partial t} + \sum_{\alpha=1}^{n} A_\alpha \phi = f, \tag{5.16}$$

$$\phi = g \quad \text{for} \quad t = 0$$

on the interval $t_{j-1} \leq t \leq t_{j+1}$, based on the two-cycle method. Consider the weak approximation schemes in the differential form.

Let

$$\frac{\partial \phi_\alpha}{\partial t} + A_\alpha \phi_\alpha = 0 \quad (\alpha = 1, 2, \ldots, n-1),$$

$$\frac{\partial \phi_n}{\partial t} + A_n \phi_n = f + \frac{\tau}{2} A_n f \tag{5.17}$$

on $t_{j-1} \leq t \leq t_j$, and let

$$\frac{\partial \phi_{n+1}}{\partial t} + A_n \phi_{n+1} = f - \frac{\tau}{2} A_n f,$$

$$\frac{\partial \phi_{n+\alpha}}{\partial t} + A_{n-\alpha+1}\phi_{n+\alpha} = 0 \quad (\alpha = 2, 3, \ldots, n) \tag{5.18}$$

on $t_j \leq t \leq t_{j+1}$. Assume that

$$\phi_1(t_{j-1}) = \phi(t_{j-1}), \quad \phi_{\alpha+1}(t_{j-1}) = \phi_\alpha(t_j) \quad (\alpha = 1, 2, \ldots, n) \tag{5.19}$$

and

$$\phi_{\alpha+1}(t_j) = \phi_\alpha(t_{j+1}) \quad (\alpha = n+1, n+2, \ldots, 2n). \tag{5.20}$$

Use now the Crank–Nicholson scheme to solve Equations (5.17)–(5.19) on $t_{j-1} \leq t \leq t_{j+1}$, letting $f = f^j$. We obtain System (4.36).

Together with System (5.17), (5.18), let us consider the following system:

$$\frac{\partial \phi_1}{\partial t} + A_1 \phi_1 = 0,$$

$$\ldots\ldots\ldots\ldots\ldots \tag{5.21}$$

$$\frac{\partial \phi_n}{\partial t} + A_n \phi_n = 0$$

on the interval $t_{j-1} \leq t \leq t_j$,

$$\frac{\partial \phi_{n+1}}{\partial t} = f \tag{5.22}$$

on $t_{j-1} \leq t \leq t_{j+1}$, and

$$\frac{\partial \phi_{n+2}}{\partial t} + A_n \phi_{n+2} = 0,$$

$$\dots \dots \dots \dots \dots \dots \dots \dots \tag{5.23}$$

$$\frac{\partial \phi_{2n+1}}{\partial t} + A_1 \phi_{2n+1} = 0$$

on the interval $t_j \leq t \leq t_{j+1}$. The initial conditions for (5.21) are

$$\phi_1(t_{j-1}) = \phi(t_{j-1}), \quad \phi_\alpha(t_{j-1}) = \phi_{\alpha-1}(t_j) \quad (\alpha = 2, 3, \dots, n), \tag{5.24}$$

for (5.22)

$$\phi_{n+1}(t_{j-1}) = \phi_n(t_j) \tag{5.25}$$

and for (5.23)

$$\phi_\alpha(t_j) = \phi_{\alpha-1}(t_{j+1}) \quad (\alpha = n+2, n+3, \dots, 2n+1). \tag{5.26}$$

The approximation and stability of the schemes obtained guarantee convergence (see Section 1.4).

4.6. Methods of Solving Equations of the Hyperbolic Type

A large class of problems of mathematical physics is related to the equations of hyperbolic type; the origins of the corresponding numerical methods are found in the fundamental contribution by Courant, Friedrichs, and Lewy [7]. A sizable amount of research was later undertaken by Ladyzenskaya [2], Godunov [2], Samarskii [3], Konovalov [15], and others. Recently a number of effective algorithms for hyperbolic-type problems have been constructed and applied in the theory of oscillations, theory of elasticity, etc. for the case of multidimensional regions. For smooth operators, solutions and input data, these algorithms are based on special splitting-up algorithms which will be discussed presently.

4.6.1. The Stabilization Method

Consider the problem

$$\frac{\partial^2 \phi}{\partial t^2} + A\phi = f \quad \text{in} \quad D \times D_t,$$

$$\phi = p, \frac{\partial \phi}{\partial t} = q \quad \text{in} \quad D \quad \text{for} \quad t = 0. \tag{6.1}$$

Methods of Solving Equations of the Hyperbolic Type

We will assume that the operator A is time invariant and that the functions p and q have properties which guarantee a sufficient smoothness of the solutions. Assume further that the operator A is positive-definite; that is,

$$(A\phi, \phi) \geq \gamma^2(\phi, \phi). \tag{6.2}$$

Let us recall that for positive-definite operators A we have

$$\gamma^2 = \alpha\left(\frac{A^* + A}{2}\right),$$

where α is the smallest eigenvalue of the operator $(A + A^*)/2$.

Consider the difference approximation of Equation (6.1) in the form

$$\frac{\phi^{j+1} - 2\phi^j + \phi^{j-1}}{\tau^2} + A\phi^j = f^j. \tag{6.3}$$

It is not difficult to show that on smooth solutions the difference Scheme (6.3) approximates the original Equation (6.1) with an accuracy of τ^2. Let us now complement Equation (6.3) with the initial data. In order not to destroy the approximation order, consider the initial conditions in the following form

$$\phi^0 = p, \qquad \phi^1 = \left(E - \frac{\tau^2}{2}A\right)p + \tau q + \frac{\tau^2}{2}f^0. \tag{6.4}$$

The second of the above relations has been obtained by expanding the solution of (6.1) into a Taylor series around $t = 0$, and subsequently eliminating the derivatives using the equation and the initial conditions from (6.1).

The problem specification (6.3) and (6.4) is thus complete. At this point it is necessary to investigate the stability of Scheme (6.3). To do this, we will use the spectral method.

Let the eigenfunctions u_n and u_n^*, and the eigenvalues $\lambda_n > 0$ correspond to the spectral problems

$$Au = \lambda u, \qquad A^*u^* = \lambda u^*. \tag{6.5}$$

Suppose further that $\{u_n\}$ form a basis. The solution of the equation will be sought in the form

$$\phi^j = \sum_n \phi_n^j u_n, \tag{6.6}$$

where

$$\phi_n^j = (\phi^j, u_n^*).$$

Substituting the Fourier Series (6.6) into (6.3) and subsequently taking the scalar product with u_n^*, we obtain the following equations for the Fourier coefficients:

$$\frac{\phi_n^{j+1} - 2\phi_n^j + \phi_n^{j-1}}{\tau^2} + \lambda_n \phi_n^j = f_n^j. \tag{6.7}$$

The general solution of the homogeneous equation which corresponds to (6.7) will be sought as a power function:

$$\phi_n^j = \eta_n^j. \tag{6.8}$$

(Let us emphasize that the letter j in the left side of the above relation stands for an index, while on the right it indicates the power.) We substitute next (6.8) into (6.7) and take $f_n^j = 0$; the result is the characteristic equation for η_n:

$$\eta_n^2 - 2\left(1 - \frac{\tau^2 \lambda_n}{2}\right)\eta_n + 1 = 0. \tag{6.9}$$

If we assume

$$\left|1 - \frac{\tau^2 \lambda_n}{2}\right| \leqslant 1, \tag{6.10}$$

it is easy to see that the roots of Equation (6.9) are complex conjugate with the amplitude one:

$$|\eta_n| = 1. \tag{6.11}$$

From (6.10) we have

$$\tau^2 \leqslant \frac{4}{\lambda_n} \quad (n = 1, 2, \ldots). \tag{6.12}$$

Clearly, (6.12) holds true for all λ_n, if τ is such that

$$\tau \leqslant \frac{2}{\sqrt{\beta(A)}}, \tag{6.13}$$

where $\beta(A)$ is an upper bound of the spectrum of A. For symmetric operators $\beta(A) = \|A\|$ and consequently

$$\tau \leqslant \frac{2}{\sqrt{\|A\|}}. \tag{6.14}$$

Let us now turn to the implicit difference schemes. Consider

$$\frac{\phi^{j+1} - 2\phi^j + \phi^{j-1}}{\tau^2} + A\frac{\phi^{j+1} + \phi^{j-1}}{2} = f^j. \tag{6.15}$$

The above scheme is of second-order accuracy in τ, and together with (6.4) it approximates Problem (6.1) up to the second power in τ. The characteristic equation for (6.15) has the form

$$\eta_n^2 - \frac{2}{1 + \frac{\tau^2 \lambda_n}{2}} \eta_n + 1 = 0, \tag{6.16}$$

and consequently

$$\eta_n = \frac{1}{1+\frac{\tau^2\lambda_n}{2}} \pm \sqrt{\left(\frac{1}{1+\frac{\tau^2\lambda_n}{2}}\right)^2 - 1}. \qquad (6.17)$$

Hence we can see that for any τ an n

$$|\eta_n| = 1. \qquad (6.18)$$

Scheme (6.15) is absolutely stable (see Richtmyer, Morton [3]). Consider now the case where

$$A = \sum_{\alpha=1}^{n} A_\alpha, \qquad (6.19)$$

with all A_α nonnegative. In order to obtain an approximate solution of (6.1) we exploit in this case the following difference approximation:

$$B\frac{\phi^{j+1} - 2\phi^j + \phi^{j-1}}{\tau^2} + A\phi^j = f^j, \qquad (6.20)$$

where

$$B = \prod_{\alpha=1}^{n}\left(E + \frac{\tau^2}{2}A_\alpha\right). \qquad (6.21)$$

From (6.20) and (6.21) we have that (6.20) is a second-order approximation of (6.1). Since (6.20) can be written as

$$\frac{\phi^{j+1} - 2\phi^j + \phi^{j-1}}{\tau^2} + B^{-1}A\phi^j = B^{-1}f^j, \qquad (6.22)$$

the Fourier analysis will imply the stability of (6.20), (6.21), provided

$$\tau \leqslant \frac{2}{\sqrt{\beta(B^{-1}A)}}. \qquad (6.23)$$

In this fashion the problem of choosing the parameter τ satisfying the stability condition has been reduced to computation of the largest eigenvalue for the problem

$$Au = \lambda Bu, \qquad (6.24)$$

under the assumption that all eigenvalues of $B^{-1}A$ are positive. This problem can be solved by the Lyusternik iterative process.

Let us form the realization of the difference scheme corresponding to Equation (6.20):

$$\left(E + \frac{\tau^2}{2} A_1\right)\xi^{j+(1/n)} = -A\phi^j + f^j,$$

$$\left(E + \frac{\tau^2}{2} A_2\right)\xi^{j+(2/n)} = \xi^{j+(1/n)}, \quad (6.25)$$

$$\dots\dots\dots\dots\dots\dots\dots\dots\dots\dots\dots\dots$$

$$\left(E + \frac{\tau^2}{2} A_n\right)\xi^{j+1} = \xi^{j+[(n-1)/n]},$$

$$\phi^{j+1} = 2\phi^j - \phi^{j-1} + \tau^2 \xi^{j+1}.$$

This problem is solved sequentially for $j = 1, 2, \ldots$ using the initial data of (6.4). Scheme (6.20) is a splitting-up scheme.

4.6.2. Reduction of the Wave Equation to an Evolution Problem

Computer-oriented constructions of absolutely stable second-order difference approximations for hyperbolic equations have led eventually to the necessity of developing special splitting-up methods similar to those for evolution problems.

The basic idea behind the formal reduction of a hyperbolic problem to an evolution problem will be illustrated on a simple example of membrane oscillations, with periodic boundary conditions relative the square $D = \{0 \leq x \leq 1, 0 \leq y \leq 1\}$:

$$\frac{\partial^2 \phi}{\partial t^2} = \frac{\partial}{\partial x} a^2 \frac{\partial \phi}{\partial x} + \frac{\partial}{\partial y} a^2 \frac{\partial \phi}{\partial y} \quad \text{in} \quad D \times D_t,$$

$$\phi = p, \ \frac{\partial \phi}{\partial t} = q \quad \text{in} \quad D \quad \text{for} \quad t = 0, \quad (6.26)$$

where $a^2 = a^2(x, y)$ is the square of the propagation velocity of the disturbances, and $p = p(x, y)$ and $q = q(x, y)$ are given functions. Periodic solutions are known to be sufficiently smooth for our purposes in all arguments x, y, and t.

First of all let us rewrite the wave Equation (6.26) as a system of equations. There results

$$\frac{\partial u}{\partial t} - a \frac{\partial \phi}{\partial x} = 0,$$

$$\frac{\partial v}{\partial t} - a \frac{\partial \phi}{\partial y} = 0 \quad \text{in} \quad D \times D_t, \quad (6.27)$$

$$\frac{\partial \phi}{\partial t} - \left(\frac{\partial au}{\partial x} + \frac{\partial av}{\partial y}\right) = 0.$$

Methods of Solving Equations of the Hyperbolic Type

Let us take the following initial data for the functions u, v and ϕ:

$$u = u^0(x, y), \qquad v = v^0(x, y), \qquad \phi = p(x, y) \quad \text{for} \quad t = 0. \tag{6.28}$$

The functions u^0, v^0 can be chosen more or less arbitrarily as long as they satisfy the relation

$$\frac{\partial au^0}{\partial x} + \frac{\partial av^0}{\partial y} = q(x, y). \tag{6.29}$$

Introduce next the matrix A and the vector ϕ:

$$A = \begin{Vmatrix} 0 & 0 & -a\dfrac{\partial}{\partial x} \\ 0 & 0 & -a\dfrac{\partial}{\partial y} \\ -\dfrac{\partial}{\partial x}a & -\dfrac{\partial}{\partial y}a & 0 \end{Vmatrix}, \qquad \phi = \begin{vmatrix} u \\ v \\ \phi \end{vmatrix}.$$

With this notation we can rewrite System (6.27) and the initial data of (6.28) as follows:

$$\frac{\partial \phi}{\partial t} + A\phi = 0 \quad \text{in} \quad D,$$

$$\phi = \phi^0 \quad \text{in} \quad D \quad \text{for} \quad t = 0, \tag{6.30}$$

where ϕ^0 has the components u^0, v^0, ϕ^0.

In order to investigate the properties of the symmetric operator A, let us form the functional

$$(A\phi, \phi) = -\int_D \left[\frac{\partial}{\partial x}(au\phi) + \frac{\partial}{\partial y}(av\phi) \right] dD = -\int_S au_n \phi \, dS. \tag{6.31}$$

Here u_n is the component of the vector $u = ui + vj$, which is orthogonal to the surface S. Because of the periodicity of a, ϕ (and also the derivatives of the solution), the values of u_n at the boundary points of the square D, which are symmetric relative to its center, are equal in magnitude and of opposite sign. Hence the surface integral in (6.31) becomes zero and we obtain the condition

$$(A\phi, \phi) = 0. \tag{6.32}$$

Note that if we require that the membrane be fixed in a frame ($\phi = 0$) rather than the periodicity conditions, then even in this case (6.32) holds true, as can be seen from (6.31). This remark is correct for an arbitrary region D. Condition (6.32) guarantees the uniqueness of the solution for the problem.

Indeed, taking the scalar product of (6.30) with ϕ and exploiting Relation (6.32), we obtain

$$\frac{d}{dt}\|\phi\|^2 = 0, \qquad (6.33)$$

where

$$\|\phi\| = \left\{\int_D (u^2 + v^2 + \phi^2)\, dD\right\}^{1/2}.$$

Assume that

$$u^0 = 0, \qquad v^0 = 0, \qquad \phi^0 = 0.$$

Then

$$\|\phi^0\| = 0. \qquad (6.34)$$

Solving Equation (6.33) with the initial conditions of (6.34), we obtain

$$\|\phi\| = \|\phi^0\| = 0.$$

This means that

$$u = v = \phi = 0$$

at all times, which proves uniqueness.

Let us now turn to the problem of formulating the splitting-up method for Problem (6.30). To this end let us introduce the following matrices:

$$A_1 = \begin{Vmatrix} 0 & 0 & -a\dfrac{\partial}{\partial x} \\ 0 & 0 & 0 \\ -\dfrac{\partial}{\partial x}a & 0 & 0 \end{Vmatrix}, \quad A_2 = \begin{Vmatrix} 0 & 0 & 0 \\ 0 & 0 & -a\dfrac{\partial}{\partial y} \\ 0 & -\dfrac{\partial}{\partial y}a & 0 \end{Vmatrix}.$$

Clearly

$$A = A_1 + A_2. \qquad (6.35)$$

In analogy with the foregoing one can show that

$$(A_1\phi, \phi) = 0, \qquad (A_2\phi, \phi) = 0. \qquad (6.36)$$

Therefore, on each interval $t_j \le t \le t_{j+1}$ Problem (6.30) can be solved using one of the splitting-up methods discussed in Section 4.3: either the stabilization method, or the predictor-corrector method, or, finally, the component-by-component splitting-up method. Note at this point that if we consider a multidimensional wave equation rather than two dimensions, then it would be desirable to use the component-by-component splitting, since it leads to an absolutely stable, second-order difference approximation under the minimal requirements regarding the definiteness of Equation (6.36)-type operators.

Let us now perform the reduction of Problem (6.30) on each of the intervals $t_j \leq t \leq t_{j+1}$, for example using the component-by-component splitting-up method. We have

$$\frac{\phi^{j+1/2} - \phi^j}{\tau} + A_1 \frac{\phi^{j+1/2} + \phi^j}{2} = 0,$$
$$\frac{\phi^{j+1} - \phi^{j+1/2}}{\tau} + A_2 \frac{\phi^{j+1} + \phi^{j+1/2}}{2} = 0.$$
(6.37)

Going back to the scalar form, these equations become

$$\frac{u^{j+1/2} - u^j}{\tau} = a \frac{\partial}{\partial x}\left(\frac{\phi^{j+1/2} + \phi^j}{2}\right),$$
$$\frac{v^{j+1/2} - v^j}{\tau} = 0,$$
$$\frac{\phi^{j+1/2} - \phi^j}{\tau} = \frac{\partial}{\partial x}\left(a \frac{u^{j+1/2} + u^j}{2}\right)$$
(6.38)

and

$$\frac{u^{j+1} - u^{j+1/2}}{\tau} = 0,$$
$$\frac{v^{j+1} - v^{j+1/2}}{\tau} = a \frac{\partial}{\partial y}\left(\frac{\phi^{j+1} + \phi^{j+1/2}}{2}\right),$$
$$\frac{\phi^{j+1} - \phi^{j+1/2}}{\tau} = \frac{\partial}{\partial y}\left(a \frac{v^{j+1} + v^{j+1/2}}{2}\right).$$
(6.39)

Taking into account that $v^{j+1/2} = v^j$, and $u^{j+1/2} = u^{j+1}$, System (6.38) can be somewhat simplified by writing

$$\frac{u^{j+1} - u^j}{\tau} = a \frac{\partial}{\partial x}\left(\frac{\phi^{j+1/2} + \phi^j}{2}\right),$$
$$\frac{\phi^{j+1/2} - \phi^j}{\tau} = \frac{\partial}{\partial x}\left(a \frac{u^{j+1} + u^j}{2}\right),$$
(6.40)

$$\frac{v^{j+1} - v^j}{\tau} = a \frac{\partial}{\partial y}\left(\frac{\phi^{j+1} + \phi^{j+1/2}}{2}\right),$$
$$\frac{\phi^{j+1} - \phi^{j+1/2}}{\tau} = \frac{\partial}{\partial y}\left(a \frac{v^{j+1} + v^j}{2}\right).$$
(6.41)

System (6.40) is solved for u^{j+1} and $\phi^{j+1/2}$, and (6.41) for v^{j+1} and ϕ^{j+1}.

Difference approximations with respect to x and y will be chosen so as to obtain eventually the absolutely stable schemes for u, v, and ϕ, which will be of second-order accuracy and which will preserve Conditions (6.36) for the finite-difference representations of A_1, A_2. Put

$$\frac{u_{k,l}^{j+1} - u_{k,l}^{j}}{\tau} = \frac{a_{k,l}}{h}\left[\left(\frac{\phi^{j+1/2} + \phi^{j}}{2}\right)_{k,l} - \left(\frac{\phi^{j+1/2} + \phi^{j}}{2}\right)_{k-1,l}\right],$$

$$\frac{\phi_{k,l}^{j+1/2} - \phi_{k,l}^{j}}{\tau} = \frac{1}{h}\left[a_{k+1,l}\left(\frac{u^{j+1} + u^{j}}{2}\right)_{k+1,l} - a_{k,l}\left(\frac{u^{j+1} + u^{j}}{2}\right)_{k,l}\right].$$
(6.42)

In the first of the above equations we use a *backward* difference, while a *forward* difference is used in the second one. Similarly for (6.41)

$$\frac{v_{k,l}^{j+1} - v_{k,l}^{j}}{\tau} = \frac{a_{k,l}}{h}\left[\left(\frac{\phi^{j+1} + \phi^{j+1/2}}{2}\right)_{k,l} - \left(\frac{\phi^{j+1} + \phi^{j+1/2}}{2}\right)_{k,l-1}\right],$$

$$\frac{\phi_{k,l}^{j+1} - \phi_{k,l}^{j+1/2}}{\tau} = \frac{1}{h}\left[a_{k,l+1}\left(\frac{v^{j+1} + v^{j}}{2}\right)_{k,l+1} - a_{k,l}\left(\frac{v^{j+1} - v^{j}}{2}\right)_{k,l}\right].$$
(6.43)

The indices k and l in (6.42) and (6.43) run through positive integers up to $N - 1$.

Consider the case where, for instance, we have the following condition on the boundary of the region D:

$$\phi = 0 \quad \text{on} \quad \partial D \times D_t. \tag{6.44}$$

Let us project (6.44) on the net $D_h \times D_\tau$; we have

$$\phi_{0,l}^{j} = \phi_{N,l}^{j} = 0 \quad \text{and} \quad \phi_{k,0}^{j} = \phi_{k,N}^{j} = 0. \tag{6.45}$$

both for integer and fractional indices j.

Note that (6.42) together with (6.45) can be used to compute $u_{k,l}^{j+1}$ and $v_{k,l}^{j+1}$ at all boundary net points, provided $\phi_{k,l}^{j}$, $\phi_{k,l}^{j+1/2}$ and $\phi_{k,l}^{j+1}$ are known.

Next, let us eliminate the variables $u_{k,l}^{j+1}$ and $v_{k,l}^{j+1}$ from (6.42) and (6.43) to obtain the difference equations for $\phi_{k,l}$. To this end let us introduce the auxiliary variables

$$\phi_{k,l}^{j+1/4} = \tfrac{1}{2}(\phi_{k,l}^{j+1/2} + \phi_{k,l}^{j}), \qquad \phi_{k,l}^{j+3/4} = \tfrac{1}{2}(\phi_{k,l}^{j+1} + \phi_{k,l}^{j+1/2}). \tag{6.46}$$

Equations (4.42) and (6.43) can now be written in the form

$$\mu_{k+1,l}^{2}(\phi_{k+1,l}^{j+1/4} - \phi_{k,l}^{j+1/4}) - \mu_{k,l}^{2}(\phi_{k,l}^{j+1/4} - \phi_{k-1,l}^{j+1/4}) - \phi_{k,l}^{j+1/4} = -f_{k,l}^{j+1/4},$$
$$\mu_{k,l+1}^{2}(\phi_{k,l+1}^{j+3/4} - \phi_{k,l}^{j+3/4}) - \mu_{k,l}^{2}(\phi_{k,l}^{j+3/4} - \phi_{k,l-1}^{j+3/4}) - \phi_{k,l}^{j+3/4} = -f_{k,l}^{j+3/4},$$
(6.47)

where

$$\mu_{k,l} = \frac{\tau a_{k,l}}{2h}$$

$$f_{k,l}^{j+1/4} = 2\phi_{k,l}^{j} + 2(\mu_{k+1,l}u_{k+1,l}^{j} - \mu_{k,l}u_{k,l}^{j}), \tag{6.48}$$

$$f_{k,l}^{j+3/4} = 2\phi_{k,l}^{j+1/2} + 2(\mu_{k,l+1}v_{k,l+1}^{j} - \mu_{k,l}v_{k,l}^{j}).$$

We have thus obtained the following algorithm for the numerical solution of (6.30).

First, determine the initial fields of functions $u_{k,l}^0$, $v_{k,l}^0$, and $\phi_{k,l}^0$, with $u_{k,l}^0$, $v_{k,l}^0$ satisfying the discrete analog of Condition (6.29). Then using the first of the formulae of (6.48), find $f_{k,l}^{j+1/4}$ and solve the first of the difference Equations (6.47) under Condition (6.45) on the boundary of D^h. Next compute

$$\phi_{k,l}^{j+1/2} = 2\phi_{k,l}^{j+1/4} - \phi_{k,l}^j,$$

where $\phi_{k,l}^{j+1/4}$ are found with the help of Relations (6.46). Solve the second equation of Equations (6.47) using Condition (6.45). After this exploit (6.46) to obtain

$$\phi_{k,l}^{j+1} = 2\phi_{k,l}^{j+3/4} - \phi_{k,l}^{j+1/2}.$$

The quantities $\phi_{k,l}^{j+1}$ are further used for finding $u_{k,l}^{j+1}$ and $v_{k,l}^{j+1}$, with the help of the first relations from (6.42) and (6.43). Thus the algorithm is complete.

Finally, it is to be noted that this method is quite easily generalized to more complicated equations of hyperbolic type; usually it leads to absolutely stable, second-order approximation schemes with minimal requirements on the operators A_α.

Chapter 5

Numerical Methods for Some Inverse Problems

In the current literature the term *inverse problems* designates various kinds of problems of mathematical physics. Broad classes of inverse problems have been studied by Levitan [16], Lavrentiev, Romanov and Vasilyev [16], and others.

We will consider two types of inverse problems. The first type involves determining past states of a process. For example, the problem of finding the initial distribution of temperature on an object, given the current field of temperatures. In the second type of problems we try to identify the coefficients of the operator with a known structure in terms of information provided by some functionals of the solution. An instance of this is the inverse problem for the Sturm–Liouville equation, in which it is necessary to determine the coefficients of a second-order differential equation using the properties of the spectral function corresponding to a certain boundary-value problem.

Inverse problems of mathematical physics are often *ill-posed* (in the classical sense): small perturbations in the observed functionals may result into large changes in the corresponding solutions. The notion of a *well-posed* problem as well as an example of an ill-posed problem were given by Hadamard at the beginning of the century.

For a long time ill-posed problems of mathematical physics were considered uninteresting and drew little attention. Intensive research into these problems has been triggered by a need to interpret geophysical data. Those who have significantly contributed to the theory of ill-posed problems of mathematical physics (in the classical sense of Hadamard) include Tikhonov [16], Lavrentiev [2, 16], John [16], Ivanov [16], Turchin [16], and others.

It has been shown that for ill-posed problems to possess stable solutions relative to data perturbations one has to impose certain additional restrictions on the admissible solutions. These problems have become known as *conditionally well-posed*.

A need for approximate solutions of conditionally well-posed problems with inaccurate data has led to the notion of a *regularization family*: along with the conditionally well-posed problem in question we construct a parametrized family of well-posed problems (the regularization family), which has the property that for a parameter approaching its limit the corresponding sequence of solutions of these problems approaches the solution of the conditionally well-posed problem. It has been shown that, depending on the accuracy of the data, the parameter (regularization parameter) can be chosen in such a way that the approximate solution of the corresponding problem from the regularization family will turn out to solve approximately the original conditionally well-posed problem.

Broad classes of conditionally well-posed problems have been studied by the authors we have just listed. We will constrain ourselves to one class of conditionally well-posed inverse problems, namely those connected with the evolution equations. Our regularization method will be based on techniques from Fourier analysis.

5.1. Basic Definitions and Examples

Consider a Hilbert space F. Each of its elements can be represented in the form of a Fourier series relative to a complete biorthogonal system of functions $\{u_n\}$ and $\{u_n^*\}$. Thus ($f \in F$)

$$f = \sum_{n=1}^{\infty} f_n u_n, \qquad (1.1)$$

where

$$f_n = (f, u_n^*). \qquad (1.2)$$

Let us construct a subspace $\Phi \subset F$, generated by the elements with no more than N nonzero terms in Expansion (1.1), which correspond to the most significant components:

$$f = \sum_{n=1}^{N} f_n u_n. \qquad (1.3)$$

Of course, such a subspace is contained in the space of all functions expandable into a Fourier series.

Consider now the equation

$$A\phi = f, \qquad (1.4)$$

where A is a linear operator with an unbounded inverse. To find its solution let us use the iterations

$$\phi^{j+1} = \phi^j - \tau(A\phi^j - f), \qquad \phi^0 = 0 \qquad (1.5)$$

or,

$$\phi^{j+1} = T\phi^j + \tau f, \qquad \phi^0 = 0, \qquad (1.6)$$

where $T = E - \tau A$ is the transition operator.

Suppose that (1.4) has a unique solution belonging to the subspace Φ. Further, let the Hilbert space be endowed with the norm

$$\|T\|_F^2 = \sup_{\phi \in F} \frac{(T\phi, T\phi)}{(\phi, \phi)} = \beta(T^*T) > 1,$$

while on Φ we have

$$\|T\|_\Phi^2 = \sup_{\phi \in \Phi} \frac{(T\phi, T\phi)}{(\phi, \phi)} < 1. \tag{1.7}$$

Under such assumptions the iterative Process (1.6) will converge on the elements $\phi^j \in \Phi$, while diverging on F as a whole. In order to ensure convergence, we require that at each step the approximate solution belong to Φ: $\phi^j \in \Phi$. This can be easily implemented in a constructive manner.

Suppose that at some step $\phi^j \in \Phi$. Using the recursive Relation (1.6) for ϕ^{j+1} we note that $T\phi^j$ may possibly lie outside Φ. In order that ϕ^j be from Φ we must have that the terms in the expression

$$\phi^j = \sum_{n=1}^{\infty} \phi_n^j u_n$$

whose indices are bigger than N be zero. From an algorithmic standpoint this can be conveniently achieved by computing only the first N Fourier coefficients $\phi_k^j = (\phi^j, u_k^*)$ ($k = 1, 2, \ldots, N$) and subsequently forming the finite sum

$$\phi^j = \sum_{k=1}^{N} \phi_k^j u_k. \tag{1.8}$$

If we continue with this procedure throughout, then all the test-functions ϕ^j—the approximate solutions of the problem—belong to Φ. The norm of the transition operator restricted to Φ is smaller than unity, and the process will thus converge to an element of Φ.

The above arrangement is by no means the only one ensuring that the computational algorithm will not take us outside the given subspace.

Let us next turn to conditionally well-posed problems of mathematical physics. Consider again Equation (1.4) with a symmetric positive-definite operator A. Assume further that $f \in \Phi$ and $\phi \in \Phi$, where Φ is a subspace of the Hilbert space F.

The problem (1.4) will be called conditionally well-posed if its solution belongs to the subspace Φ and if

$$\|\phi\|_\Phi \leq M_0 \|f\|_\Phi, \quad \phi \in \Phi, \tag{1.9}$$

where M_0 is a constant and

$$\|f\|_\Phi < \infty. \tag{1.10}$$

The solution of (1.4) is formally written as

$$\phi = A^{-1} f. \tag{1.11}$$

Its norm on the elements of the subspace Φ can be estimated as follows:

$$\|\phi\|_\Phi \leq \|A^{-1}\|_\Phi \|f\|_\Phi. \tag{1.12}$$

Basic Definitions and Examples

Hence we have that in order for the problem to be well-posed it is enough to require that

$$\|A^{-1}\|_\Phi \leq M. \tag{1.13}$$

The following significant detail is of interest: if A is symmetric, then

$$\|A^{-1}\|_\Phi = \sup_{\phi \in \Phi} \frac{(A^{-1}\phi, \phi)}{(\phi, \phi)}; \qquad \|A^{-1}\|_F = \sup_{\phi \in F} \frac{(A^{-1}\phi, \phi)}{(\phi, \phi)}. \tag{1.14}$$

It is easily seen that for $\Phi \subset F$

$$\|A^{-1}\|_\Phi \leq \|A^{-1}\|_F = 1/[\alpha(A)].$$

Thus the question of whether the problem is well-posed or not may be equivalently answered by both (1.9) and (1.13). Moreover, if the constant M is given, the main problem in arranging the computational algorithm comes down to constructing the subspace Φ. This involves not only the very solution ϕ but also the whole sequence of its approximations ϕ^j.

Let us next investigate another feature of conditionally well-posed problems, namely the accuracy of input data. Here we minimally have to deal with the error resulting from the difference approximation of (1.4) or the error caused by the inaccuracies in f and the coefficients of A.

Denote by $\bar{A}$, $\bar{f}$, and $\bar{\phi}$ the exact operator, vector, and solution correspondingly, so that

$$\bar{A}\bar{\phi} = \bar{f},$$

and assume

$$A = \bar{A} + \delta A, \qquad f = \bar{f} + \delta f. \tag{1.15}$$

Assume we know *a priori* that

$$\|\delta A\|_\Phi \leq \varepsilon_1; \tag{1.16}$$

$$\|\delta f\|_\Phi \leq \varepsilon_2. \tag{1.17}$$

In solving (1.4) we eventually have to arrange the iterative process in such a way that the resulting approximations belong to the subspace Φ. If A is a positive matrix, one can use various iterations which converge to (1.4) as we have seen in Chapter 3. In this case the subspace of approximate solutions Φ coincides with the Hilbert (Euclidean) space F. One must say that this circumstance is one of the conveniences provided by the positivity assumption. Taking the corresponding choice of τ, the iterative process

$$\phi^{j+1} = \phi^j - \tau(A\phi^j - f), \qquad \phi^0 = 0 \tag{1.18}$$

converges and may be optimized, for instance, by properly terminating the process (see Section 3.7), say, at the j_0th step, and with the *a priori* information from Equations (1.16) and (1.17) taken into account.

Suppose next that the operator A in (1.4) is symmetric and its spectrum contains both positive and negative eigenvalues. This situation is typical for conditionally well-posed problems.

If the original matrix A is not symmetric, the Gauss transformation allows one to replace the problem by the one in which the operator is symmetric and positive semidefinite.

Under the conditions stated, Process (1.18) can be shown to diverge. Indeed, let

$$\phi = \sum_n \phi_n u_n, \qquad f = \sum_n f_n u_n, \qquad (1.19)$$

where $\{u_n\}$ is a complete eigenfunction system of A. Substituting (1.19) into (1.18) and taking the inner product with u_n, we obtain the following recursive relation for the Fourier coefficients:

$$\phi_n^{j+1} = \phi_n^j - \tau(\lambda_n \phi_n^j - f_n), \qquad \phi_n^0 = 0$$

For the residual $\xi^j = A\phi^j - f$ we then have

$$\xi_n^{j+1} = (1 - \tau\lambda_n)\xi_n^j, \qquad \xi_n^0 = -f_n. \qquad (1.20)$$

Hence

$$\xi_n^j = -(1 - \tau\lambda_n)^j f_n. \qquad (1.21)$$

Consequently

$$\xi^j = -\sum_n (1 - \tau\lambda_n)^j f_n u_n. \qquad (1.22)$$

The iterative Process (1.18) converges if

$$\lim_{j \to \infty} \xi^j = 0.$$

If the operator A had for its eigenvalues only positive numbers from the interval

$$\alpha(A) \leq \lambda_n(A) \leq \beta(A),$$

then Process (1.19) could be stabilized by a choice of τ from

$$0 < \tau < \frac{2}{\beta} \qquad (1.23)$$

In our case, however, the matrix A may have negative eigenvalues. If we take τ from Interval (1.23), then all the residual components corresponding to the positive eigenvalues λ will diminish at the rate T_n^j as the iterations progress ($T_n = (1 - \tau\lambda_n)^j < 1$, j indicating the power). As far as the components corresponding to the negative eigenvalues are concerned, we have

$$T_n^j = (1 - \tau\lambda_n)^j > 1,$$

so that these residual components will grow, resulting in a divergent process.

Thus Iterations (1.18), in which the test functions ϕ^j span the whole Hilbert space, diverge.

The two-step residual method (see Section 3.2.1) may serve as an example of a process which does not take the approximating sequence outside Φ:

$$\phi^{j+1} = \phi^j - \tau_j(A\phi^j - f) - \gamma_j A^*(A\phi^j - f).$$

Let us briefly comment on a practical computational approach to conditionally well-posed problems. These problems usually lead to systems of linear equations which involve ill-conditioned matrices of a general structure. As a rule, they can be solved using the multistep residual methods (see Section 3.2.1), thus providing a fast convergence. Another possibility is to use the method of conjugate gradients preceded by the symmetrization of the equations via the Gauss transformation. This latter procedure will be described later in connection with the inverse evolution problems (see Section 5.3). In any case, the iterations should be interrupted as soon as the norm of the residual becomes approximately equal to the *a priori* input data error, that is $\|\xi^j\| \approx \varepsilon_1 + \varepsilon_2$. As we have pointed out in Section 3.6, we have at this point the highest attainable accuracy of the solution.

5.2. Fourier Series Method for Inverse Evolution Problems

Let A be a time invariant, positive matrix with a real spectrum in the interval $\alpha(A) \leq \lambda \leq \beta(A)$, and let the vector function ϕ solve the following Cauchy problem:

$$\frac{d\phi}{dt} - A\phi = 0 \quad (0 \leq t \leq t_0), \tag{2.1}$$

$$\phi = g \quad \text{for} \quad t = 0,$$

where g is a given vector. Consider the spectral problems

$$Au = \lambda u; \quad A^*u^* = \lambda u^*. \tag{2.2}$$

and suppose they define biorthogonal bases $\{u_n\}$, $\{u_n^*\}$. Writing

$$\phi = \sum_n \phi_n u_n, \quad g = \sum_n g_n u_n, \tag{2.3}$$

substituting into (2.1), and multiplying the result by u_n^* (in the sense of inner product), we obtain the following system of ordinary differential equations for the Fourier coefficients:

$$\frac{d\phi_n}{dt} - \lambda_n \phi_n = 0,$$

$$\phi_n = g_n \quad \text{for} \quad t = 0 \tag{2.4}$$

$$(n = 1, 2, \ldots, N).$$

Solving each of the above equations separately yields

$$\phi_n = g_n e^{\lambda_n t} \quad (n = 1, 2, \ldots, N), \tag{2.5}$$

and hence the solution of (2.1) can be represented in the form

$$\phi(t) = \sum_{n=1}^{N} g_n e^{\lambda_n t} u_n. \tag{2.6}$$

Thus, our solution is given in terms of a Fourier sum, each term of which grows exponentially in time, the rate of growth being dependent on the corresponding eigenvalue λ_n.

Suppose we are interested in a solution which is physically meaningful. Thus consider a similar problem as (2.1), but now well-posed:

$$\frac{d\phi}{dt} - A\phi = 0 \quad (0 \leqslant t \leqslant t_0),$$
$$\phi = h \quad \text{for} \quad t = t_0. \tag{2.7}$$

Similarly as above we obtain

$$\phi = \sum_{n=1}^{N} h_n e^{-\lambda_n (t_0 - t)} u_n. \tag{2.8}$$

Let us require, that for $t = 0$ the solution of (2.8) coincides with the vector g from (2.1). From this we obtain the relation between the Fourier coefficients of h and g:

$$g_n = h_n e^{-\lambda_n t_0}. \tag{2.9}$$

In this manner we have derived a rather simple formula for g in terms of h:

$$g = \sum_{n=1}^{N} h_n e^{-\lambda_n t_0} u_n. \tag{2.10}$$

Moreover, small errors in h (or h_n) do not cause large errors in g. Our problem, however, is just the opposite to the above: what we have at our disposal is the information regarding g, while h is to be computed in terms of g by the formula

$$h = \sum_{n=1}^{N} g_n e^{\lambda_n t_0} u_n. \tag{2.11}$$

If we had accurate information about the function g and could compute without error, the function h could be reconstructed via Equation (2.11) without difficulty. In our situation, however, the function g is known only approximately, with the corresponding error bound given *a priori*. At the same time the numerical procedures are subject to round-off errors in the computer. In view of these two limitations the computation of h by Equation (2.11) becomes more difficult.

Suppose first of all that the system of eigenvectors u_n is known and that the Fourier expansion of the initial data may be used to draw useful information including sufficiently accurate error estimates for the Fourier components g_n.

If the problem is of a statistical nature and can be repeated many times, well-developed correlation techniques allow a significant increase in the accuracy of the data entering g_n, even if the error of a single measurement drastically excedes the relevant information. In any case, the preliminary processing of the observation data may serve to draw conclusions regarding the systematic (or random, if only one single measurement is available) error in g_n. Thus we have that for any n

$$g_n = \bar{g}_n(1 + \delta_n),$$

where g_n denotes the exact value (not known *a priori*), and δ_n is a relative inaccuracy which we consider as known.

The error estimate δ_n is usually small for the low-frequency components, but grows quickly with the frequency as a rule. Therefore, starting with some index, the coefficients g_n describe in fact the errors in the input data. Returning now back to Equation (2.11), it follows that the highest-frequency components grow at the highest exponential rate. Hence if we do not take notice and leave these parasitic frequencies in, we may obtain a wrong result: these components contain no practical useful information, and being multiplied by large factors $\exp(\lambda_n t_0)$ they may sometimes cause an irrepairable damage to the solution of the problem. The primary task is therefore to determine the information content of the coefficients g_n.

Suppose that using the *a priori* information it has been concluded that the relative inaccuracies in the first n_0 coefficients g_n is smaller than η, i.e., $\delta_n < \eta$, where η is the largest admissible error. Algorithm (2.11) then becomes similar to the one we have already considered when constructing the subspace Φ for Problem (2.1). The only thing we need is to drop the parasitic components in (2.11):

$$h = \sum_{n=1}^{n_0} g_n e^{\lambda_n t_0} u_n. \tag{2.12}$$

Choosing n_0 may become a problem in the case where the operators A and A^* are complicated enough so that the corresponding spectral problem can not be solved in a simple manner.

To illustrate this situation consider the algorithm based on the iterative method formulated in Section 1.1. The eigenfunctions u_n and u_n^* of major significance (and the corresponding eigenvalues) can be determined via the orthogonalization algorithm, the essence of which is as follows: Assuming the largest eigenvalue has already been found, the functions u_1 and u_1^*, corresponding to the smallest eigenvalue $\alpha(A)$, are obtained by the following iterations (see Section 1.1):

$$\psi_1^{(n+1)} = B \frac{\psi_1^{(n)}}{\|\psi_1^{(n)}\|}, \qquad \psi_1^{(0)} = g, \tag{2.13}$$

and

$$\psi_1^{*(n+1)} = B^* \frac{\psi_1^{*(n)}}{\|\psi_1^{*(n)}\|}, \qquad \psi_1^{*(0)} = g, \tag{2.14}$$

where
$$B = \beta(A)E - A, \qquad B^* = \beta(A)E - A^*. \tag{2.15}$$
Then
$$u_1 = \frac{1}{\sqrt{c_1}} \lim_{n \to \infty} \psi_1^{(n)}, \qquad u_1^* = \frac{1}{\sqrt{c_1}} \lim_{n \to \infty} \psi_1^{*(n)}, \tag{2.16}$$
and the normalizing constant c_1 is chosen from the condition
$$(u_1, u_1^*) = 1.$$

Having chosen the eigenfunctions u_1 and u_1^*, the eigenvalue $\alpha(A) = \lambda_1$ may be determined as the limit
$$\alpha(A) = \lim_{n \to \infty} \frac{\|A\psi_1^{(n)}\|}{\|\psi_1^{(n)}\|}. \tag{2.17}$$

The computation of the next eigenfunctions in order of increasing eigenvalues follows the same pattern:
$$\psi_k^{(n+1)} = B \frac{\psi_k^{(n)}}{\|\psi_k^{(n)}\|}, \qquad \psi_k^{(0)} = g - \sum_{k'=1}^{k-1} (g, u_{k'}^*)u_{k'}, \tag{2.18}$$
and
$$\psi_k^{*(n+1)} = B \frac{\psi_k^{*(n)}}{\|\psi_k^{*(n)}\|}, \qquad \psi_k^{*(0)} = g - \sum_{k'=1}^{k-1} (g, u_{k'})u_{k'}^*. \tag{2.19}$$
Then
$$u_k = \frac{1}{\sqrt{c_k}} \lim_{n \to \infty} \psi_k^{(n)}; \qquad u_k^* = \frac{1}{\sqrt{c_k}} \lim_{n \to \infty} \psi_k^{*(n)};$$
$$\lambda_k = \lim_{n \to \infty} \frac{\|A\psi_k^{(n)}\|}{\|\psi_k^{(n)}\|}. \tag{2.20}$$

Note that by the special choice we took for the initial approximations, the elements $\psi_k^{(0)}$ and $\psi_k^{*(0)}$ have zeros as their first $(k - 1)$ components. The lowest-frequency nonzero components in the iterative processes are those with the index k. Unfortunately this can only happen if we are able to compute without errors. Since any computer can only operate with words of restricted lengths, the round-off errors will make the first $(k - 1)$ components nontrivial, and in fact they will steadily increase with progressing iterations. In order that the error does not become large at the terminal step, it is recommended to orthogonalize the quantities $\psi_k^{(n)}$ and $\psi_k^{*(n)}$ several times in the course of computations (especially for large numbers k). Convergence can be accelerated using various well-developed linear-algebraic approaches (Faddeev and Faddeeva [8], Gavurin [9], and others).

5.3. Inverse Evolution Problems with Time-Varying Operators

Consider the evolution problem

$$\frac{d\phi}{dt} - A(t)\phi = 0, \qquad 0 \leq t \leq t_0, \tag{3.1}$$

$$\phi = g \quad \text{for} \quad t = 0,$$

with $A > 0$ depending on time. We assume as before that (3.1) resulted by reducing a problem of mathematical physics to a system of ordinary differential equations. Since the Fourier method is no longer applicable in this case, (3.1) must be solved by numerical methods.

Let us consider one of the possible algorithms. In correspondence to (3.1) let us define the following *model* [a problem in a sense close to (3.1)]:

$$\frac{d\bar{\phi}}{dt} - \bar{A}\bar{\phi} = 0, \qquad 0 \leq t \leq t_0, \tag{3.2}$$

$$\bar{\phi} = g \quad \text{for} \quad t = 0.$$

where the operator $\bar{A} > 0$ is now time invariant, has a positive spectrum $\alpha(\bar{A}) \leq \lambda(\bar{A}) \leq \beta(\bar{A})$, and in some sense is close to the operator $A(t)$. To be specific, let

$$A(t) = \bar{A} + \delta A(t), \tag{3.3}$$

where

$$\|\delta A(t)\| \ll \|\bar{A}\| \tag{3.4}$$

for any t from the interval $0 \leq t \leq t_0$. Equations (3.2) will be used for obtaining a necessary *a priori* information to be used later in designing the numerical algorithm for solving the original Problem (3.1).

Using the methods for time-invariant problems from Section (3.2), we first determine the eigenvectors u_n and u_n^* ($n = 1, \ldots, m$) with the major information content (from the viewpoint of the input data errors) and the corresponding eigenvalues λ_n. The remaining eigenvectors are not needed, since the Fourier components g_n ($n = m + 1, \ldots, N$) are to be dropped as the numerical error involved exceeds (sometimes quite considerably) the useful information they may carry. Thus

$$g = \sum_{n=1}^{m} g_n u_n, \tag{3.5}$$

where

$$g_n = (g, u_n^*).$$

As a result, the solution $\bar{\phi}$ corresponding to the model of (3.2) can be written in the form

$$\bar{\phi}(t) = \sum_{n=1}^{m} g_n e^{\lambda_n t} u_n, \qquad (3.6)$$

$0 \leq t \leq t_0$.

Let us try to solve the model Problem (3.2) numerically. For this consider, for instance, the following difference scheme of second-order accuracy in $\Delta t = \tau$:

$$\frac{\bar{\phi}^{j+1} - \bar{\phi}^j}{\tau} - \bar{A}\frac{\bar{\phi}^{j+1} + \bar{\phi}^j}{2} = 0, \qquad \bar{\phi}^0 = g \qquad (3.7)$$

$$(j = 1, 2, \ldots, j_0).$$

Assume we know the set of eigenelements u_n and u_n^*. (This assumption is made only for theoretical purposes, in order to obtain an *a priori* information regarding the solution.) Using the Fourier method, we obtain

$$\bar{\phi}^j = \sum_{n=1}^{N} \bar{\phi}_n^j u_n, \qquad (3.8)$$

and combining further with (3.7) we have

$$\bar{\phi}_n^{j+1} = \frac{1 + \dfrac{\tau \lambda_n}{2}}{1 - \dfrac{\tau \lambda_n}{2}} \bar{\phi}_n^j, \qquad \bar{\phi}_n^0 = g_n \qquad (3.9)$$

$$(j = 1, 2, \ldots, j_0).$$

Hence

$$\bar{\phi}_n^j = \left(\frac{1 + 1\dfrac{\tau \lambda_n}{2}}{1 - \dfrac{\tau \lambda_n}{2}}\right)^j g_n. \qquad (3.10)$$

Thus

$$\bar{\phi}^j = \sum_{n=1}^{N} T_n^j g_n u_n, \qquad (3.11)$$

where

$$T_n = \frac{1 + \dfrac{\tau \lambda_n}{2}}{1 - \dfrac{\tau \lambda_n}{2}}.$$

Let τ be such that the denominator in (3.11) does not become zero for any n. Then

$$\tau < \frac{2}{\beta(\bar{A})} \tag{3.12}$$

Note that this condition is compatible with the emphasis we place on the components with the highest information content.

A formal analysis of (3.11) shows that $T_n > 1$ for all n, and that the high-frequency components which correspond to the large indices n grow quickly in amplitude as n increases. For these indices we have therefore $T_n \gg 1$ and in particular $T_n^j \gg 1$. In the course of processing the input data g we have dropped all the Fourier components in (3.5) beginning with the index $n = m + 1$; hence it may seem at first that we are guaranteed that the Fourier sum

$$g = \sum_{n=1}^{m} g_n u_n$$

generates a solution with the same number of components, i.e.,

$$\bar{\phi}^j = \sum_{n=1}^{m} T_n^j g_n u_n. \tag{3.13}$$

This would be indeed the case if the computations were made with the infinite accuracy. Since this is not the case, we will immediately observe the components g_n with $n > m$. Although small, they have large weights (proportional to $T_n^j \gg 1$), and eventually they may corrupt significantly the desired solution. In order to avoid the catastrophic growth of errors in the high-frequency Fourier components, we need to find a construction which would automatically take the elements from F into certain subspace Φ.

Let us define Φ as follows. An element will be considered to belong to Φ if its last $(N - m)$ components in the Fourier sum do not grow faster in amplitude than several amplitudes of the last informative component indexed by m. With Φ constructed this way, the round-off errors will not grow on its elements faster than the amplitudes of the mth component. This guarantees that the computational scheme is well-posed.

Lavrentiev [2, 26] and Lions and Lattes [16] have suggested a modification of (3.2) by taking $\bar{A} = \bar{A} - \varepsilon \bar{A}^2$ instead of $\bar{A}$. In this case we have

$$\frac{d\bar{\phi}_\varepsilon}{dt} - \bar{A}\bar{\phi}_\varepsilon = -\varepsilon \bar{\phi}^2 \bar{\phi}_\varepsilon \quad (0 \leq t \leq t_0), \tag{3.14}$$

$$\bar{\phi}_\varepsilon = g \quad t = 0,$$

where ε is arbitary at the moment. This parameter will be chosen from the requirement that the solution stay in Φ. For simplicity assume that $\bar{A} = \bar{A}^*$ and consider the difference scheme

$$\frac{\bar{\phi}_\varepsilon^{j+1} - \bar{\phi}_\varepsilon^j}{\tau} - (\bar{A} - \varepsilon \bar{A}^2)\frac{\bar{\phi}_\varepsilon^{j+1} + \bar{\phi}_\varepsilon^j}{2} = 0, \tag{3.15}$$

$$\bar{\phi}_\varepsilon^0 = g.$$

The above equation will be solved using the Fourier series expansion along the eigenfunctions of the operator $\bar{A}$. We obtain

$$\bar{\phi}_\varepsilon = \sum_{n=1}^{N} \frac{1 + \frac{\tau\lambda_n}{2} - \varepsilon\frac{\tau\lambda_n^2}{2}}{1 - \frac{\tau\lambda_n}{2} + \varepsilon\frac{\tau\lambda_n^2}{2}} g_n u_n. \tag{3.16}$$

In order to choose ε, let us require that the relative error in the mth component, due to the operator $\varepsilon\bar{A}^2$, does not exceed η (usually $\eta < 1$ depending on the ratio between the useful information and the inaccuracies which can not be accounted for (the noise) in the mth component). The above requirement implies

$$\frac{\tau\lambda_m}{2} = \frac{\eta\varepsilon\tau\lambda_m^2}{2}. \tag{3.17}$$

Hence

$$\varepsilon = \frac{1}{\eta\lambda_m}. \tag{3.18}$$

In this fashion we have determined an important *a priori* quantity which will be needed later. It is easily seen that for ε given by (3.18) the amplitudes of all the components with indices $n > m$ will increase with time at a rate not higher than T_m^j.

We will have occasion to use yet another *a priori* quantity. Consider

$$\bar{\phi}^j = \sum_{n=1}^{m} g_n e^{\tau\lambda_n j} u_n \tag{3.19}$$

and

$$\bar{\phi}_\varepsilon^j = \sum_{n=1}^{N} g_n T_n^j(\varepsilon) u_n, \tag{3.20}$$

where

$$T_n(\varepsilon) = \frac{1 + \frac{\tau\lambda_n}{2} - \varepsilon\frac{\tau\lambda_n^2}{2}}{1 - \frac{\tau\lambda_n}{2} + \varepsilon\frac{\tau\lambda_n^2}{2}}.$$

Since $\bar{\phi}_\varepsilon^j$ belongs to Φ, we will not introduce a large error by replacing it with

$$\bar{\phi}_\varepsilon^j = \sum_{n=1}^{m} g_n T_n^j(\varepsilon) u_n, \tag{3.21}$$

where we limit ourselves only to the first m members in the series. Using the above form, the solution can be found in a constructive fashion (the system of functions u_n and u_n^*, $n = 1, \ldots, m$, has already been found). Next

Inverse Evolution Problems with Time-Varying Operators

we compute $\bar{\phi}^j$ and $\bar{\phi}_\varepsilon^j, j = 1, \ldots, j_0$, using (3.19) and (3.21). Let us introduce the new vectors

$$\bar{\phi} = \begin{vmatrix} \bar{\phi}^1 \\ \bar{\phi}^2 \\ \cdots \\ \bar{\phi}^{j_0} \end{vmatrix}, \quad \bar{\phi}_\varepsilon = \begin{vmatrix} \bar{\phi}_\varepsilon^1 \\ \bar{\phi}_\varepsilon^2 \\ \cdots \\ \bar{\phi}_\varepsilon^{j_0} \end{vmatrix}$$

and compute the norm

$$\|\bar{\phi} - \bar{\phi}_\varepsilon\| = \delta. \tag{3.22}$$

The above norm is the last *a priori* quantity we need. The other two, τ and ε, are given by (3.12) and (3.18).

Let us formulate the numerical algorithm for solving the original Problem (3.1). Taking into account the above analysis, let us construct the following approximation problem:

$$\frac{\phi^{j+1} - \phi^j}{\tau} - (A_j - \varepsilon A_j^2)\frac{\phi^{j+1} + \phi^j}{2} = 0, \quad \phi^0 = g, \tag{3.23}$$

$$\tau = \frac{2}{\beta(\bar{A})}; \quad \varepsilon = \frac{1}{\eta\lambda_m(\bar{A})}. \tag{3.24}$$

Let

$$\phi = \begin{vmatrix} \phi^1 \\ \phi^2 \\ \cdots \\ \phi^{j_0} \end{vmatrix}, \quad f = \begin{vmatrix} -R_0 g \\ 0 \\ \cdots \\ 0 \end{vmatrix},$$

$$\Lambda = \begin{Vmatrix} -S_0 & 0 & 0 & 0 & \cdots & 0 & 0 \\ R_1 & -S_1 & 0 & 0 & \cdots & 0 & 0 \\ 0 & R_2 & -S_2 & 0 & \cdots & 0 & 0 \\ 0 & 0 & R_3 & -S_3 & \cdots & 0 & 0 \\ \cdots & & & & & & \\ 0 & 0 & 0 & 0 & \cdots & R_{j_0-1} & -S_{j_0-1} \end{Vmatrix},$$

where

$$S_j = E - \frac{\tau}{2}(A_j - \varepsilon A_j^2); \quad R_j = E + \frac{\tau}{2}(A_j - \varepsilon A_j^2);$$

$$A_j = A(t_{j+1/2}).$$

The problem can then be written as

$$\Lambda\phi = f. \tag{3.25}$$

Next let us symmetrize by multiplying with Λ^*, i.e.

$$\Lambda^*\Lambda\phi = \phi^* f, \tag{3.26}$$

and formulate an interative process. In particular, we can use conveniently the conjugate gradients method, which doesn't require computation of the spectral boundaries of $\Lambda^*\Lambda$.

Having specified the successive approximations, we still have to determine the optimal number of iterations k_0, so that we reach the highest attainable accuracy under the given *a priori* conditions. Since such a number can not be computed with a high precision, we will assume that the *a priori* estimate of (3.22), obtained for the model, is also acceptable for (3.1). Thus suppose

$$\|\phi - \phi_\varepsilon\| = \delta, \qquad (3.27)$$

where ϕ is the exact solution of (3.1) at the net points and ϕ_ε solves the difference problem with the regularization operator. Under these conditions it is natural to continue the iterative Process (3.23) as long as the iteration error stays larger than the approximation error of (3.27). This can be easily implemented. Introduce the residual ξ^k according to the formula

$$\xi^h = \Lambda^*(\Lambda\phi^k - f) = \Lambda^*\Lambda(\phi^k - \phi). \qquad (3.28)$$

We have then

$$\|\xi^k\| \leqslant \|\Lambda^*\Lambda\| \|\phi^k - \phi\|. \qquad (3.29)$$

Clearly, the approximation error $\|\phi - \phi_\varepsilon\| = \delta$ is equivalent to the residual

$$\|\xi^k\| \leqslant \delta \|\Lambda^*\Lambda\|. \qquad (3.30)$$

This means that the numerical process is to be continued until the residual no longer comes close to the right-hand side of (3.30) in norm. Thus we have the following parametric estimate for k_0:

$$\|\xi^{k_0}\| \leqslant \beta(\Lambda^*\Lambda)\delta. \qquad (3.31)$$

As we have seen, inverse evolution problems require a formidable preliminary analysis of various simple models, which eventually allow one to obtain a necessary *a priori* information needed for the qualitative construction of numerical algorithms. In particular cases we may face even more complex situations. Nevertheless, the above development is sufficient to give an idea of how simple models and error analysis can be used in order to formulate numerical methods. Although the regularization process has been looked at from only one point of view, it is enough to see various possible numerical approaches to the inverse problems. Deeper considerations can be found in the monograph by Lavrentiev [2].

In conclusion let us note that the methods and ideas described above can also be applied to Cauchy problems for elliptic equations. These problems are ill-posed in the classical sense and require methods from the theory of the conditionally well-posed problems. A sizable research in this direction has been carried out by Tikhonov [16], Lavrentiev [16], Ivanov [16], and others.

5.4. Methods of Perturbation Theory for Inverse Problems

Some inverse problems can be formulated in the framework of the theory of conjugate functions and perturbation theory. This approach is beginning to play an ever increasing role in forming numerical algorithms, especially when dealing with complicated problems of mathematical physics in which it is difficult to estimate *a priori* the effects of various factors on the solution of the problem. The approach has acquired a special significance in the problem of planning an experiment, where the objective is to obtain the functionals with the highest information content.

Important results in this direction have been obtained in the theory of radiation by Fuchs [16], Usachev [16], Kadomtsev [16] and Marchuk, and Orlov [16]. The inverse problems are also actively studied in the areas of pattern recognition, identification, and optimization theory. These problems are studied in detail in Pontryagin [3], Balakrishnan [3], Lions [2], and others.

5.4.1. Some Problems of the Linear Theory of Measurements

The theory of measurements has become greatly important for the purposes of organizing information systems. The measurement techniques allow one to obtain data (functionals) regarding the process and to analyze and control the process. We will not discuss the particular elementary measurements such as voltage or current measurements on various segments of an electrical network. Instead we will be interested in complex physical phenomena and processes which are to be understood and quantitatively evaluated within a required accuracy. Such problems are emerging all the time, especially in new branches of technology. It is impossible to design methods for measuring the coefficient of neutron multiplication in the reactor if the physical process of the chain reaction and the diffusion of neutrons are not clearly understood in detail, or if we do not know the equations describing the behavior of the nuclear reactor under various changing conditions.

There is no doubt that the measurement methods as well as the measuring instruments themselves improve considerably as the theory of the physical process develops. As a rule, advances in the theory and experiment are accompanied by new or improved measurement methods.

A question arises whether one could not formulate at this time more or less general approaches to measurement methods as applied to various processes with the possibility of a formal mathematical description of the algorithm. It turns out that such an approach can be indeed formulated at least for problems with linear operators. This particular class of problems will be in fact the topic of our discussion in what follows.

It may be conjectured that the theory of variation measurements of physical quantities can be based on perturbation theory. The essence of the matter is as follows: Suppose we are studying a complicated physical process using an instrument with known physical characteristics. Its readings are in relation with the field of physical quantities under investigation, and are functionals on this field. In most cases, however, the experimenter is interested not in the field itself, but rather in its deviations caused by some (usually small) perturbations. This requires that the measurements be made with enough accuracy as to permit the observations of the field deviations from some "standard" state. Thus, assume that this first requirement on the instrument is met, so that we have at our disposal sufficiently accurate measurements of the deviations from the standard. We now ask whether this information will suffice for a satisfactory interpretation of the experiment and whether we can reconstruct the information about the perturbed state of the system with a sufficient accuracy. Unfortunately this question is usually very difficult to answer. The reason is that the problem of reconstructing the information about the field of a physical quantity using measurements is, as a rule, an ill-posed problem of mathematical physics.

In order to sidestep this essential difficulty of data processing, it is necessary that the deviations of the instrument reading are related from the very beginning to the deviations in the physical parameters of the process under consideration. In this case the error in the characteristic considered becomes proportional to the error in the deviation of the instrument reading (the variation of the functional), and hence we can use the maximal information provided by the instrument for interpretation purposes. Taking the above point of view we now develop the theory as based on the results of Marchuk and Orlov [16].

5.4.2. Conjugate Functions and the Notion of Value

Consider a function $\phi(x)$ satisfying the equation

$$L\phi(x) = q(x), \tag{4.1}$$

where L is some linear operator and $q(x)$ is a source distribution in the medium. By x we represent all the variables of the problem (the time and space coordinates, energy, velocity, and direction). We also assume that the operator L and the function ϕ are real, and that $\phi \in \Phi$.

In order to be specific, we will assume that the process is related to a diffusion or to the propagation of a substance; the results, however, reach far beyond the scope of these types of problems.

Let us introduce a Hilbert space of functions endowed with the inner product

$$(g, h) = \int g(x)h(x)\, dx, \tag{4.2}$$

where the integration is performed over the region D of definition of g and h.

Various physical problems are usually solved with the objective of evaluating eventually certain functionals of the flow $\phi(x)$. Any quantity which is linearly related to $\phi(x)$ can be represented in the form of this inner product. For instance, while observing some process, we may account for the characteristic $\Sigma(x)$ of the measuring instrument by taking

$$J_\Sigma[\phi] = \int \phi(x)\Sigma(x)\,dx = (\phi, \Sigma). \tag{4.3}$$

Hence, we will consider physical quantities which can be expressed as linear functionals of $\phi(x)$:

$$J_p[\phi] = (\phi, p),$$

where p designates the physical process we are interested in. Along with L consider its adjoint L^*, defined by

$$(g, Lh) = (h, L^*g), \tag{4.4}$$

for any g and h, and introduce (formally at this moment) the nonhomogeneous adjoint equation

$$L^*\phi_p^* = p(x), \tag{4.5}$$

where the function $p(x)$ is yet to be determined, and $\phi_p^* \in \Phi^*$. The original Equation (4.1) will be called fundamental. Take for g and h in (4.4) the solutions ϕ and ϕ_p^* of (4.1) and (4.5). We obtain

$$(\phi_p^*, L\phi) = (\phi, L^*\phi_p^*) \tag{4.6}$$

or, using (4.1) and (4.5) again,

$$(\phi_p^*, q) = (\phi, p), \tag{4.7}$$

In other words, $J_q[\phi_p^*] = J_p[\phi]$. Hence, in order to find $J_p[\phi]$, we can proceed in two ways: either we solve (4.1) and use the formula

$$J_p[\phi] = (\phi, p), \tag{4.8}$$

or we solve (4.5) and take

$$J_p[\phi] = J_q[\phi_p^*] = (\phi^*, q). \tag{4.9}$$

Thus each linear functional can be put into a correspondence with a function $\phi_p^*(x)$ satisfying (4.5), where the free element in this equation is chosen to be $p(x)$—the function characterizing the process we are interested in.

Suppose that at the point x_0 of the medium there is a "source of unit power":

$$q(x) = \delta(x - x_0). \tag{4.10}$$

Since

$$(\phi(x), \delta(x - x_0)) = \phi(x_0), \tag{4.11}$$

we have

$$J_p[\phi] = J_{q=\delta(x-x_0)}[\phi_p^*] = \phi_p^*(x_0). \quad (4.12)$$

Consequently, the conjugate function $\phi_p^*(x)$ describes the dependence of the functional $J_p[\phi] = (\phi, p)$ on the location of the source.

Imagine a physical system (or instrument) which involves measurements of a certain linear functional $J_p[\phi]$ of the solution, related for instance to the density of the particles in the medium. Assume there occur emissions of a certain number of particles at some point (or, conversely, absorption of these particles). Then the measured value of $J_p[\phi]$ will correspondingly increase or decrease, and this change will depend on the location of the emission (or absorption) point. As can be seen from the foregoing, this dependence is described by the conjugate function $\phi_p^*(x)$, which in turn satisfies (4.5). Consequently, the conjugate function $\phi_p^*(x)$ describes the effect of depositing the particles at a given point. The function $\phi_p^*(x)$ can be called a *value* of the substance at the point x, relative the functional $J_p[\phi] = (\phi, p)$. (The term value fits well in problems of the theory of radiation. A more suitable term may well be found for another application.)

The interpretation of the conjugate function $\phi_p^*(x)$ as the value of the substance also helps to clarify the exposition of perturbation theory for arbitrary functionals $J_p[\phi]$. Indeed, suppose we change the number of particles in an element of volume Δx, surrounding the point x, by an increment of ΔN. The corresponding change in J_p can then be expressed by the formula

$$\delta J_p = \delta N \phi_p^*(x). \quad (4.13)$$

If the system under consideration is subject to some small changes in the parameters, so that the operator L becomes $L + \delta L$, then the number of particles in each element x changes correspondingly by the amount of $\Delta N = -\Delta x \delta L \phi$. The overall change of J_p is thus given by

$$\delta J_p = -\int \phi_p^*(x) \, \delta L \phi(x) \, dx. \quad (4.14)$$

A rigorous derivation of this result will be given later.

Relation (4.13) facilitates the measurements of the value distribution function of the system. It can be accomplished by changing in a certain manner the number of particles at various points x, while measuring the corresponding changes in the quantity L_p. The notion of a value can also be useful in the theory of measuring instruments. An instrument is usually designed to measure a single variable J_p. Hence for each instrument we can introduce a well-defined value function $\phi_p^*(x)$, which need be calculated only once. If the distributions of the substance and of its value are known, Relation (4.14) can be used in the following two ways:

First, it can be used for determining the quantities δL, i.e., various characteristics of the mutual interaction between the particles and the matter, by taking measurements of δJ_p for various changes in the parameters of δL.

Methods of Perturbation Theory for Inverse Problems

For instance, we can measure sections of the interacting neutrons for various figures (shapes) by accommodating these figures into an instrument and evaluating $\delta\Sigma = \delta L$ using the increments of J_p.

Secondly, (4.14) allows one to correct the effect of various perturbing factors of the instrument on the functional J_p.

Finally, the notion of value makes it possible to derive the equations for ϕ_p^* directly from its physical meaning; the procedure is the same as when deriving the equation for the flow of neutrons using the law of preserving neutrons.

The above formulas can also be used to prove the reciprocity theorem for the Green functions $G(x, x_0)$ and $G^*(x, x_1)$ corresponding to the fundamental and adjoint equations respectively. Indeed, the function $G(x, x_0)$ satisfies (4.1) for $q(x) = \delta(x - x_0)$, and $G^*(x, x_1)$ satisfies (4.5) for $p(x) = \delta(x - x_1)$. Substituting $\phi(x) = G(x, x_0)$, $\phi_p^*(x) = G^*(x, x_1)$ and the above equations for q and p into Formula (4.7), we obtain

$$G(x_1, x_0) = G^*(x_0, x_1), \tag{4.15}$$

which is the reciprocity theorem.

5.4.3. Perturbation Theory for Linear Functionals

If the properties of the medium interacting with the field are subject to changes, i.e. of the operator in (4.1) become

$$L' = L + \delta L,$$

then both the field $\phi(x)$ itself and the functional $J_p[\phi]$ are also changing:

$$\phi(x) \to \phi'(x), \qquad J_p[\phi] \to J_p' = J_p + \delta J_p.$$

Let us find the relation between the increments δL and δJ_p. The perturbed system is described by the equation

$$L'\phi' = (L + \delta L)\phi' = q. \tag{4.16}$$

The conjugate function of the unperturbed system corresponding to J_p is given by the equation

$$L^*\phi_p^* = p. \tag{4.17}$$

Take the inner products of (4.16) and ϕ^*, and (4.17) and ϕ'. Then subtracting, we obtain

$$(\phi_p^*, L'\phi') - (\phi', L^*\phi_p^*) = (\phi_p^*, \delta L \phi') \tag{4.18}$$

on one hand, and

$$(\phi_p^*, q) - (\phi', p) = J_p[\phi] - J_p[\phi'] = -\delta J_p \tag{4.19}$$

on the other hand [in agreement with (4.7)].

A comparison of (4.18) and (4.19) gives the following relation for the increment of the functional:

$$\delta J_p = -(\phi_p^*, \delta L \phi'). \tag{4.20}$$

If instead of (4.16) and (4.17) we consider the perturbed adjoint equation

$$(L^* + \delta L^*)\phi_p^{*\prime} = p \tag{4.21}$$

and the unperturbed fundamental equation (4.1) respectively, a similar procedure yields

$$\delta J_p = -(\phi, \delta L^* \phi_p^{*\prime}), \tag{4.22}$$

which is, of course, equivalent to (4.20).

Note an important aspect of applying the formulas of perturbation theory. These formulas are written in a form applicable to the variation of the functional, the admissible inaccuracy of the variation usually being in the limits of several percent. Therefore the calculation of the variations indicated does not require precise knowledge of the fundamental and adjoint problems; it is enough to use their approximate solutions.

If the perturbation of L (and L^* for that matter) is small enough, so that the functions ϕ and ϕ_p^* are not seriously distorted, then in (4.20) and (4.22) one can put $\phi' \approx \phi$, $\phi^{*\prime} \approx \phi^*$, and obtain the following mutually equivalent formulas:

$$\delta J_p = -(\phi_p^*, \delta L \phi), \tag{4.23}$$

$$\delta J_p = -(\phi, \delta L^* \phi_p^*). \tag{4.24}$$

Besides their direct application in estimating various effects and their use in measurement analysis, the obtained formulas of perturbation theory have yet another quite important application.

In theoretical considerations as well as practical computations the original complicated problem is often approximated by a simplified model. For this it is clearly necessary that the replacement leave unchanged certain characteristics of the system which are fundamental for the process under consideration. As an example consider the approximation of a time-varying differential equation by a time invariant equation. An instance of this approach is the method of efficient boundary conditions: essentially, we replace the true boundary conditions by simpler ones, but such that they result in correct values of a certain functional.

The formulas of perturbation theory obtained above facilitate formulations of quite general approaches to various problems. Consider a system characterized by the operator L, and assume that the most significant quantity for our purposes is the functional $J_p[\phi]$. If the sought simplified model can be characterized by the operator $L' = L + \delta L$, then for J_p not to be affected by the replacement of the true system with the model it is enough that

$$\delta J_p = -(\phi_p^*, [L' - L]\phi') = 0, \quad \text{i.e.} \quad (\phi_p^*, L'\phi') = (\phi_p^*, L\phi'). \tag{4.25}$$

Methods of Perturbation Theory for Inverse Problems

For more quantities J_{p_1}, J_{p_2}, etc. we similarly obtain conditions of the type of (4.25) with the solutions $\phi_{p_1}^*$, $\phi_{p_2}^*$, etc.

Condition (4.25) does not determine the desired model uniquely. But being the necessary condition, it may still help in finding the model in conjunction with other conditions. In particular, if the model operator L' involves one or more parameters, (4.25) can serve to find their values. (The form of L' can be determined from physical considerations.)

5.4.4. Numerical Methods for Inverse Problems and Design of Experiment

Suppose that we have at our disposal a set of functionals (measurements) J_{p_i} ($i = 1, \ldots, n$). Suppose that the measurements are essentially different (for instance, the measurements are taken at different points of the domain of the solution using the same single instrument; or, several instruments are used to register different characteristics of the considered phenomenon). For the sake of simplicity, the random errors are assumed to be already removed from the measurements by preliminary data processing.

Let us put each J_{p_i} in correspondence with the respective value function for the unperturbed problem, i.e. for the model in which the operator L and its domain are considered to be known. Solving now a total of n different problems

$$L^*\phi_{p_i}^* = p_i \qquad (i = 1, 2, \ldots, n), \tag{4.26}$$

we find the value functions $\phi_{p_i}^*$. Having done this, let us solve the single fundamental problem with the model ("unperturbed") operator L, adjoint to L^*:

$$L\phi = f. \tag{4.27}$$

We will assume that $\phi \in \Phi$ and $\phi^* \in \Phi^*$, where Φ and Φ^* are the domains of L and L^* respectively. Next consider the total of n formulas of the theory of small perturbations

$$(\phi_{p_i}^*, \delta L \phi) = -\delta J_{p_i} \qquad (i = 1, 2, \ldots, n), \tag{4.28}$$

where δL designates the difference between L' and L. Suppose that L is known:

$$L = \sum_{k=1}^{m} [\alpha_k A_k + B_k(\beta_k C_k)], \tag{4.29}$$

where A_k, B_k, and C_k are elementary linear operators (for example differentiation, or integration, or their various combinations); the coefficients $\alpha_k(x)$ and $\beta_k(x)$ are to be determined (their rough approximations in the unperturbed problem are usually known).

Our problem now is to reconstruct the coefficients α_k' and β_k' appearing in the expression

$$L' = \sum_{k=1}^{m} [\alpha_k' A_k + B_k(\beta_k' C_k)]. \tag{4.30}$$

Using (4.29) and (4.30) we get

$$\delta L' = \sum_{k=1}^{m} [\delta\alpha_k A_k + B_k(\delta\beta_k C_k)], \qquad (4.31)$$

where

$$\delta\alpha_k = \alpha'_k - \alpha_k, \qquad \delta\beta_k = \beta'_k - \beta_k.$$

We substitute (4.31) into (4.28), and under the corresponding assumptions obtain the following system of equations:

$$\sum_{k=1}^{m} [(\phi^*_{p_i}, \delta\alpha_k A_k \phi) + (B^*_k \phi^*_{p_i}, \delta\beta_k C_k \phi)] = -\delta J_{p_i}$$

$$(i = 1, 2, \ldots, n). \qquad (4.32)$$

The next procedure is to parametrize the variations $\delta\alpha_k$ and $\delta\beta_k$. To begin with, consider the simplest case where $\delta\beta_k = 0$ and $\delta\alpha_k$ are constant. Under these conditions System (4.32) becomes a problem of linear algebra

$$\sum_{k=1}^{m} \delta\alpha_k(\phi^*_{p_i}, A_k \phi) = -\delta J_{p_i} \qquad (i = 1, 2, \ldots, n). \qquad (4.33)$$

Here $(\phi^*_{p_i}, A_k \phi)$ are elements of a matrix which can be calculated.

Let y be a vector with the components $\delta\alpha_k$, let F be another vector with the components $-\delta J_{p_i}$, and finally let $a_{ik} = (\phi^*_{p_i}, A_k \phi)$ be the entries of a matrix Λ. Then

$$\Lambda y = F. \qquad (4.34)$$

If the number of functionals n equals the number of variations of the coefficients α_k to be determined, then $\delta\alpha_k$ can be found, in principal, from (4.34). If $n > m$, (4.34) is overdefined, and its solution (if it exists) can usually be found with the help of the least-squares method under the assumption that the quadratic functional below achieves its minimum on y:

$$\|\Lambda y - F\|^2 = \min. \qquad (4.35)$$

The minimizing vector y is sometimes referred to as a *quasisolution* of (4.34). If $n = m$, System (4.34) can be solved by methods we have discussed in Chapter 3 in connection with the analysis of numerical methods with inaccurate data.

In the case where $\delta\alpha_k(x)$ and $\delta\beta_k(x)$ are functions, the inverse problem can be solved by various parametric methods, the essence of which consists of the following: It is assumed that based on an *a priori* analysis of the behavior of physical parameters (usually as a result of statistical and correlation analysis) we have found a certain complete orthogonal system of functions

Methods of Perturbation Theory for Inverse Problems

$u_{k,l}(x)$ and $v_{k,l}(x)$ such that for a small number $n(k)$ they give a reasonable approximation of α_k and β_k:

$$\delta\alpha_k(x) = \sum_{l=1}^{n(k)} a_{k,l} u_{k,l}(x);$$
$$\delta\beta_k(x) = \sum_{l=1}^{n(k)} b_{k,l} v_{k,l}(x),$$
(4.36)

where $a_{k,l}$ and $b_{k,l}$ are yet to be determined.

Let us substitute the expressions (4.36) into (4.32). There results

$$\sum_{k=1}^{m} \sum_{l=1}^{n(k)} [a_{k,l}(\phi^*_{p_i}, u_{k,l} A_k \phi) + b_{k,l}(B^*_k \phi^*_{p_i}, v_{k,l} C_k \phi)] = -\delta J_{p_i}$$
(4.37)

$(i = 1, 2, \ldots, n)$.

Using a suitable ordering, let us relabel the coefficients $a_{k,l}$ and $b_{k,l}$ as y_j ($j = 1, 2, \ldots$) and introduce further a matrix Λ so that the equation

$$\Lambda y = F$$

is equivalent to (4.37). In this fashion we again obtain a problem in linear algebra, from which we eventually find $a_{k,l}$ and $b_{k,l}$, and consequently $\delta\alpha_k$ and $\delta\beta_k$.

So far we have only considered the case where the solution of the model is close to reality; thus one could interchange ϕ' and ϕ and use the theory of small perturbations. If the unperturbed (model) state of the process differs from that of the true process, the above algorithm can be taken only as a first approximation of the solution to the inverse problem. Once the variations $\delta\alpha_k$ and $\delta\beta_k$ are found, one can modify the coefficients α_k and β_k and find

$$\alpha'_k = \alpha_k + \delta\alpha_k,$$
$$\beta'_k = \beta_k + \delta\beta_k.$$

Next one has to solve the "perturbed" problem

$$L'\phi' = f$$
(4.38)

with the operator

$$L' = \sum_{k=1}^{m} [\alpha'_k A_k + B_k(\beta'_k C_k)],$$

and then switch to a new approximation in the solution of the inverse problem, replacing (4.32) by a more general perturbation formula

$$\sum_{k=1}^{m} [(\phi^*_{p_i}, \delta\alpha_k A_k \phi') + B^*_k \phi^*_{p_i}, \delta\beta_k C_k \phi')] = -\delta J_{p_i}$$
(4.39)

$(i = 1, 2, \ldots, n)$.

One must repeat the computational cycle in order to improve the accuracy of the variations $\delta\alpha_k$ and $\delta\beta_k$. This is called a *second approximation* in solving the inverse problem. It is understood that the above process can be continued. The successive approximations may be shown to converge, depending on specific information about the elementary operators A_k and the domain of the operators L, L^*.

Let us illustrate our algorithm with a simple example, namely

$$-\frac{d}{dx}\beta(x)\frac{d\phi'}{dx} + \alpha(x)\phi' = f(x), \quad \phi'(0) = \phi'(1) = 0, \quad (4.40)$$

where the unknown coefficients $\alpha(x)$ and $\beta(x)$ are *a priori* assumed to be, for instance, continuous on $0 \leq x \leq 1$, and approximately equal to $\bar{\alpha}$ and $\bar{\beta}$, i.e.,

$$\alpha(x) = \bar{\alpha} + \delta\alpha, \quad \beta(x) = \bar{\beta} + \delta\beta(x). \quad (4.41)$$

Of course, if the values of $\bar{\alpha}(x)$ and $\bar{\beta}(x)$ can be specified with more precision using an *a priori* information, then the approximation by the constants $\bar{\alpha}$ and $\bar{\beta}$ becomes void.

By a preliminary analysis we eventually conclude that $\delta\alpha(x)$ and $\delta\beta(x)$ may be represented in the form of the finite sums

$$\delta\alpha(x) = \sum_{l=1}^{n} a_l u_l(x), \quad \delta\beta(x) = \sum_{l=1}^{n} b_l v_l(x), \quad (4.42)$$

where $\{u_k(x)\}$ and $\{v_k(x)\}$ form complete systems of orthonormal functions (for example trigonometric polynomials, or Legendre polynomials).

Let $p_1(x), p_2(x), \ldots, p_n(x)$ be the measurement characteristics, so that each of the instruments registers the functional

$$J'_{p_i}[\phi'] = \int_0^1 p_i(x)\phi'(x)\,dx \quad (i = 1, 2, \ldots, n). \quad (4.43)$$

The functions p_i can be viewed as the characteristics of the instrument.

Consider next the unperturbed problem (model) corresponding to Problem (4.40):

$$-\frac{d}{dx}\bar{\beta}\frac{d\phi}{dx} + \bar{\alpha}\phi = f, \quad \phi(0) = \phi(1) = 0. \quad (4.44)$$

Along with (4.44) consider also the adjoint problems which correspond to the model chosen:

$$-\frac{d}{dx}\bar{\beta}\frac{d\phi^*_{p_i}}{dx} + \bar{\alpha}\phi^*_{p_i} = p_i(x), \quad \phi^*_{p_i}(0) = \phi^*_{p_i}(1) = 0 \quad (4.45)$$

$$(i = 1, 2, \ldots, n).$$

According to the general theory, we have

$$J_{p_i}[\phi] = \int_0^1 p_i(x)\phi(x)\,dx = \int_0^1 f(x)\phi^*_{p_i}(x)\,dx. \quad (4.46)$$

Assume now that the model Problems (4.44) and (4.45) have been solved, and let us find the variations δJ_{p_i} from the formula

$$\delta J_{p_i} = J'_{p_i} - J_{p_i} \quad (i = 1, 2, \ldots, n), \tag{4.47}$$

where J'_{p_i} is the measurement with the characteristic p_i (see (4.43), where ϕ' is unknown); J_{p_i} can be expressed by means of any of the formulas in (4.46). The measurement accuracy must be such as to guarantee that the variation δJ_{p_i} can be evaluated.

Consider now the formulas of the theory of small perturbations (4.33):

$$A = E; \quad B = -\frac{d}{dx}; \quad C = \frac{d}{dx}.$$

Taking into account the boundary conditions for $\phi^*_{p_i}$ and ϕ, we obtain

$$\int_0^1 \left(\delta\alpha \phi \phi^*_{p_i} + \delta\beta \frac{d\phi}{dx} \frac{d\phi^*_{p_i}}{dx} \right) dx = -\delta J_{p_i}. \tag{4.48}$$

Substituting (4.42) into (4.48), we obtain

$$\sum_{l=1}^n \left(a_l \int_0^1 u_l \phi \phi^*_{p_i} dx + b_l \int_0^1 v_l \frac{d\phi}{dx} \frac{d\phi^*_{p_i}}{dx} dx \right) = -\delta J_{p_i} \tag{4.49}$$

$$(i = 1, 2, \ldots, n).$$

Solving this system, we find the coefficients a_k and b_k, and based on the representation (4.42) we obtain the first approximation for α' and β'. These quantities can be made more precise by the successive approximation method described earlier. In the same manner we can formulate and solve more complicated inverse problems, including the problem of determining the source perturbations δf.

Here we have to design a rather complicated experiment. The problem can be formulated as follows: Consider a family of measurements (which are practically realizable). We are required to choose a measurement from this family with the highest information content, having in mind the specific inverse problem of reconstructing the required characteristics of the medium (the coefficients of the equation). In the general framework of optimization this problem turns out to be very difficult. Nevertheless, we can consider certain particular approaches to its solution.

Thus, let us suppose that prior to the execution of the experiment we have constructed a model of the unperturbed problem, which is subsequently used to describe the linear functionals of the solution; using the *a priori* information about the measurement accuracy we then make conclusions regarding the necessary measurement accuracy of the functionals. Suppose that the necessary measurement accuracy requirements on δJ_{p_i} are met. Next we consider various families of measurements and choose those which give the best conditioning of the matrix Λ. The system of linear equations obtained thereby can be easily inverted; thus we have produced a sort of optimal design of experiment (of course, we have omitted economy considerations, which sometimes may play a decisive role). One may fail to achieve the high requirements regarding the accuracy of the measurements of the functionals J_{p_i} (chosen so as to maximize the conditioning of the matrix B). In this case we face a more complicated problem of experiment planning: we are constrained by a given accuracy compatible with the instrument technology. This belongs to another class of problems, namely optimization problems with constraints.

We have not considered here the problems of statistical processing of the empirical data. These problems are dealt with in sufficient detail in the relevant literature, and in any case, they do not add any major complication to the theory of inverse problems (see Marchuk and Drobyshev [16]).

Chapter 6

The Simplest Problems of Mathematical Physics

We will consider a number of simple and at the same time typical problems of mathematical physics, and use them to illustrate the fundamental methods of numerical mathematics.

6.1. The Poisson Equation

Let us attempt to illustrate the numerical approaches to some problems related to elliptic equations.

6.1.1. The Dirichlet Problem for the One-Dimensional Poisson Equation

To begin with, let us consider the Dirichlet problem for the one-dimensional Poisson equation

$$-\frac{d^2\phi}{dx^2} = f \quad (0 < x < 1), \tag{1.1}$$

$$\phi(0) = a, \quad \phi(1) = b,$$

where $f(x)$ represents the sources, and a and b are some given constants.

Let us divide the interval $0 \leq x \leq 1$ into N equal subintervals $x_k \leq x \leq x_{k+1}$ of length $h = x_k - x_{k-1}$. Writing down the second-order approximation of (1.1) with respect to h, we obtain

$$\frac{-\phi_{k-1} + 2\phi_k - \phi_{k+1}}{h^2} = f_k \quad (k = 1, 2, \ldots, N-1), \tag{1.2}$$

$$\phi_0 = a, \quad \phi_N = b.$$

In agreement with the general principles of constructing difference schemes, we need to eliminate first the boundary points from the difference Equation (1.2). This can be done with the help of the boundary conditions. Of course, the operator of the problem changes in the process, and we obtain the following problem:

$$\frac{2\phi_1 - \phi_2}{h^2} = \frac{a}{h^2} + f_1,$$

$$\frac{-\phi_{k-1} + 2\phi_k - \phi_{k+1}}{h^2} = f_k \quad (k = 2, 3, \ldots, N-2), \tag{1.3}$$

$$\frac{-\phi_{N-2} + 2\phi_{N-1}}{h^2} = \frac{b}{h^2} + f_{N-1}.$$

Thus we have a system of $N - 1$ equations with the unknowns $\phi_1, \phi_2, \ldots, \phi_{N-1}$. It can be written in the matrix form

$$A\phi = g, \tag{1.4}$$

where

$$A = \frac{1}{h^2} \begin{Vmatrix} 2 & -1 & 0 & 0 & \cdots & 0 & 0 \\ -1 & 2 & -1 & 0 & \cdots & 0 & 0 \\ 0 & -1 & 2 & -1 & \cdots & 0 & 0 \\ \cdots & \cdots & \cdots & \cdots & \cdots & \cdots & \cdots \\ 0 & 0 & 0 & 0 & \cdots & -1 & 2 \end{Vmatrix},$$

$$\phi = \begin{Vmatrix} \phi_1 \\ \phi_2 \\ \vdots \\ \phi_{N-1} \end{Vmatrix}, \quad g = \begin{Vmatrix} g_1 \\ g_2 \\ \vdots \\ g_{N-1} \end{Vmatrix},$$

$$g_k = \begin{cases} f_1 + \dfrac{a}{h^2} & (k = 1), \\ f_k & (k = 2, 3, \ldots, N - 2), \\ f_{N-1} + \dfrac{b}{h^2} & (k = N - 1). \end{cases}$$

It is not difficult to verify that for $\phi \neq 0$

$$(A\phi, \phi) > 0.$$

In order to solve (1.3), we use the factorization method, described in Section 3.8. The result is

$$\beta_{k+1} = \frac{1}{2 - \beta_k},$$

$$z_{k+1} = \beta_{k+1}(z_k + h^2 g_k) \quad (k = 1, 2, \ldots, N - 1), \tag{1.5}$$

$$\phi_k = \beta_{k+1}\phi_{k+1} + z_{k+1} \quad (k = N - 1, N - 2, \ldots, 1).$$

In order to obtain the initial conditions β_2 and z_2, let us specialize the first of Equations (1.4) to $k = 1$:

$$\phi_1 = \beta_2 \phi_2 + z_2,$$

and choose β_2 and z_2 so that the above equation becomes identical with the first equation in (1.3). Thus

$$\beta_2 = \tfrac{1}{2}, \quad z_2 = (a + f_1 h^2)/2. \tag{1.6}$$

Initial conditions for ϕ_k are taken as

$$\phi_N = 0. \tag{1.7}$$

The Poisson Equation

6.1.2. The One-Dimensional Neumann Problem

Consider now the Neumann problem

$$-\frac{d^2\phi}{dx^2} = f,$$

$$\frac{d\phi}{dx} = a \quad \text{for} \quad x = 0, \tag{1.8}$$

$$\frac{d\phi}{dx} = b \quad \text{for} \quad x = 1,$$

where a and b are given constants.

Let us integrate the above equation over the entire domain of definition of the solution, while making use of the boundary conditions. There results

$$a - b = \int_0^1 f(x)\,dx, \tag{1.9}$$

This relation represents a necessary condition for solvability of (1.8). If $a = b$, it says that the total number of sources sums to zero, i.e., each emission source is counter balanced by an absorption.

In order to obtain the second-order accuracy difference analog for the problem, the solution of the problem (which must be sufficiently smooth) is extended to the additional intervals of length h adjoining its domain $0 \leq x \leq 1$ from the right and left. This means that we are considering an auxiliary net region $x_1 = -h$, $x_0 = 0$, $x_k = kh$, $x_N = 1$, $x_{N+1} = 1 + h$. Let us define on this net the approximation of the problem as follows:

$$\frac{-\phi_{k-1} + 2\phi_k - \phi_{k+1}}{h^2} = f_k \quad (k = 0, 1, \ldots, N),$$

$$\frac{\phi_1 - \phi_{-1}}{2h} = a, \quad \frac{\phi_{N+1} - \phi_{N-1}}{2h} = b. \tag{1.10}$$

As a preliminary work, let us remove the boundary conditions by eliminating ϕ_{-1} and ϕ_{N+1} from (1.10). Solving for ϕ_{-1}, ϕ_{N+1}, we obtain

$$\phi_{-1} = \phi_1 - 2ha; \quad \phi_{N+1} = \phi_{N-1} + 2hb. \tag{1.11}$$

Substituting into (1.10), we obtain

$$\frac{\phi_0 - \phi_1}{h^2} = \frac{f_0}{2} - \frac{a}{h},$$

$$\frac{-\phi_{k-1} + 2\phi_k - \phi_{k+1}}{h^2} = f_k \quad (k = 1, \ldots, N-1), \tag{1.12}$$

$$\frac{-\phi_{N-1} + \phi_N}{h^2} = \frac{f_N}{2} + \frac{b}{h}.$$

Introduce the matrix

$$A = \frac{1}{h^2} \begin{Vmatrix} 1 & -1 & 0 & \cdots & 0 & 0 \\ -1 & 2 & -1 & \cdots & 0 & 0 \\ 0 & -1 & 2 & \cdots & 0 & 0 \\ \cdots & \cdots & \cdots & \cdots & \cdots & \cdots \\ 0 & 0 & 0 & \cdots & 2 & -1 \\ 0 & 0 & 0 & \cdots & -1 & 1 \end{Vmatrix} \quad (1.13)$$

and the vectors

$$\phi = \begin{vmatrix} \phi_0 \\ \phi_1 \\ \phi_2 \\ \cdots \\ \phi_N \end{vmatrix}, \quad g = \begin{vmatrix} g_0 \\ g_1 \\ g_2 \\ \cdots \\ g_N \end{vmatrix}, \quad g_k = \begin{cases} \dfrac{f_0}{2} - \dfrac{a}{h} & (k = 0), \\ f_k & (k = 1, 2, \ldots, N-1), \\ \dfrac{f_N}{2} + \dfrac{b}{h} & (k = N). \end{cases}$$

We can then write

$$A\phi = g. \quad (1.14)$$

The first question to be resolved when analyzing the matrix A is that regarding its definiteness. It is well known that in the Neumann difference problem the matrix A is singular, since its smallest eigenvalue is zero. This can be seen rather easily by recognizing that the spectral problem

$$Au = \lambda u$$

has for its eigenvector the vector u_0 with identical components, the corresponding eigenvalue being $\lambda = 0$.

Naturally, if A is singular, one has to impose additional constraints on the sources $\{g_k\}_{k=0}^N$ so as to guarantee the solvability, similarly as in (1.9). It is known that for the solvability of (1.14) it is necessary to ensure that g is orthogonal to u_0. This can be achieved by removing the common factor from the components g_k, namely by replacing g_k with $g_k - \bar{g}$, where

$$\bar{g} = \sum_{k=0}^{N} \frac{g_k}{(N+1)}.$$

Equation (1.14) is then solved by the factorization method. Note that an actual realization of the method may involve fractions of the type "0/0," which should be replaced by an arbitrary constant.

6.1.3. The Two-Dimensional Poisson Equation

Consider a Poisson equation in two dimensions

$$-\left(\frac{\partial^2 \phi}{\partial x^2} + \frac{\partial^2 \phi}{\partial y^2}\right) = f \quad \text{in} \quad D, \tag{1.15}$$

$$\phi = g \quad \text{on} \quad \partial D,$$

where the domain D is the square $0 \leq x \leq 1$, $0 \leq y \leq 1$, and $g(x, y)$ is a sufficiently smooth function defined on the boundary of D.

Let us choose a grid of net points in D, defined by coordinate lines $x = x_k$, $y = y_l$. Consider the corresponding difference equation

$$\frac{-\phi_{k-1,l} - \phi_{k+1,l} - \phi_{k,l-1} - \phi_{k,l+1} + 4\phi_{k,l}}{h^2} = f_{k,l},$$

$$\phi_{0,l} = a_l, \qquad \phi_{N,l} = b_l, \tag{1.16}$$

$$\phi_{k,0} = c_k, \qquad \phi_{k,N} = a_k$$

$$(k, l = 1, 2, \ldots, N - 1),$$

where

$$a_l = g_{0,l}, \qquad b_l = g_{N,l}, \qquad c_k = g_{k,0}, \qquad d_k = g_{k,N}.$$

Next we eliminate the boundary conditions in (1.16) and write the result in the matrix form. For this purpose let us introduce the following matrix A and the vectors $\{\phi_l; g_l\}_{l=1}^{N-1}$:

$$A = \frac{1}{h^2} \begin{Vmatrix} 2 & -1 & 0 & -0 & \cdots & 0 & 0 \\ -1 & 2 & -1 & -0 & \cdots & 0 & 0 \\ 0 & -1 & 2 & -1 & \cdots & 0 & 0 \\ \multicolumn{7}{c}{\dotfill} \\ 0 & 0 & 0 & 0 & \cdots & -1 & 2 \end{Vmatrix}, \quad \phi_l = \begin{Vmatrix} \phi_{1,l} \\ \phi_{2,l} \\ \phi_{3,l} \\ \vdots \\ \phi_{N-1,l} \end{Vmatrix},$$

$$g_l = \begin{Vmatrix} f_{1,l} - \dfrac{a_l}{h^2} \\ f_{2,l} \\ f_{3,l} \\ \vdots \\ f_{N-1,l} + \dfrac{b_l}{h^2} \end{Vmatrix}.$$

Define

$$B = h^2 A + 2E,$$

and rewrite (1.16) in the form

$$\frac{1}{h_2}(-\phi_{l-1} + B\phi_l - \phi_{l+1}) = g_l \qquad (l = 1, 2, \ldots, N-1), \qquad (1.17)$$

assuming

$$\phi_0 = c, \qquad \phi_N = d, \qquad (1.18)$$

where

$$c = \begin{vmatrix} c_1 \\ c_2 \\ \cdots \\ c_{N-1} \end{vmatrix}, \qquad d = \begin{vmatrix} d_1 \\ d_2 \\ \cdots \\ d_{N-1} \end{vmatrix},$$

Eliminating the boundary Conditions (1.18) from (1.17), we obtain

$$\frac{1}{h^2}(B\phi_1 - \phi_2) = g_1 + \frac{1}{h^2}c,$$

$$\frac{1}{h^2}(-\phi_{l-1} + B\phi_l - \phi_{l+1}) = g_l \qquad (l = 2, 3, \ldots, N-2), \qquad (1.19)$$

$$\frac{1}{h^2}(-\phi_{N-2} + B\phi_{N-1}) = g_{N-1} + \frac{1}{h^2}d.$$

Let us rewrite System (1.19) using the block matrix and block vector notation:

$$\Lambda = \frac{1}{h^2} \begin{Vmatrix} B & -E & 0 & 0 & \cdots & 0 & 0 \\ -E & B & -E & 0 & \cdots & 0 & 0 \\ 0 & -E & B & -E & \cdots & 0 & 0 \\ \cdots & \cdots & \cdots & \cdots & \cdots & \cdots & \cdots \\ 0 & 0 & 0 & 0 & \cdots & -E & B \end{Vmatrix}, \qquad \phi = \begin{Vmatrix} \phi_1 \\ \phi_2 \\ \vdots \\ \phi_{N-1} \end{Vmatrix},$$

$$F = \begin{Vmatrix} \frac{1}{h^2}c + g_1 \\ g_2 \\ \vdots \\ \frac{1}{h^2}d + g_{N-1} \end{Vmatrix},$$

The Poisson Equation

where E is the unit matrix. System (1.19) then becomes

$$\Lambda\phi = F. \tag{1.20}$$

Write next the matrix Λ as a sum of two matrices,

$$\Lambda = \Lambda_1 + \Lambda_2$$

where

$$\Lambda_1 = \begin{vmatrix} A & 0 & 0 & \cdots & 0 \\ 0 & A & 0 & \cdots & 0 \\ 0 & 0 & A & \cdots & 0 \\ \hdotsfor{5} \\ 0 & 0 & 0 & \cdots & A \end{vmatrix},$$

$$\Lambda_2 = \frac{1}{h^2} \begin{vmatrix} 2E & -E & 0 & 0 & \cdots & 0 \\ -E & 2E & -E & 0 & \cdots & 0 \\ 0 & -E & 2E & -E & \cdots & 0 \\ \hdotsfor{6} \\ 0 & 0 & 0 & 0 & \cdots & 2E \end{vmatrix}.$$

It is not difficult to verify that

$$\Lambda_1\phi = \begin{vmatrix} A\phi_1 \\ A\phi_2 \\ \vdots \\ A\phi_{N-1} \end{vmatrix}, \quad \Lambda_2\phi = \frac{1}{h^2} \begin{vmatrix} 2\phi_1 - \phi_2 \\ \vdots \\ -\phi_{l-1} + 2\phi_l - \phi_{l+1} \\ \vdots \\ -\phi_{N-2} + 2\phi_{N-1} \end{vmatrix}. \tag{1.21}$$

Introduce new vectors as follows:

$$(\Lambda_1\phi)_l = A\phi_l \quad (l = 1, 2, \ldots, N-1);$$

$$(\Lambda_2\phi)_k = \begin{cases} \dfrac{1}{h^2}(2\phi_1 - \phi_2)_k & (l = 1), \\[6pt] \dfrac{1}{h^2}(-\phi_{l-1} + 2\phi_l - \phi_{l+1})_k & (l = 2, 3, \ldots, N-2), \\[6pt] \dfrac{1}{h^2}(-\phi_{N-2} + 2\phi_{N-1})_k & (l = N-1). \end{cases} \tag{1.22}$$

We see that in this notation $(\Lambda_1 \phi)_l$ and $(\Lambda_2 \phi)_k$ are components of the vectors $\Lambda_1 \phi$ and $\Lambda_2 \phi$, while having the following convenient representations:

$$(\Lambda_1 \phi)_l = \frac{1}{h^2} \begin{vmatrix} 2\phi_{1,l} - \phi_{2,l} \\ \vdots \\ -\phi_{k-1,l} + 2\phi_{k,l} - \phi_{k+1,l} \\ \vdots \\ -\phi_{N-2,l} + 2\phi_{N-1,l} \end{vmatrix}$$

$(l = 2, \ldots, N-2)$,

$$(\Lambda_2 \phi)_k = \frac{1}{h^2} \begin{vmatrix} 2\phi_{k,1} - \phi_{k,2} \\ \vdots \\ -\phi_{k,l-1} + 2\phi_{k,l} - \phi_{k,l+1} \\ \vdots \\ -\phi_{k,N-2} + 2\phi_{k,N-1} \end{vmatrix}$$

$(l = 1, 2, \ldots, N-1; k = 1, 2, \ldots, N-1)$.

Similarly, let us represent F in the form

$$F = \begin{vmatrix} F_1 \\ F_2 \\ \vdots \\ F_{N-1} \end{vmatrix}, \quad F_1 = \begin{vmatrix} f_{1,1} + \dfrac{c_1}{h^2} + \dfrac{a_1}{h^2} \\ f_{2,1} + \dfrac{c_2}{h^2} \\ f_{3,1} + \dfrac{c_3}{h^2} \\ \vdots \\ f_{N-1,1} + \dfrac{c_{N-1}}{h} + \dfrac{b_1}{h^2} \end{vmatrix}, \quad F_l = \begin{vmatrix} f_{1,l} + \dfrac{a_l}{h^2} \\ f_{2,l} \\ f_{3,l} \\ \vdots \\ f_{N-1,l} + \dfrac{b_l}{h^2} \end{vmatrix},$$

$(l = 2, \ldots, N-2)$,

$$F_{N-1} = \begin{vmatrix} f_{1,N-1} + \dfrac{d_1}{h^2} + \dfrac{a_{N-1}}{h^2} \\ f_{2,N-1} + \dfrac{d_2}{h^2} \\ f_{3,N-1} + \dfrac{d_3}{h^2} \\ \vdots \\ f_{N-1,N-1} + \dfrac{d_{N-1}}{h^2} + \dfrac{b_{N-1}}{h^2} \end{vmatrix}.$$

The Poisson Equation

As a result we obtain the component version of the problem (k, l are fixed):

$$(\Lambda_1\phi)_{k,l} + (\Lambda_2\phi)_{k,l} = F_{k,l}. \tag{1.23}$$

Finally we obtain the vector matrix form of (1.16)

$$(\Lambda_1 + \Lambda_2)\phi = F, \tag{1.24}$$

by taking $l = 1, 2, \ldots, N - 1$ in (1.23).

One can easily verify, that each component of (1.24) corresponds to a difference equation from (1.16) with the boundary conditions already accounted for.

In order to form the algorithm, let us now compute the upper and lower spectral boundaries of the operator Λ.

Considering the form of the eigenfunctions of Λ (see Section 1.1.7), that is,

$$u_{mp}^{kl} = \sin m\pi kh \sin p\pi lh \qquad (k, l = 1, 2, \ldots, N - 1), \tag{1.25}$$

we find that

$$\lambda_{mp} = \frac{4}{h^2}\left(\sin^2 \frac{m\pi h}{2} + \sin^2 \frac{p\pi h}{2}\right). \tag{1.26}$$

Hence

$$\alpha = \frac{8}{h^2}\sin^2 \frac{\pi h}{2}, \qquad \beta = \frac{8}{h^2}\cos^2 \frac{\pi h}{2}, \tag{1.27}$$

where

$$\alpha = \alpha(\Lambda), \qquad \beta = \beta(\Lambda).$$

Let us use one of the iterative methods considered in Chapter 3:

$$\phi^{j+1} = \phi^j - \tau_j(\Lambda\phi^j - F), \tag{1.28}$$

or, equivalently,

$$\phi^{j+1} = \phi^j - \tau_j \xi^j,$$
$$\xi^j = \Lambda\phi^j - F.$$

The iterative process is to be continued until we have the inequality

$$\|\phi^j - \phi\| \leq \varepsilon,$$

where ε is an *a priori* constant. This estimate takes place if

$$\|\xi^j\| \leq \alpha(\Lambda)\varepsilon. \tag{1.29}$$

Generally speaking, the rate of convergence can be improved by replacing Iterations (1.28) by successive approximations with the Chebyshev acceleration:

$$\phi^{j+1} = \phi^j - \tau_j B^{-1}(\Lambda\phi^j - F), \tag{1.30}$$

where

$$B = (E + \sigma\Lambda_1)(E + \sigma\Lambda_2); \qquad \sigma = \frac{2}{\sqrt{\alpha\beta}}; \qquad \alpha = \alpha(\Lambda); \beta = \beta(\Lambda).$$

Let us find the spectral boundaries of the operator $B^{-1}\Lambda$. Since the matrices Λ_1 and Λ_2 have a common basis, the eigenvalues λ_1 and λ_2 of the matrices Λ_1 and Λ_2 correspondingly are related to the eigenvalues of the problem

$$B^{-1}\Lambda u = \lambda(B^{-1}\Lambda)u$$

by means of the following formula:

$$\lambda(B^{-1}\Lambda) = \frac{\lambda_1 + \lambda_2}{\left(1 + \dfrac{2\lambda_1}{\sqrt{\alpha\beta}}\right)\left(1 + \dfrac{2\lambda_2}{\sqrt{\alpha\beta}}\right)}, \tag{1.31}$$

$$\frac{\alpha}{2} \leq \lambda_1 \leq \frac{\beta}{2}, \quad \frac{\alpha}{2} \leq \lambda_2 \leq \frac{\beta}{2}.$$

Expression (1.31) can be written as

$$\lambda(B^{-1}\Lambda) = \frac{\sqrt{\alpha\beta}}{2} f(x, y), \tag{1.32}$$

where

$$x = \frac{2\lambda_1}{\sqrt{\alpha\beta}}, \quad y = \frac{2\lambda_2}{\sqrt{\alpha\beta}}; \quad f(x, y) = \frac{x + y}{(1 + x)(1 + y)}. \tag{1.33}$$

Thus, in order to define the spectral boundaries of the matrix $B^{-1}\Lambda$, it is enough to find the maximum and minimum of the function $f(x, y)$ on the square $\sqrt{\alpha/\beta} \leq x, y \leq \sqrt{\beta/\alpha}$. Looking at the derivatives of this function, i.e.,

$$\frac{\partial f}{\partial x} = \frac{1 - y}{1 + y} \cdot \frac{1}{(1 + x)^2}, \quad \frac{\partial f}{\partial y} = \frac{1 - x}{1 + x} \cdot \frac{1}{(1 + y)^2}, \tag{1.34}$$

it is not difficult to show, that the maximum of $f(x, y)$ is attained at two of the corners:

$$\max_{\sqrt{\alpha/\beta} \leq x, y \leq \sqrt{\beta/\alpha}} f(x, y) = f\left(\sqrt{\frac{\alpha}{\beta}}, \sqrt{\frac{\beta}{\alpha}}\right) = f\left(\sqrt{\frac{\beta}{\alpha}}, \sqrt{\frac{\alpha}{\beta}}\right), \tag{1.35}$$

while the other two corners yield the minimum:

$$\min_{\sqrt{\alpha/\beta} \leq x, y \leq \sqrt{\beta/\alpha}} f(x, y) = f\left(\sqrt{\frac{\alpha}{\beta}}, \sqrt{\frac{\alpha}{\beta}}\right) = f\left(\sqrt{\frac{\beta}{\alpha}}, \sqrt{\frac{\beta}{\alpha}}\right). \tag{1.35a}$$

From this and from (1.32) it follows that

$$\alpha(B^{-1}\Lambda) = \frac{\alpha}{\left(1 + \sqrt{\dfrac{\alpha}{\beta}}\right)^2}, \quad \beta(B^{-1}\Lambda) = \frac{1}{2}\sqrt{\frac{\alpha}{\beta}} \cdot \frac{(\alpha + \beta)}{\left(1 + \sqrt{\dfrac{\alpha}{\beta}}\right)^2}. \tag{1.36}$$

The Poisson Equation

Asymptotically, this means that for $\beta(\Lambda) \gg \alpha(\Lambda)$ we have

$$p(B^{-1}\Lambda) = \frac{\beta(B^{-1}\Lambda)}{\alpha(B^{-1}\Lambda)} = \frac{1}{2} \cdot \frac{\alpha + \beta}{\sqrt{\beta\alpha}} \simeq \frac{1}{2}\sqrt{\frac{\beta}{\alpha}} = \frac{1}{2}[p(\Lambda)]^{1/2}, \quad (1.37)$$

and

$$s = \frac{2}{\sqrt{p(B^{-1}\Lambda)}} = 2^{3/2} p^{-1/4}(\Lambda). \quad (1.38)$$

Consider now the realization scheme of the iterative Process (1.30):

$$\xi^j = \Lambda\phi^j - F, \quad (E + \sigma\Lambda_1)\xi^{j+1/2} = \xi^j,$$

$$(E + \sigma\Lambda_2)\xi^{j+1} = \xi^{j+1/2}, \quad \phi^{j+1} = \phi^j - \tau_j\xi^{j+1}, \quad (1.39)$$

where the choice of τ_j depends on the particular optimization method.

Using the component representation, the second and third equations of System (1.39) are handled as follows: we first solve the problem

$$(1 + 2\sigma)\xi^{j+1/2}_{1,l} - \sigma\xi^{j+1/2}_{2,l} = \xi^j_{1,l},$$

$$-\sigma\xi^{j+1/2}_{k-1,l} + (1 + 2\sigma)\xi^{j+1/2}_{k,l} - \sigma\xi^{j+1/2}_{k+1,l} = \xi^j_{k,l}, \quad (1.40)$$

$$-\sigma\xi^{j+1/2}_{N-2,l} + (1 + 2\sigma)\xi^{j+1/2}_{N-1,l} = \xi^j_{N-1,l}$$

for a fixed $l = 1, \ldots, N - 1$, and then, for a fixed $k = 1, \ldots, N - 1$:

$$(1 + 2\sigma)\xi^{j+1}_{k,1} - \sigma\xi^{j+1}_{k,2} = \xi^{j+1/2}_{k,1},$$

$$-\sigma\xi^{j+1}_{k,l-1} + (2\sigma + 1)\xi^{j+1}_{k,l} - \sigma\xi^{j+1}_{k,l-1} = \xi^{j+1/2}_{k,l}, \quad (1.41)$$

$$-\sigma\xi^{j+1}_{k,N-2} + (1 + 2\sigma)\xi^{j+1}_{k,N-1} = \xi^{j+1/2}_{k,N-1}.$$

To do this, we can use the factorization method.

In conclusion let us note, that the two-dimensional Neumann problem can be similarly reduced to Problem (1.24), which differs from the one we are considering only in the number of components of the solution to be determined: it will not be $(N - 1)^2$ as in the Dirichlet problem, but $(N + 1)^2$.

An important difference between the realization schemes of the difference analog of the Neumann problem and the Dirichlet problem is that in the Neumann problem $\alpha(\Lambda) = 0$. Therefore the right-hand side F, as well as the approximate solution ϕ^j must be orthogonal to the vector with identical components at each step. This means that before a new iteration step can be executed we must eliminate the constant factor from the vector ϕ^j.

With the orthogonalization procedure of this kind it is permissible to take the smallest nonzero eigenvalue for the lower spectral boundary.

Since the Neumann problem for the difference analog of the Laplace operator is known to have the spectrum given by the formula

$$\lambda_{mp} = \frac{4}{h^2}\left(\sin^2\frac{m\pi h}{2} + \sin^2\frac{p\pi h}{2}\right), \tag{1.42}$$

the spectral boundaries can be taken as

$$\alpha^*(\Lambda) = \frac{8}{h^2}\sin^2\frac{\pi h}{2}, \qquad \beta(\Lambda) = \frac{8}{h^2}. \tag{1.43}$$

The parameter σ is given in this case by

$$\sigma = \sqrt{\frac{2}{\alpha^*\beta}}, \tag{1.44}$$

and instead of (1.29) we now have

$$\|\xi^j\| < \alpha^*(\Lambda)\varepsilon \tag{1.45}$$

as the index for terminating the iterative process. Here the differences end.

More general iterative processes for the Neumann problem in the form of (1.30) can be treated similarly.

6.1.4. A Problem of Boundary Conditions

The above approaches to the Poisson equation allow one to make an important and fairly general conclusion with regard to forming efficient algorithms for solving boundary-value problems of mathematical physics. It mainly concerns the problem of boundary conditions. In the above we have shown how to eliminate the boundary conditions imposed on the solution of the problem at hand and how to modify them in view of the difference analog of the problem. There is a deeper sense in such an approach: if we can remove the boundary conditions from consideration by means of difference methods, we may then proceed without worrying about them any further: they are automatically taken care of by the modified difference equations. This is of importance for stationary problems, and especially for nonstationary problems in which the boundary conditions require careful treatment. In particular, this was the reason why we did not consider in Chapter 5 the splitting-up methods for nonstationary problems (in the differential formulation). This would have required additional work as to the compatibility of the boundary conditions and the decomposed system. From our point of view it is much simpler first to put the original problem of mathematical physics into correspondence with a system of difference equations (with respect to the spatial variables) and then to eliminate the boundary conditions using the difference analogs of the boundary conditions, the accuracy of which matches that of the difference equations. Having done this, one can next proceed by approximating the equations in time using the splitting-up

The Poisson Equation

method or another algorithm. This approach allows one to sidestep the compatibility problem for the boundary conditions as mentioned above, which would run through all stages of forming the numerical algorithm when using splitting-up schemes.

Let us now turn to the following fact, which is closely related to the above problem. Construction of difference schemes for problems of mathematical physics may in some cases benefit from a Fourier series representation of the solution along the eigenelements of the operator from the corresponding difference problem. We have already used this approach many times when analyzing the properties of numerical algorithms. In order to actually apply this method it is necessary, however, that the difference problem be closed with respect to the homogeneous boundary conditions. Nonhomogeneous boundary conditions can be made into homogeneous ones by an auxiliary transformation, which may be conveniently implemented by a method we now will illustrate with a simple example provided by Problem (1.1). Thus, consider the system of difference Equations (1.2). Let us extend the domain of definition of the solution x_k, $k = 1, \ldots, N - 1$ by adding two more net points $k = 0$ and $k = N$, and defining

$$\phi_0 = 0 \qquad \phi_N = 0.$$

Of course, this extension carries only a formal character and has no relation whatsoever to the actual values of the solution at the points $x = 0$ and $x = 1$, which are defined by the nonhomogeneous boundary Conditions (1.1). This means that once the problem has been solved, we have to go back and restrict the domain by eliminating the two points. This method permits one to consider the following equivalent problem:

$$\frac{-\phi_{k-1} + 2\phi_k - \phi_{k+1}}{h^2} = g_k \qquad (k = 1, 2, \ldots, N - 1), \qquad (1.46)$$

$$\phi_0 = 0, \qquad \phi_N = 0,$$

where

$$g_k = \begin{cases} \dfrac{a}{h^2} + f_1 & (k = 1), \\ f_k & (k = 2, 3, \ldots, N - 2), \\ \dfrac{b}{h^2} + f_{N-1} & (k = N - 1). \end{cases} \qquad (1.47)$$

Hence Problem (1.1) with nonhomogeneous boundary conditions has been reduced to a homogeneous problem, ready for the Fourier method.

In the case of nonhomogeneous Neumann Problem (1.10) the domain of the solution is to be extended by adjoining the net points x_{-1} and x_{N+1}.

Eventually we arrive at the problem

$$\frac{-\phi_{k-1} + 2\phi_k - \phi_{k+1}}{h^2} = g_k \quad (k = 0, 1, 2, \ldots, N), \tag{1.48}$$

$$\phi_{-1} = \phi_1, \quad \phi_{N+1} = \phi_{N-1},$$

where

$$g_k = \begin{cases} f_0 - \dfrac{2a}{h} & (k = 0), \\ f_k & (k = 1, 2, \ldots, N-1), \\ f_N + \dfrac{2b}{h} & (k = N). \end{cases} \tag{1.49}$$

Having found the solution of (1.48) and (1.49), we may drop the auxiliary components ϕ_{-1} and ϕ_{N+1}.

One can tackle multidimensional stationary and nonstationary problems in a similar manner.

If necessary, Equations (1.46)–(1.49) can be handled the same way as the nonhomogeneous Problems (1.1) and (1.10), i.e. by eliminating the boundary values. Again, we eventually finish up with Problems (1.3) and (1.12).

In what follows, we therefore will not consider the nonhomogeneous boundary conditions, since the above described algorithms permit one to reduce them to homogeneous conditions or to exclude them completely from consideration.

6.2. The Heat Equation

The problem of heat conduction is a typical nonstationary problem of mathematical physics and occupies a prominent place in history. It was the heat equation which stimulated many results of principal importance in numerical mathematics and which led to constructions of first-class algorithms for problems of mathematical physics. These studies go back to the classical work by O'Brien and Hyman and Kaplan [7], where the convergence problem of approximate solutions to the exact ones were taken up. By now there is a whole series of monographs and original papers devoted solely to the numerical treatment of the heat equation (for references see, for instance, Saulyev [3]). Thus, we will restrict our attention to some of the methods which have found the broadest applications.

6.2.1. The One-Dimensional Problem of Heat Conduction

To begin with, consider a simple problem of heat propagation in a homogeneous bar of finite length, heated by internal sources located on the

The Heat Equation

boundary. The problem can be described as follows:

$$\frac{1}{c^2} \cdot \frac{\partial \phi}{\partial t} = \frac{\partial^2 \phi}{\partial x^2} + f(x, t), \tag{2.1}$$

$$\phi(0, t) = a(t), \qquad \phi(1, t) = b(t), \qquad \phi(x, 0) = \phi^0(x),$$

where f, a, b and ϕ^0 are given sufficiently smooth functions; $c^2 = \text{const}$; the variables x and t run through the domain of definition of the solution $D \times D_t = \{0 \leq x \leq 1, 0 \leq t \leq T\}$.

We will solve (2.1) using the finite-difference method. Thus, let us first approximate (2.1) with respect to x with second-order accuracy in $h = \Delta x$. Divide the interval $0 \leq x \leq 1$ into N subintervals of equal length $h = 1/N$ and denote the points of the grid by x_k ($k = 1, \ldots, N$). We arrive at the following problem, similar to (1.2):

$$\frac{1}{c^2} \cdot \frac{d\phi_k}{dt} + \frac{-\phi_{k-1} + 2\phi_k - \phi_{k+1}}{h^2} = f_k(t), \tag{2.2}$$

$$\phi_0 = a(t), \qquad \phi_N = b(t), \qquad \phi_k = \phi_k^0 \quad \text{for} \quad t = 0$$

$$(k = 1, 2, \ldots, N-1),$$

or, using the vector-matrix notation,

$$\frac{1}{c^2} \cdot \frac{d\phi}{dt} + A\phi = g, \tag{2.3}$$

$$\phi = \phi^0 \quad \text{for} \quad t = 0,$$

where A is a positive matrix, and $\phi(t)$ and $g(t)$ are vector functions which have been defined for an arbitrary t in Problem (1.4).

Using the ideas from Section 6.1.4, Equations (2.2) can be rewritten as

$$\frac{1}{c^2} \cdot \frac{d\phi_k}{dt} + \frac{-\phi_{k-1} + 2\phi_k - \phi_{k+1}}{h^2} = g_k, \tag{2.4}$$

$$\phi_0 = 0, \qquad \phi_N = 0, \qquad \phi_k = \phi_k^0 \quad \text{for} \quad t = 0,$$

where g_k is defined by (1.47).

As we have pointed out in Section 6.1.4, the solution of (2.4) makes sense only at the net points $x_1, \ldots, x_{N-1}$.

Let us now consider the system of ordinary differential equations of (2.4). Integrate each of them with respect to time over the interval $t_j \leq t \leq t_{j+1}$:

$$\frac{\phi_k^{j+1} - \phi_k^j}{\tau} = \frac{\bar{\phi}_{k-1}^j - 2\bar{\phi}_k^j + \bar{\phi}_{k+1}^j}{h^2} + \bar{g}_k^j, \tag{2.5}$$

where

$$\bar{\phi}_k^j = \frac{1}{\Delta t} \int_{t_j}^{t_{j+1}} \phi_k \, dt, \qquad \bar{g}_k^j = \frac{1}{\Delta t} \int_{t_j}^{t_{j+1}} g_k \, dt. \tag{2.6}$$

Here we have adopted the notation
$$\phi_k^j = \phi_k(t_j), \qquad \tau = c^2 \Delta t.$$

The kind of difference equations that we will obtain depends on a particular approximation in (2.6). Consider the following three simple interpolations:

$$\frac{1}{\Delta t}\int_{t_j}^{t_{j+1}} \phi_k \, dt = \begin{cases} \phi_k^j, \\ \phi_k^{j+1}, \\ \frac{1}{2}(\phi_k^j + \phi_k^{j+1}); \end{cases} \qquad \frac{1}{\Delta t}\int_{t_j}^{t_{j+1}} g_k \, dt = \begin{cases} g_k^j, \\ g_k^{j+1}, \\ \frac{1}{2}(g_k^{j+1} + g_k^j). \end{cases}$$

Correspondingly, they result into the following most widely used difference schemes:

the explicit triangular scheme (∴)
$$\frac{\phi_k^{j+1} - \phi_k^j}{\tau} = \frac{\phi_{k-1}^j - 2\phi_k^j + \phi_{k+1}^j}{h^2} + g_k^j; \tag{2.7}$$

the implicit triangular scheme (∵)
$$\frac{\phi_k^{j+1} - \phi_k^j}{\tau} = \frac{\phi_{k-1}^{j+1} - 2\phi_k^{j+1} + \phi_{k+1}^{j+1}}{h^2} + g_k^{j+1}; \tag{2.8}$$

the Crank–Nicholson scheme (∷)
$$\frac{\phi_k^{j+1} - \phi_k^j}{\tau} = \frac{\phi_{k-1}^{j+1} - 2\phi_k^{j+1} + \phi_{k-1}^{j+1}}{2h^2}$$
$$+ \frac{\phi_{k-1}^j - 2\phi_k^j + \phi_{k+1}^j}{2h^2} + \frac{g_k^{j+1} + g_k^j}{2}. \tag{2.9}$$

The systems of Equations (2.7)–(2.9) must be adjoined in addition by the boundary conditions
$$\phi_0^j = 0, \qquad \phi_N^j = 0. \tag{2.10}$$

The triangular scheme (∴) can be solved explicitly for the unknown ϕ^{j+1}:
$$\phi^{j+1} = \phi^j + \mu(\phi_{k-1}^j - 2\phi_k^j + \phi_{k+1}^j) + \tau g_k^j,$$
$$\phi_0^j = 0, \qquad \phi_N^j = 0, \qquad \mu = \frac{\tau}{h^2}. \tag{2.11}$$

The implicit triangular scheme (∵) has a more complicated realization; we have to solve eventually the following difference equation:
$$-\phi_{k-1}^{j+1} + \left(2 + \frac{1}{\mu}\right)\phi_k^{j+1} - \phi_{k+1}^{j+1} = \frac{1}{\mu}\phi_k^j + h^2 g_k^{j+1},$$
$$\phi_0^{j+1} = 0, \qquad \phi_N^{j+1} = 0. \tag{2.12}$$

The Heat Equation

Finally, the Crank–Nicholson scheme ($\genfrac{}{}{0pt}{}{\cdot\,\cdot\,\cdot}{\cdot\,\cdot\,\cdot}$) has the following realization:

$$-\xi_{k-1}^{j+1} + \frac{2(1+\mu)}{\mu}\xi_k^{j+1} - \xi_{k+1}^{j+1} = \frac{2}{\mu}\phi_k^j + h^2 g_k^{j+1/2},$$

$$\xi_0^{j+1} = 0, \qquad \xi_N^{j+1} = 0, \qquad \phi_k^{j+1} = 2\xi_k^{j+1} - \phi_k^j, \qquad (2.13)$$

where

$$g_k^{j+1/2} = \tfrac{1}{2}(g_k^{j+1} + g_k^j).$$

Problems (2.12) and (2.13) can be efficiently solved by the factorization method.

Let us next investigate the stability of the difference schemes of Equations (2.7)–(2.9), assuming (2.10). To this end, let us expand the solution into a Fourier series with respect to the complete system of functions $\{\sin n\pi kh\}$, where $h = 1/N$ [the functions clearly satisfy (2.10)]. Let

$$\phi_k^j = \sum_{n=1}^{N-1} \Phi_n^j \sin n\pi kh, \qquad g_k^j = \sum_{n=1}^{N-1} G_n^j \sin n\pi kh, \qquad (2.14)$$

where

$$\Phi_n^j = \frac{1}{q_n} \sum_{k=1}^{N-1} \phi_k^j \sin n\pi kh, \qquad G_n^j = \frac{1}{q_n} \sum_{k=1}^{N-1} g_k^j \sin n\pi kh.$$

$$q_n = \sum_{k=1}^{N-1} \sin^2 n\pi kh.$$

Let us substitute (2.14) into (2.7)–(2.9), multiply the results by $\sin(l n\pi h)$, and then sum them up in l. As a result we obtain the recursive relations for the Fourier coefficients: for the scheme ($\genfrac{}{}{0pt}{}{\cdot\,\cdot\,\cdot}{\cdot\,\cdot\,\cdot}$)

$$\frac{\Phi_n^{j+1} - \Phi_n^j}{\tau} + \lambda_n \Phi_n^j = G_n^j, \qquad (2.15)$$

for the scheme ($\genfrac{}{}{0pt}{}{\cdot\,\cdot\,\cdot}{\cdot\,\cdot\,\cdot}$)

$$\frac{\Phi_n^{j+1} - \Phi_n^j}{\tau} + \lambda_n \Phi_n^{j+1} = G_n^{j+1}, \qquad (2.16)$$

for the scheme ($\genfrac{}{}{0pt}{}{\cdot\,\cdot\,\cdot}{\cdot\,\cdot\,\cdot}$)

$$\frac{\Phi_n^{j+1} - \Phi_n^j}{\tau} + \lambda_n \frac{\Phi_n^{j+1} + \Phi_n^j}{2} = \frac{G_n^{j+1} + G_n^j}{2}, \qquad (2.17)$$

where

$$\lambda_n = \frac{4}{h^2} \sin^2 \frac{n\pi h}{2}. \qquad (2.18)$$

Solving (2.15)–(2.17), we get correspondingly

$$\Phi_n^{j+1} = (1 - \tau\lambda_n)\Phi_n^j + \tau G_n^j,$$

$$\Phi_n^{j+1} = \frac{1}{1 + \tau\lambda_n}\Phi_n^j + \frac{\tau}{1 + \tau\lambda_n}G_n^{j+1},$$

$$\Phi_n^{j+1} = \frac{1 - \frac{\tau}{2}\lambda_n}{1 + \frac{\tau}{2}\lambda_n}\Phi_n^j + \frac{\tau}{2}\cdot\frac{G_n^{j+1} + G_n^j}{1 + \frac{\tau}{2}\lambda_n}. \tag{2.19}$$

Note that

$$\frac{4}{h^2}\sin^2\frac{\pi h}{2} \leqslant \lambda_n \leqslant \frac{4}{h^2}\cos^2\frac{\pi h}{2} < \frac{4}{h^2}. \tag{2.20}$$

For $(\pi h/2) \ll 1$ we have the asymptotic relation

$$\pi^2 \leqslant \lambda_n \leqslant \frac{4}{h^2}. \tag{2.21}$$

Taking (2.20) into account, we conclude that for

$$\tau \leqslant \frac{h^2}{2\cos^2\frac{\pi h}{2}} \tag{2.22}$$

the triangular scheme $(\therefore)$ is stable, since for all n

$$|1 - \tau\lambda_n| < 1. \tag{2.23}$$

Instead of (2.22) one can use the sufficient condition

$$\tau \leqslant \frac{h^2}{2}. \tag{2.24}$$

In the cases of the implicit triangular scheme and the Crank–Nicholson scheme we have correspondingly the inequalities

$$\left|\frac{1}{1 + \tau\lambda_n}\right| < 1, \qquad \left|\frac{1 - \frac{\tau}{2}\lambda_n}{1 + \frac{\tau}{2}\lambda_n}\right| < 1,$$

which hold true for any n and $\tau > 0$. This means that, while the scheme $(\therefore)$ requires Condition (2.22) for its stability, the schemes $(\cdot\!\cdot\!\cdot)$ and $(\cdot\!\cdot\!\cdot)$ are absolutely stable. The Crank–Nicholson scheme has been used most extensively.

In the case of the Neumann problem for the heat equation, the operator A and the vector functions ϕ and g can be determined from (1.14). As a result we obtain equations similar to (2.5) and (2.6), which can be solved without any difficulty. At this point let us make the following remark: In the Neumann problem for the Laplace equation we have "filtered" away from the approximate solution the parasitic component (frequency) corresponding to the

The Heat Equation

eigenvalue $\lambda = 0$. This can no longer be done in the problem of heat conduction; we need to keep this frequency, since it describes the overall increase or decrease in the temperature of the bar due to the external sources.

6.2.2. The Two-Dimensional Problem of Heat Conduction

The problem of heat conduction in two dimensions can be formulated as follows:

$$\frac{1}{c^2}\frac{\partial \phi}{\partial t} - \Delta \phi = f \quad \text{in} \quad D \times D_t$$

$$\phi = g \quad \text{or} \quad \frac{\partial \phi}{\partial n} = g \quad \text{on} \quad \partial D \times D_t \tag{2.25}$$

$$\phi = s(x, y) \quad \text{in} \quad D \quad \text{for} \quad t = 0,$$

where $D \equiv \{0 \leq x \leq 1, 0 \leq y \leq 1\}$.

First of all let us reduce this problem to its finite-difference version in the variables (x, y). The manipulations similar to those used in the two-dimensional Dirichlet problem lead to the following representation:

$$\frac{1}{c^2} \cdot \frac{d\phi}{dt} + (\Lambda_1 + \Lambda_2)\phi = F, \tag{2.26}$$

$$\phi = s \quad \text{for} \quad t = 0,$$

where the matrices Λ_1, Λ_2 and the vector functions ϕ and F are defined the same way as in Problem (1.24), the only difference being that now the components of ϕ and F depend on t. Note that if the heat equation is supplemented with the Dirichlet boundary conditions, we have

$$\Lambda_1 > 0, \qquad \Lambda_2 > 0, \tag{2.27}$$

whereas in the case of Neumann conditions

$$\Lambda_1 \geqslant 0, \qquad \Lambda_2 \geqslant 0. \tag{2.28}$$

Divide the interval $0 \leq t \leq T$ into subintervals by means of the points t_j. Putting $\tau = c^2 \Delta t$, let us approximate (2.26) using the component splitting-up method. Then $(t_{j-1} \leq t \leq t_{j+1})$

$$\left(E + \frac{\tau}{2}\Lambda_1\right)\phi^{j-1/2} = \left(E - \frac{\tau}{2}\Lambda_1\right)\phi^{j-1},$$

$$\left(E + \frac{\tau}{2}\Lambda_2\right)(\phi^j - \tau f^j) = \left(E - \frac{\tau}{2}\Lambda_2\right)\phi^{j-1/2},$$

$$\left(E + \frac{\tau}{2}\Lambda_2\right)\phi^{j+1/2} = \left(E - \frac{\tau}{2}\Lambda_2\right)(\phi^j + \tau f^j), \tag{2.29}$$

$$\left(E + \frac{\tau}{2}\Lambda_1\right)\phi^{j+1} = \left(E - \frac{\tau}{2}\Lambda_1\right)\phi^{j+1/2}.$$

Note that the right-hand sides above can be computed in terms of components by the explicit schemes. Therefore the system of equations in the component form we are dealing with can be represented by means of the three-point schemes similar to (1.39), the role of σ and ξ_k^j being now played by the parameter $\tau/2$ and the right-hand sides in (2.29).

The Neumann conditions can be handled similarly.

6.3. The Wave Equation

The hyperbolic-type equations play an important role in applications, and the corresponding numerical methods are developed with a sufficient completeness. A distinguishing feature of the hyperbolic equations is the fact that the domains of their solutions are bounded by characteristic cones. Thus, the part of the domain $D \times D_t$, which lies outside such a cone, has no influence on the solution. Another feature is that the hyperbolic problems admit the existence of the solutions which are not smooth. This latter circumstance is of particular concern to numerical analysts.

Consider the following one-dimensional problem, describing small oscillations of a homogeneous string:

$$\frac{1}{c^2} \cdot \frac{\partial^2 \phi}{\partial t^2} = \frac{\partial^2 \phi}{\partial x^2} + f(x, t),$$

$$\phi(0, t) = a(t), \qquad \phi(1, t) = b(t), \qquad (3.1)$$

$$\phi(x, 0) = p(x), \qquad \frac{\partial \phi}{\partial t}(x, 0) = q(x),$$

where c designates the velocity of propagation of excitations along the string; $a(t)$, $b(t)$, $f(t, x)$, $p(x)$ and $q(x)$ are given functions.

The theory of string oscillations is fairly complete. Here we are concerned with the numerical aspects of the problem. For this purpose let us first transform (3.1) to a system of ordinary differential equations in the time variable, by making use of the difference relation in the x variable. The method described in Section 6.2 fits well for our present problem. Thus, we obtain the following approximation of (3.1):

$$\frac{1}{c^2} \cdot \frac{d^2 \phi_k}{dt^2} + \frac{-\phi_{k-1} + 2\phi_k - \phi_{k+1}}{h^2} = g_k(t),$$

$$\phi_0 = 0, \qquad \phi_N = 0, \qquad (3.2)$$

$$\phi_k = p_k, \qquad \frac{d\phi_k}{dt} = q_k \quad \text{for} \quad t = 0,$$

where, similarly as in (2.4), the function $g_k(t)$ is defined by (1.47).

As pointed out earlier, the solution $\phi_k(t)$ of (3.2) is only meaningful at the net points $k = 1, 2, \ldots, N - 1$. At the points $k = 0$ and $k = N$ the solution

The Wave Equation   231

is defined by the boundary conditions from (3.1); the homogeneous conditions in (3.2) are the results of a special extension of the domain of definition of the solution, and do not approximate the function $\phi(x, t)$ at x_0 and x_N. Note that Problem (3.2) has an accuracy of second order with respect to $h = \Delta x$.

In order to obtain next the finite-difference approximation of (3.2) with respect to the time variable, let us introduce a system of netpoints $t = t_i$ (where $t_{i+1} = t_i + \Delta t$), and consider the following two most useful schemes, the explicit "cross" scheme $(\cdot\!:\!\cdot)$

$$\frac{\phi_k^{j+1} - 2\phi_k^j + \phi_k^{j-1}}{\tau^2} = \frac{\phi_{k-1}^j - 2\phi_k^j + \phi_{k+1}^j}{h^2} + g_k^j,$$

$$\phi_0^j = 0, \qquad \phi_N^j = 0, \qquad \phi_k^0 = p_k, \qquad (3.3)$$

$$\phi_k^1 = p_k + \Delta t q_k + \frac{\tau^2}{2}\left(\frac{p_{k-1} - 2p_k + p_{k+1}}{h^2} + q_k^0\right),$$

where $\tau = c\Delta t$, and the implicit scheme $(\cdot\!\vdots\!\cdot)$, which is analoguous to the Crank–Nicholson scheme:

$$\frac{\phi_k^{j+1} - 2\phi_k^j + \phi_k^{j-1}}{\tau^2} = \frac{\phi_{k-1}^{j+1} - 2\phi_k^{j+1} + \phi_{k+1}^{j+1}}{2h^2}$$

$$+ \frac{\phi_{k-1}^{j-1} - 2\phi_k^{j-1} + \phi_{k+1}^{j-1}}{2h^2} + g_k^j, \qquad (3.4)$$

$$\phi_0^{j+1} = 0, \qquad \phi_N^{j+1} = 0, \qquad \phi_k^0 = p_k,$$

$$\phi_k^1 = \phi_k^0 + \Delta t q_k + \frac{\tau^2}{2}\left(\frac{p_{k-1} - 2p_k + p_{k+1}}{h^2} + q_k^0\right).$$

It is not difficult to verify that for sufficiently smooth solutions $\phi(x, t)$ the difference schemes (3.3) and (3.4) approximate the original problem with second-order accuracy with respect to both h and τ.

Problems (3.3) and (3.4) will be solved using the method of Fourier series with respect to the eigenfunctions

$$u_n(k) = \sin n\pi kh \qquad (k, n = 1, 2, \ldots, N - 1),$$

as in (2.14). There result the following recursive relations for the Fourier coefficients: for the explicit "cross" scheme we have

$$\frac{\Phi_n^{j+1} - 2\Phi_n^j + \Phi_n^{j-1}}{\tau^2} + \frac{4}{h^2}\sin^2\frac{n\pi h}{2}\,\Phi_n^j = G_n^j,$$

$$\Phi_n^0 = P_n, \quad \Phi_n^1 = P_n + \Delta t Q_n + \frac{2\tau^2}{h^2}\sin^2\frac{n\pi h}{2}P_n + \frac{\tau^2}{2}G_n^0, \qquad (3.5)$$

where Φ_n, G_n, P_n, and Q_n are the Fourier coefficients of the elements ϕ_k, g_k, p_k, and q_k respectively. For the implicit scheme $(\genfrac{}{}{0pt}{}{\cdot\,\cdot\,\cdot}{\cdot\,\cdot\,\cdot})$

$$\frac{\Phi_n^{j+1} - 2\Phi_n^j + \Phi_n^{j-1}}{\tau^2} + \frac{4}{h^2} \sin^2 \frac{n\pi h}{2} \cdot \frac{\Phi_n^{j+1} + \Phi_n^{j-1}}{2} = G_n^j,$$

$$\Phi_n^0 = P_n, \quad \Phi_n^1 = P_n + \Delta t Q_n + \frac{2\tau^2}{h^2} \sin^2 \frac{n\pi h}{2} P_n + \frac{\tau^2}{2} G_n^0. \tag{3.6}$$

Our next step is establishing a criterion for countable stability for the problems with smooth input data. For that let us investigate the behavior of the linearly independent solutions of the homogeneous Problems (3.5) and (3.6), depending on j. The solutions will be sought in the form

$$\Phi_n^j = A_n \eta_n^j, \tag{3.7}$$

where A_n and η_n are constants (note that η_n^j in the above equation means the jth power of η_n). Putting $G_n^j = 0$ in (3.5) and (3.6) and substituting then (3.7) into these equations, we obtain the following expressions for η_n: in the case of the "cross" scheme $(\cdot\genfrac{}{}{0pt}{}{\cdot}{\cdot}\cdot)$

$$\eta_n^2 - 2(1 - \mu_n^2)\eta_n + 1 = 0, \tag{3.8}$$

and in the case of the implicit scheme $(\genfrac{}{}{0pt}{}{\cdot\,\cdot\,\cdot}{\cdot\,\cdot\,\cdot})$

$$\eta_n^2 - \frac{2}{1 + \mu_n^2} \eta_n + 1 = 0, \tag{3.9}$$

where

$$\mu_n^2 = 2\frac{\tau^2}{h^2} \sin^2 \frac{n\pi h}{2}.$$

Solving the quadratic Equations (3.8) and (3.9) we obtain

$$\eta_n = 1 - \mu_n^2 \pm \sqrt{(1 - \mu_n^2)^2 - 1}; \tag{3.10}$$

and

$$\eta_n = \frac{1}{1 + \mu_n^2} \pm \sqrt{\left(\frac{1}{1 + \mu_n^2}\right)^2 - 1}. \tag{3.11}$$

Let us look first at the solution corresponding to the "cross" scheme. It is easy to see that if

$$\mu_n^2 = 2\frac{\tau^2}{h^2} \sin^2 \frac{n\pi h}{2} < 2, \tag{3.12}$$

then

$$|\eta_{n_i}| = 1, \quad |\eta_{n_1} - \eta_{n_2}| > C\tau, \tag{3.13}$$

and hence the difference scheme is stable. It is necessary that Condition (3.12) holds true for all $n = 1, 2, \ldots, N - 1$. The latter is clearly true if τ and h are related as written below:

$$\frac{\tau^2}{h^2} < \frac{1}{\sin^2 \frac{\pi h n}{2}} \tag{3.14}$$

or, more simply, if the two parameters satisfy the Courant condition:

$$\frac{\tau}{h} < 1. \tag{3.15}$$

It is not difficult to verify that in the case of the implicit scheme ($\vdots$) we have

$$|\eta_n| = 1 \tag{3.16}$$

for any $\tau > 0$ and any n. This means that the given scheme is absolutely stable (see Richtmyer and Morton [3]).

In the same fashion one may consider small oscillations of a membrane. In this case we obtain the equation

$$\frac{1}{c^2} \frac{\partial^2 \phi}{\partial t^2} = \frac{\partial^2 \phi}{\partial x^2} + \frac{\partial^2 \phi}{\partial y^2} + f(x, y, t) \quad \text{in} \quad D \times D_t \tag{3.17}$$

with the boundary condition

$$\phi = g \quad \text{on} \quad \partial D \times D_t, \tag{3.18}$$

and the initial conditions

$$\phi = p, \quad \frac{\partial \phi}{\partial t} = q \quad \text{in} \quad D \quad \text{for} \quad t = 0.$$

Assuming the input data sufficiently smooth, the solution of this problem in the square D can be found by the finite-difference method, in the same manner as in the case of the string. The method of Fourier series is similar to that we have used earlier for the two-dimensional difference analog of the heat equation. Rather than repeating this route again, we will now turn to a more general problem of the hyperbolic type.

6.4. The Equation of "Motion"

The equation of motion has become a subject of considerable interest among scientists, mainly because of its significant applications. For example, one of the important building elements of the hydrodynamical process—the propagation of particles along a trajectory—is governed by this equation.

The development of numerical methods for the equation of motion has been triggered by the needs of hydrodynamics and aerodynamics. In the last couple of decades the numerical methods in hydrodynamics have been enriched by a number of interesting ideas due to Neuman and Richtmyer [7], Dorodnitsyn [3], Godunov [4], Lax [6, 7], Babenko and Rusanov [12], and many others.

Recent stimuli for the development of efficient numerical methods for the equation of motion are coming from the problems of weather forecasting and from ocean dynamics. Thanks to works by Kurihara and Holloway [4], Bryan [4], Marchuk [4], and others, there are now universal, efficient algorithms for solving such problems.

The propagation problem of a substance along the particle trajectories may be considered to belong among the simpler problems of mathematical physics. It can be described by the following equation:

$$\frac{d\phi}{dt} = 0,$$

where

$$\frac{d}{dt} = \frac{\partial}{\partial t} + u\frac{\partial}{\partial x} + v\frac{\partial}{\partial y} + w\frac{\partial}{\partial z}$$

and u, v, and w are the components of the velocity vector $ui + vj + wk$, so that

$$u = \frac{dx}{dt}, \quad v = \frac{dy}{dt}, \quad w = \frac{dz}{dt}.$$

The above equation is further restrained by imposing additional conditions, the simplest of which can be written as

$$\phi(x, y, z, 0) = f(x, y, z),$$

assuming the unbounded region.

A similar problem arises in the framework of the general algorithm for numerical solution of equations of hydrodynamics, the theory of radiation and many others. Because of this fact we will discuss thoroughly the possibilities of numerical treatment of such problems.

6.4.1. The Simplest Equations of Motion

When solving problems of hydrodynamics, hydrothermodynamics, weather forecasting, ocean dynamics, and others, we often have to deal with equations describing the transfer of a substance along trajectories. A simple equation of this kind is as follows:

$$\frac{\partial \phi}{\partial t} + u\frac{\partial \phi}{\partial x} = 0 \quad \text{in} \quad D \times D_t,$$

$$\phi = f(x) \quad \text{in} \quad D \quad \text{for} \quad t = 0, \tag{4.1}$$

where u is a given velocity and $f(x)$ represents the initial distribution of ϕ. The region D is taken as the whole real line, and $D_t = \{0 \le t \le T\}$. Assume that $\phi(x, t)$ and $f(x)$ are periodic in x, with the period 2π. If $u =$ const, then the problem (4.1) has an immediate solution

$$\phi(x, t) = f(x - ut), \qquad (4.2)$$

assuming f is differentiable. Solution (4.2) describes the propagation of the initial perturbation along the characteristic

$$x - ut = \text{const}.$$

This means that $\phi(x, t) =$ const on an arbitrary line $x - ut =$ const.

Thus Problem (4.1) defines for $u > 0$ a process of propagation of the perturbation in the direction of the growing x. These well-known facts should be kept in mind when constructing difference analogs of Problem (4.1). If the velocity $u = u(x, t)$ is changing, an analytic solution of (4.1) becomes a problem. In this case it is usually necessary to resort to numerical methods based on difference approximations.

Consider a simple difference scheme with $u =$ const and, to be specific, take $u > 0$. We then have the explicit scheme

$$\frac{\phi_k^{j+1} - \phi_k^j}{\tau} + u \frac{\phi_k^j - \phi_{k-1}^j}{\Delta x} = 0, \qquad (4.3)$$

and the implicit scheme

$$\frac{\phi_k^{j+1} - \phi_k^j}{\tau} + u \frac{\phi_k^{j+1} - \phi_{k-1}^{j+1}}{\Delta x} = 0. \qquad (4.4)$$

Both schemes are of first-order accuracy in Δx and τ. Indeed, assuming that the solution $\phi(x, t)$ and the initial distribution $f(x)$ are sufficiently smooth, we can expand into the Taylor series in the vicinity of $x = x_k, t = t_j$:

$$\phi(x, t) = (\phi)_k^j + (\phi_t)_k^j (t - t_j) + (\phi_x)_k^j (x - x_k) + \cdots \qquad (4.5)$$

Substituting (4.5) into (4.3) we obtain

$$\frac{\partial \phi}{\partial t} + u \frac{\partial \phi}{\partial x} = \frac{u \Delta x}{2} \cdot \frac{\partial^2 \phi}{\partial x^2} - \frac{\tau}{2} \cdot \frac{\partial^2 \phi}{\partial t^2}. \qquad (4.6)$$

In (4.6) we have dropped the higher-order terms. From (4.1) we have

$$\frac{\partial^2 \phi}{\partial t^2} = u^2 \frac{\partial^2 \phi}{\partial x^2}. \qquad (4.7)$$

Then Expression (4.6) becomes

$$\frac{\partial \phi}{\partial t} + u \frac{\partial \phi}{\partial x} = \frac{u \Delta x - \tau u^2}{2} \cdot \frac{\partial^2 \phi}{\partial x^2} \quad \text{for} \quad x = x_k, t = t_j. \qquad (4.8)$$

The above procedure has been suggested by Zhukov†. Some analysis of (4.8) shows, that for

$$\frac{u\tau}{\Delta x} < 1$$

Relation (4.8) can be treated as the heat equation with the domain $x_{k-1} \leqq x \leqq x_k, t_j \leqq t \leqq t_{j+1}$. Assuming that the terms we have dropped in (4.8) are small, we obtain the equation

$$\frac{\partial \phi}{\partial t} + u \frac{\partial \phi}{\partial x} = \mu \frac{\partial^2 \phi}{\partial x^2},$$

where

$$\mu = \frac{u\Delta x - u^2\tau}{2}$$

is a so-called *coefficient of artificial* or "*countable*" *viscosity*. Note, by the way, that if

$$\frac{u\tau}{\Delta x} = 1,$$

then $\mu = 0$, so that all the terms dropped are zero, and the explicit Scheme (4.3) has an infinite order of approximation with respect to Δx and τ.

In particular, it is to be noted that if

$$\frac{u\tau}{\Delta x} > 1$$

we get the equation

$$\frac{\partial \phi}{\partial t} + u \frac{\partial \phi}{\partial x} = -|\mu| \frac{\partial^2 \phi}{\partial x^2}. \qquad (4.9)$$

It is easy to see, that for the initial condition

$$\phi = \phi^0(x) \quad \text{for} \quad t = 0 \qquad (4.10)$$

Equation (4.9) results in an ill-posed problem (in the sense of Hadamard). The solution of this problem is unstable with respect to small variations in the initial data. Thus any difference equations for problems like (4.1) must be formed so as to satisfy the well-posedness condition

$$\mu = \frac{u\tau}{\Delta x} \leqslant 1.$$

Next let us investigate the problem of countable stability of Scheme (4.3). To this end consider first the spectral problem

$$(A^h \omega)_k \equiv u \frac{\omega_k - \omega_{k-1}}{\Delta x} = \lambda \omega_k \qquad (4.11)$$

---

† See the monograph by Rozhdestvenskii and Yanenko [2]. This type of analysis was further developed in a number of papers by Yanenko and Shokin [7].

The Equation of "Motion"

on the infinite net interval $D_h = (-\infty < x_k < \infty)$. The solution of (4.11), which is bounded in D_h, is of the form

$$\omega_k = e^{ikp\Delta x}, \tag{4.12}$$

where p is an arbitrary integer. Substituting (4.12) into (4.11), we obtain the following eigenvalue:

$$\lambda_p = \frac{u}{\Delta x}\left(2\sin^2\frac{p\Delta x}{2} + i\sin p\Delta x\right). \tag{4.13}$$

Let us write (4.3) in the operator form

$$\frac{\phi^{j+1} - \phi^j}{\tau} + A^h\phi^j = 0. \tag{4.14}$$

The solution of Equation (4.14) will be sought in the form

$$\phi^j = \sum_{p=-\infty}^{\infty} \phi_p^j e^{ikp\Delta x},$$

where ϕ_p^j is the Fourier coefficient of ϕ^j. We have

$$\frac{\phi_p^{j+1} - \phi_p^j}{\tau} + \lambda_p \phi_p^j = 0.$$

Hence

$$\phi_p^{j+1} = T_p \phi_p^j, \tag{4.16}$$

where $T_p = 1 - \tau\lambda_p$ is the *transition factor* of the Fourier coefficients.

Let us find a condition under which the Fourier coefficients do not grow in modulus. For that we may take

$$|1 - \tau\lambda_p| \leq 1. \tag{4.17}$$

This inequality takes place if

$$\frac{u\tau}{\Delta x} \leq 1.$$

Indeed,

$$|1 - \tau\lambda_p|^2 = \left(1 - \frac{2u\tau}{\Delta x}\sin^2\frac{p\Delta x}{2}\right)^2 + \left(\frac{u\tau}{\Delta x}\right)^2 \sin^2 p\Delta x$$

$$= 1 - 4\frac{u\tau}{\Delta x}\sin^2\frac{p\Delta x}{2} + \left(\frac{u\tau}{\Delta x}\right)^2\left(4\sin^4\frac{p\Delta x}{2} + \sin^2 p\Delta x\right)$$

$$= 1 - 4\frac{u\tau}{\Delta x}\sin^2\frac{p\Delta x}{2} + 4\left(\frac{u\tau}{\Delta x}\right)^2\sin^2\frac{p\Delta x}{2}\left(\sin^2\frac{p\Delta x}{2} + \cos^2\frac{p\Delta x}{2}\right)$$

$$= 1 - 4\sin^2\frac{p\Delta x}{2}\left(\frac{u\tau}{\Delta x}\right)\left(1 - \frac{u\tau}{\Delta x}\right) \geq 0.$$

If $u\tau/\Delta x > 1$, then $|1 - \tau\lambda_p| > 1$. For $u\tau/\Delta x \leq 1$ we have correspondingly $|1 - \tau\lambda_p| \leq 1$.

In this manner we arrive at the condition for countable stability of the difference scheme. It is easy to see that in the given case the stability criterion we have established coincides with the condition of well-posedness (4.8).

Let us next discuss the implicit difference Scheme (4.4). Using the above method, one can again show that (4.4) is of first-order accuracy in Δx and τ. With the help of the Taylor expansion around $x = x_k$ and $t = t_j$ we get the "asymptotic equation"

$$\frac{\partial \phi}{\partial t} + u \frac{\partial \phi}{\partial x} = \mu \frac{\partial^2 \phi}{\partial x^2}, \qquad (4.18)$$

where

$$\mu = \frac{u\Lambda x + u^2 \tau}{2}.$$

Already here we can observe the fundamental difference between (4.8) and (4.18). In the latter equation the coefficient of countable viscosity is always positive. Therefore Equation (4.18) with corresponding sufficiently smooth initial data is always well-posed. It is not difficult to show that (4.4) is stable for any ratio of the steps, i.e., it is absolutely stable, since the transition factor of every Fourier coefficient is equal to

$$T_p = \frac{1}{1 + \tau \lambda_p}.$$

Hence it follows that

$$|T_p| \leqq 1.$$

In addition to (4.3) and (4.4) there are yet some other interesting and very convenient difference schemes, for instance

$$\frac{\phi_k^{j+1} - \phi_k^j}{\tau} + u \frac{\phi_{k+1}^{j+1} - \phi_{k-1}^{j+1}}{2\Delta x} = 0; \qquad (4.19)$$

or

$$\frac{\phi_k^{j+1} - \phi_k^j}{\tau} + u \frac{\phi_{k+1}^{j+1/2} - \phi_{k-1}^{j+1/2}}{2\Delta x} = 0, \qquad (4.20)$$

where

$$\phi_k^{j+1/2} = \tfrac{1}{2}(\phi_k^{j+1} + \phi_k^j).$$

It is not difficult to show, that Scheme (4.19) is a first-order approximation in τ, and second order in Δx. The differential equation corresponding to this scheme will have the form of (4.18), where

$$\mu = u^2 \tau / 2.$$

The Equation of "Motion"

The stability is determined by the transition operators for the Fourier coefficients. We obtain the spectral problem

$$(A^h\omega)_k \equiv u\,\frac{\omega_{k+1} - \omega_{k-1}}{2\Delta x} = \lambda\omega_k, \tag{4.21}$$

and we seek its solution in the form of (4.12). There results

$$\lambda_p = i\,\frac{u}{\Delta x}\sin p\Delta x. \tag{4.22}$$

From (4.19) it is immediate that the Fourier coefficients must satisfy

$$\frac{\phi_p^{j+1} - \phi_p^j}{\tau} + \lambda_p \phi_p^{j+1} = 0,$$

or

$$\phi_0^{j+1} = T_p \phi^j, \tag{4.23}$$

where

$$T_p = \frac{1}{1 + \tau\lambda_p}.$$

Taking into account (4.22), we get

$$T_p = \frac{1}{1 + i\,\dfrac{u\tau}{\Delta x}\sin p\Delta x}$$

and consequently

$$|T_p| = \frac{1}{\sqrt{1 + \left(\dfrac{u\tau}{\Delta x}\right)^2 \sin^2 p\Delta x}} \leqslant 1.$$

The absolute countable stability of (4.19) follows from this.

The most interesting scheme in applications is that of Crank–Nicholson (4.20). It is not difficult to verify that this scheme is a second-order approximation in τ and Δx and that it is not dissipative. This means that we have $\mu = 0$ in the differential equation (4.18) and that the terms we have dropped from consideration have the order of $\tau\Delta x$, τ^2 and Δx^2. As far as countable stability is concerned, we have

$$T_p = \frac{1 - i\,\dfrac{u\tau}{2\Delta x}\sin p\Delta x}{1 + i\,\dfrac{u\tau}{2\Delta x}\sin p\Delta x}.$$

Hence

$$|T_p| = 1,$$

and the scheme is absolutely stable. Let us remark that if, in (4.3), we replace the form

$$u \frac{\phi_k^j - \phi_{k-1}^j}{\Delta x}$$

of the difference expression for $u(\partial \phi/\partial x)$ by the form

$$u \frac{\phi_{k+1}^j - \phi_k^j}{\Delta x},$$

then the resulting difference scheme (with $u > 0$) becomes unstable, regardless of the relation between the steps. The proof is trivial.

In conclusion, let us consider yet another interesting numerical method for Problem (4.1), based on the so called "running-count" scheme. The scheme has been introduced by Landau, Meyman and Khalatnikov [4], and has the following form:

$$\frac{\phi_k^{j+1} - \phi_k^j}{\tau} + \frac{u}{\Delta x} \left[\left(\frac{\phi_k^{j+1} + \phi_k^j}{2} \right) - \left(\frac{\phi_{k-1}^{j+1} + \phi_{k-1}^j}{2} \right) \right] = 0. \quad (4.24)$$

As can be easily shown, the scheme is a second-order approximation in τ and first order in x. It can be implemented by the following recursive realization:

$$\phi_k^{j+1} = \frac{1 - \dfrac{u\tau}{2\Delta x}}{1 + \dfrac{u\tau}{2\Delta x}} \phi_k^j + \frac{\dfrac{u\tau}{2\Delta x}}{1 + \dfrac{u\tau}{2\Delta x}} (\phi_{k-1}^{j+1} + \phi_{k-1}^j). \quad (4.25)$$

Following the stability analysis of Neuman and using the Fourier method it is not difficult to prove that (4.24) is absolutely stable.

In a similar fashion one can construct a "running-count" scheme for the multidimensional problem of motion and show its stability under the assumption that the equation has constant coefficients.

In the above we have been assuming throughout that u was a positive constant. It u is negative, we can still replace x by $(-x)$, and obtain Equation (4.1). In applications, however, we are particularly interested in the case where $u = u(x, t)$. It can be shown readily by a simple analysis that in this case we may be loosing countable stability, even when using implicit dissipative schemes. This is true, in particular, of nonlinear problems. The essence of the matter is as follows: If we expand the solution of the difference scheme and the coefficient $u(x_k, t_j)$ into the Fourier series, the products of the Fourier series will yield both slower and faster harmonic components than those interacting.

As a result of such a process it may happen, that the "energy" due to the round-off errors will be transferred from the low frequencies to the highest frequencies, and the computational process will thus become unstable, despite the fact that the given difference scheme with constant coefficients is countably stable. This kind of instability is usually labeled as *nonlinear*. Sometimes it also appears when dealing with linear problems with variable coefficients. Thus, difference schemes for nonlinear equations, or equations with variable coefficients, which would be stable with respect to arbitrary excitations, are the focal points of current development. In most cases countable instability can be negotiated with the help of dissipative difference schemes, which correspond to a certain choice of the coefficient of countable viscosity μ. Such schemes, however, turn out to be first-order approximations in τ or x or both.

Of particular interest in applications are equations of the form

$$\frac{\partial \phi}{\partial t} + \frac{\partial u\phi}{\partial x} = 0, \qquad (4.26)$$

where $u = u(x, t)$.

Below, when discussing the multidimensional equations of the type of (4.26), we will show how to obtain absolutely stable, first- (or even second-) order difference schemes for equations of such type (on certain classes of the coefficients).

6.4.2. The Two-Dimensional Equation of Motion with Variable Coefficients

Consider an ensemble of particles moving in the plane (x, y) along given trajectories. In the framework of fluid mechanics the problem can be formulated as follows:

$$\frac{\partial \phi}{\partial t} + u\frac{\partial \phi}{\partial x} + v\frac{\partial \phi}{\partial y} = 0 \quad \text{in} \quad D \times D_t, \qquad (4.27)$$

$$\phi(x, y, 0) = g \quad \text{in} \quad D$$

Here $u = u(x, y, t)$ and $v = v(x, y, t)$.

Assume that the components u, v of the velocity vector satisfy at each time instant the continuity condition

$$\frac{\partial u}{\partial x} + \frac{\partial v}{\partial y} = 0. \qquad (4.28)$$

Let D be the rectangle $\{0 \le x \le a, 0 \le y \le b\}$, and suppose that the solution of the problem, as well as the velocity components u and v, are periodic, with identical values on the opposite sides of the rectangle.

Let us rewrite the evolution Equation (4.27) in the operator form:

$$\frac{\partial \phi}{\partial t} + A\phi = 0, \qquad (4.29)$$

where

$$\phi = g \quad \text{for} \quad t = 0,$$

$$A = u\frac{\partial}{\partial x} + v\frac{\partial}{\partial y}.$$

It is not difficult to show, that A satisfies $(A\phi, \phi) = 0$ in the present case. Indeed, introducing the inner product as usual, we have

$$(A\phi, \phi) = \int_0^a dx \int_0^b dy \left(u \frac{\partial \phi}{\partial x} + v \frac{\partial \phi}{\partial y} \right) \phi. \tag{4.30}$$

Using (4.28), the integrand can be written as

$$\left(u \frac{\partial \phi}{\partial x} + v \frac{\partial \phi}{\partial y} \right) \phi = \frac{\partial u \frac{\phi^2}{2}}{\partial x} + \frac{\partial v \frac{\phi^2}{2}}{\partial y}.$$

Hence

$$(A\phi, \phi) = \int_0^a dx \int_0^b dy \left(\frac{\partial u \frac{\phi^2}{2}}{\partial x} + \frac{\partial v \frac{\phi^2}{2}}{\partial y} \right). \tag{4.31}$$

By the periodicity of the solution on the boundary it now follows that

$$(A\phi, \phi) = 0. \tag{4.32}$$

Thus A is antisymmetric and consequently permits one to construct an absolutely stable difference scheme.

Let us next attempt to split up the operator A in such a way that each of the elementary operators A_j ($j = 1, 2$) also satisfy the above condition:

$$(A_j \phi, \phi) = 0. \tag{4.33}$$

The component-by-component splitting-up difference scheme allows us in this case to obtain an absolutely stable difference scheme of second-order accuracy.

The formal decomposition of the operator A into the factors

$$A_1 = u \frac{\partial}{\partial x}, \qquad A_2 = v \frac{\partial}{\partial y} \tag{4.34}$$

does not satisfy Condition (4.33). It is not difficult to verify that

$$(A_1 \phi, \phi) = -\frac{1}{2} \int_0^a dx \int_0^b \phi^2 \frac{\partial u}{\partial x} dy, \quad (A_2 \phi, \phi) = -\frac{1}{2} \int_0^a dx \int_0^b \phi^2 \frac{\partial v}{\partial y} dy.$$

and therefore the operators A_1 and A_2 can not be taken as elementary operators for the construction of the sequential splitting-up scheme.

Let us take the operators A_1 and A_2 in the following, more complex form:

$$A_1 \phi = u \frac{\partial \phi}{\partial x} + \frac{\phi}{2} \frac{\partial u}{\partial x}, \qquad A_2 \phi = v \frac{\partial \phi}{\partial y} + \frac{\phi}{2} \frac{\partial v}{\partial y}. \tag{4.35}$$

The Equation of "Motion"   243

It is not difficult to see that we now have (4.33) holding for the above operators and that

$$(A_1 + A_2)\phi = u\frac{\partial \phi}{\partial x} + v\frac{\partial \phi}{\partial y} + \frac{\phi}{2}\left(\frac{\partial u}{\partial x} + \frac{\partial v}{\partial y}\right)$$

$$= u\frac{\partial \phi}{\partial x} + v\frac{\partial \phi}{\partial y} = A\phi.$$

i.e. $A_1 + A_2 = A$. Here we have used the fact that the coefficients u and v satisfy Equation (4.28).

Hence we have satisfied all the assumptions for the applicability of the splitting-up method and can thus write

$$\frac{\phi^{j-1/2} - \phi^{j-1}}{\tau} + \left(u^j\frac{\partial}{\partial x} + \frac{1}{2}\cdot\frac{\partial u^j}{\partial x}\right)\frac{\phi^{j-1/2} + \phi^{j-1}}{2} = 0,$$

$$\frac{\phi^j - \phi^{j-1/2}}{\tau} + \left(v^j\frac{\partial}{\partial y} + \frac{1}{2}\cdot\frac{\partial v^j}{\partial y}\right)\frac{\phi^j + \phi^{j-1/2}}{2} = 0,$$

$$\frac{\phi^{j+1/2} - \phi^j}{\tau} + \left(v^j\frac{\partial}{\partial y} + \frac{1}{2}\cdot\frac{\partial v^j}{\partial y}\right)\frac{\phi^{j+1/2} + \phi^j}{2} = 0,$$

$$\frac{\phi^{j+1} - \phi^{j+1/2}}{\tau} + \left(u^j\frac{\partial}{\partial x} + \frac{1}{2}\cdot\frac{\partial u^j}{\partial x}\right)\frac{\phi^{j+1} + \phi^{j+1/2}}{2} = 0,$$

(4.36)

where $t_{j-1} \leq t \leq t_{j+1}$.

If the functions u, v and the solution ϕ are sufficiently smooth in all their arguments, then (4.36) is a second-order approximation and is absolutely stable in the sense that

$$\|\phi^{j+1}\| = \|\phi^{j-1}\| = \cdots = \|g\|. \qquad (4.37)$$

From this example we may see how the formal splitting into the operators (4.34) may compromise the very idea of splitting; only by introducing additional considerations can we arrive at theoretically justifiable and practically efficient schemes.

With these preliminaries as an introduction, let us now turn to constructing the difference scheme for Problem (4.27). To begin with, let us consider feasible ways of approximating the operator A in the space variables x and y. As pointed out in Chapter 2, a convenient approximation of problems of mathematical physics, which conserves the additivity properties and qualitative features of the operator, can be achieved by the sequential coordinate approximation method.

Suppose that the coefficients u and v are sufficiently smooth, and consider Equation (4.27) in the form

$$\frac{\partial \phi}{\partial t} + \frac{\partial u\phi}{\partial x} + \frac{\partial v\phi}{\partial y} = 0 \quad \text{in} \quad D \times D_t,$$

$$\phi = g \quad \text{in} \quad D \quad \text{for} \quad t = 0. \tag{4.38}$$

Consider the operator A defined by the expression

$$A\phi = \frac{\partial u\phi}{\partial x} + \frac{\partial v\phi}{\partial y} - \frac{\phi}{2}\left(\frac{\partial u}{\partial x} + \frac{\partial v}{\partial y}\right). \tag{4.39}$$

Write its difference analog as follows:

$$(A^h\phi)_{k,l} = \frac{u_{k+1,l}\phi_{k+1,l} - u_{k-1,l}\phi_{k-1,l}}{2\Delta x}$$

$$+ \frac{v_{k,l+1}\phi_{k,l+1} - v_{k,l-1}\phi_{k,l-1}}{2\Delta y}$$

$$- \frac{\phi_{k,l}}{2}\left(\frac{u_{k+1,l} - u_{k-1,l}}{2\Delta x} + \frac{v_{k,l+1} - v_{k,l-1}}{2\Delta y}\right). \tag{4.40}$$

Clearly, difference Expression (4.40) approximates (4.39) with second-order accuracy in Δx and Δy, provided the function u, v, and ϕ are smooth enough. Equation (4.40) has a serious drawback, though, since the form of the operator A^h disturbs the antisymmetric structure, i.e.

$$(A^h\phi, \phi) \neq 0. \tag{4.41}$$

This means that our customary approximation becomes unsatisfactory for the purposes of constructing the computational algorithm for (4.27).

We now show that the approximation of (4.39) in the form

$$(A^h\phi)_{k,l} = \frac{u_{k+1/2,l}\phi_{k+1,l} - u_{k-1/2,l}\phi_{k-1,l}}{2\Delta x}$$

$$+ \frac{v_{k,l+1/2}\phi_{k,l+1} - v_{k,l-1/2}\phi_{k,l-1}}{2\Delta y} \tag{4.42}$$

satisfies the fundamental relation

$$(A^h\phi, \phi) = 0, \tag{4.43}$$

and approximates Expression (4.38) with second-order accuracy in Δx and Δy. Let us exploit the following approximation of the coefficients:

$$u_{k+1/2,l} = u_{k+1,l} - \frac{u_{k+1,l} - u_{k,l}}{2}; \quad v_{k,l+1/2} = v_{k,l+1} - \frac{v_{k,l+1} - v_{k,l}}{2};$$

$$u_{k-1/2,l} = u_{k-1,l} + \frac{u_{k,l} - u_{k-1,l}}{2}; \quad v_{k,l-1/2} = v_{k,l-1} + \frac{v_{k,l} - v_{k,l-1}}{2}. \tag{4.44}$$

The Equation of "Motion"   245

The substitution of these expressions into (4.42), followed by simple manipulations give

$$(A^h\phi)_{k,l} = \frac{u_{k+1,l}\phi_{k+1,l} - u_{k-1,l}\phi_{k-1,l}}{2\Delta x}$$
$$+ \frac{v_{k,l+1}\phi_{k,l+1} - v_{k,l-1}\phi_{k,l-1}}{2\Delta y}$$
$$- \frac{\phi_{k,l}}{2}\left(\frac{u_{k+1,l} - u_{k-1,l}}{2\Delta x} + \frac{v_{k,l+1} - v_{k,l-1}}{2\Delta y}\right)$$
$$- (\Delta x^2 R_{k,l} + \Delta y^2 Q_{k,l}), \tag{4.45}$$

where the quantities $R_{k,l}$ and $Q_{k,l}$ approach the limits

$$R_{k,l} \to \frac{1}{4} \cdot \frac{\partial}{\partial x}\left(\frac{\partial u}{\partial x} \cdot \frac{\partial \phi}{\partial x}\right), \quad Q_{k,l} \to \frac{1}{4} \cdot \frac{\partial}{\partial y}\left(\frac{\partial v}{\partial y} \cdot \frac{\partial \phi}{\partial y}\right)$$

as $\Delta x \to 0$ and $\Delta y \to 0$.

Assume now, that the coefficients $u_{k,l} v_{k,l}$ satisfy the following relation:

$$\frac{u_{k+1,l} - u_{k-1,l}}{2\Delta x} + \frac{v_{k,l+1} - v_{k,l-1}}{2\Delta y} = O(h^2). \tag{4.46}$$

If the coefficients u, v and the solution ϕ have bounded third derivatives with respect to x and y, it can be seen that (4.45) [with (4.46)] differs from (4.40) by the same second-order term as does (4.40) from (4.39). Hence (4.42) approximates (4.39) with second-order accuracy relative to Δx and Δy.

We will show that the operator A^h we have constructed satisfies Condition (4.43), and moreover that each of the operators A_1^h, A_2^h defined by

$$A_1^h \phi = \frac{u_{k+1/2,l}\phi_{k+1,l} - u_{k-1/2,l}\phi_{k-1,l}}{2\Delta x},$$
$$A_2^h \phi = \frac{v_{k,l+1/2}\phi_{k,l+1} - v_{k,l-1/2}\phi_{k,l-1}}{2\Delta y}, \tag{4.47}$$

also satisfy the corresponding equation

$$(A_\alpha^h \phi, \phi) = 0. \tag{4.48}$$

To this end, we introduce the inner product

$$(a, b) = \sum_k \sum_l a_{k,l} b_{k,l} \Delta x \Delta y,$$

for vector quantities a, b; with its help we may write

$$(A_1^h \phi, \phi) = \frac{1}{2}\sum_k \sum_l \Delta y \left(\sum_k u_{k+1/2,l}\phi_{k+1,l} - \sum_k u_{k-1/2,l}\phi_{k-1,l}\right)\phi_{k,l},$$
$$(A_2^h \phi, \phi) = \frac{1}{2}\sum_l \sum_k \Delta x \left(\sum_l v_{k,l+1/2}\phi_{k,l+1} - \sum_l v_{k,l-1/2}\phi_{k,l-1}\right)\phi_{k,l}. \tag{4.49}$$

After a little arrangement we obtain Equalities (4.48). Equation (4.43) follows immediately from (4.48). We have thus constructed the necessary spatial approximations. Our next task is the time reduction of the system of ordinary differential equations

$$\frac{d\phi^h}{dt} + A^h \phi^h = 0 \quad \text{in} \quad D_h \times D_t, \tag{4.50}$$

$$\phi^h = g^h \quad \text{in} \quad D_h \quad \text{for} \quad t = 0,$$

where

$$A^h = A_1^h + A_2^h,$$

ϕ^h is a vector function with the components $\phi_{k,l}$ and A_α^h satisfies condition (4.48). This means that the problem (4.50) can be solved using the splitting-up method. Dropping the unimportant index h, we obtain the following system on the interval $t_{j-1} \leq t \leq t_{j+1}$:

$$\frac{\phi^{j-1/2} - \phi^{j-1}}{\tau} + A_1^j \frac{\phi^{j-1/2} + \phi^{j-1}}{2} = 0,$$

$$\frac{\phi^j - \phi^{j-1/2}}{\tau} + A_2^j \frac{\phi^j + \phi^{j-1/2}}{2} = 0,$$

$$\frac{\phi^{j+1/2} - \phi^j}{\tau} + A_2^j \frac{\phi^{j+1/2} + \phi^j}{2} = 0, \tag{4.51}$$

$$\frac{\phi^{j+1} - \phi^{j+1/2}}{\tau} + A_1^j \frac{\phi^{j+1} + \phi^{j+1/2}}{2} = 0.$$

Thus Problem (4.27) has been reduced to a system of simple one-dimensional equations, which can be solved by the factorization method for three-point difference equations.

6.4.3. The Multidimensional Equation of Motion

Consider now the multidimensional equation, describing the hydrodynamical motion of a substance along a trajectory,

$$\frac{\partial \Phi}{\partial t} + \sum_{\alpha=1}^{n} v_\alpha \frac{\partial \Phi}{\partial x_\alpha} = 0 \quad \text{in} \quad D \times D_t, \tag{4.52}$$

$$\Phi(x, 0) = f(x) \quad \text{in} \quad D.$$

Suppose that the coefficients in (4.52) satisfy the continuity conditions

$$\sum_{\alpha=1}^{n} \frac{\partial v_\alpha}{\partial x_\alpha} = 0 \quad \text{in} \quad D \times D_t. \tag{4.53}$$

The Equation of "Motion"

Problems (4.52) and (4.53) can be rewritten in the divergence form

$$\frac{\partial \Phi}{\partial t} + \sum_{\alpha=1}^{n} \frac{\partial v_\alpha \Phi}{\partial x_\alpha} = 0 \quad \text{in} \quad D \times D_t, \tag{4.54}$$

$$\Phi(x, 0) = f(x) \quad \text{in} \quad D.$$

The form (4.54) is of primary importance when using splitting-up methods.

To begin with, let us consider the difference equation obtained for (4.54) by using the Crank–Nicholson scheme:

$$\frac{\Phi^{j+1} - \Phi^j}{\tau} + \sum_{\alpha=1}^{n} \frac{\partial u_\alpha^j \Phi^{j+1/2}}{\partial x_\alpha} = 0 \quad (t_j \leqslant t \leqslant t_{j+1}), \tag{4.55}$$

where

$$\Phi^{j+1/2} = \tfrac{1}{2}(\Phi^{j+1} + \Phi^j). \tag{4.56}$$

Also, in (4.55) we have used certain approximations of the coefficients v_α. It is natural to choose either first- or second-order accuracy with respect to τ. For instance, in order to obtain the first order we may take

$$u_\alpha^j = v_\alpha(x, t_j),$$

while the second order is obtained by the choice

$$u_\alpha^j = \frac{v_\alpha(x, t_{j+1}) + v_\alpha(x, t_j)}{2}.$$

Denote

$$A^j \Phi = \sum_{\alpha=1}^{n} A_\alpha^j \Phi, \qquad A_\alpha^j \Phi = \frac{\partial u_\alpha^j \Phi}{\partial x_\alpha}.$$

Then (4.55) can be written in the form

$$\frac{\Phi^{j+1} - \Phi^j}{\tau} + A^j \Phi^{j+1/2} = 0, \qquad \Phi^{j+1/2} = \tfrac{1}{2}(\Phi^{j+1} + \Phi^j)$$

or,

$$\left(E + \frac{\tau}{2} A^j\right) \Phi^{j+1} = \left(E - \frac{\tau}{2} A^j\right) \Phi^j. \tag{4.57}$$

Suppose (for the sake of simplicity) that the solution of the problem is periodic relative to the n-dimensional rectangle D. Define the inner product

$$(a, b) = \int_D ab\, dD.$$

Then it can be easily verified that

$$(A^j \Phi, \Phi) = 0. \tag{4.58}$$

Solve for Φ^{j+1} in Equation (4.57):

$$\Phi^{j+1} = \left(E + \frac{\tau}{2} A^j\right)^{-1} \left(E - \frac{\tau}{2} A^j\right) \Phi^j.$$

With the help of the Kellogg lemma and (4.58) we find

$$\|\Phi^{j+1}\| = \|\Phi^j\|. \tag{4.59}$$

Consider next the approximation of (4.55) in the spatial coordinates. To this end, project the region D on D_h and write (we drop from the symbols u, Φ the indices which do not change)

$$\frac{\Phi^{j+1} - \Phi^j}{\tau} + \Lambda^j \Phi^{j+1/2} = 0, \tag{4.60}$$

where

$$\Lambda^j = \sum_{\alpha=1}^{n} \Lambda^j_\alpha, \tag{4.61}$$

$$\Lambda^j_\alpha \Phi = \frac{1}{2\Delta x_\alpha} (u^j_{\alpha, k_\alpha + 1/2} \Phi_{k_\alpha + 1} - u^j_{\alpha, k_\alpha - 1/2} \Phi_{k_\alpha - 1}), \tag{4.62}$$

and the index k_α corresponds to the net points of the variable x_α. Introduce the inner product

$$(a, b) = \sum_{k_1 \cdots k_n} a_{k_1 k_2 \cdots k_n} b_{k_1 k_2 \cdots k_n} \Delta x_1 \Delta x_2 \cdots \Delta x_n. \tag{4.63}$$

It is not difficult to verify that

$$(\Lambda^j \Phi, \Phi) = 0. \tag{4.64}$$

Consequently, using (4.64), we have

$$\|\Phi^{j+1}\| = \|\Phi^j\| = \cdots = \|f\|,$$

where Φ^j solves (4.60). Thus, exploiting the approximation of A^j by Λ^j from (4.60)–(4.62), we have again obtained an absolutely stable scheme.

Let us now investigate the approximation of A^j by Λ^j. In this connection consider again the elementary operators A^j_α and Λ^j_α. Below, the coefficients in $\Lambda^j_\alpha \Phi$ will be taken as follows

$$\begin{aligned} u^j_{\alpha, k_\alpha + 1/2} &= u^j_{\alpha, k_\alpha + 1} - \tfrac{1}{2}(u^j_{\alpha, k_\alpha + 1} - u^j_{\alpha, k_\alpha}), \\ u^j_{\alpha, k_\alpha - 1/2} &= u^j_{\alpha, k_\alpha - 1} + \tfrac{1}{2}(u^j_{\alpha, k_\alpha} - u^j_{\alpha, k_\alpha - 1}). \end{aligned} \tag{4.65}$$

Then we have

$$\Lambda^j_\alpha \Phi = \frac{u^j_{\alpha, k_\alpha + 1} \Phi_{\alpha, k_\alpha + 1} - u^j_{\alpha, k_\alpha - 1} \Phi_{k_\alpha - 1}}{2\Delta x_\alpha} - \frac{\Phi_{k_\alpha}}{2} \frac{u^j_{\alpha, k_\alpha + 1} - u^j_{\alpha, k_\alpha - 1}}{2\Delta x_\alpha} - \frac{\Delta x_\alpha^2}{4} R^j_\alpha, \tag{4.66}$$

The Equation of "Motion"

where

$$R_\alpha^j = \frac{1}{\Delta x_\alpha^3}[(u_{\alpha,k_\alpha+1}^j - u_{\alpha,k_\alpha}^j)(\Phi_{k_\alpha+1} - \Phi_{k_\alpha}) - (u_{\alpha,k_\alpha}^j - u_{\alpha,k_\alpha-1}^j)(\Phi_{k_\alpha} - \Phi_{k_\alpha-1})]. \tag{4.67}$$

Recalling the assumption on the smoothness of the solution and the coefficients $u(x, t)$, we have that

$$R_\alpha^j \to \frac{\partial}{\partial x_\alpha}\left(\frac{\partial u^j}{\partial x_\alpha} \cdot \frac{\partial \Phi}{\partial x_\alpha}\right)$$

as $\Delta x_\alpha \to 0$. It follows from (4.66) that, generally speaking, the operator Λ_α^j does not approximate the operator A_α^j.

Consider the full operator Λ^j and investigate the expression $\Lambda^j \Phi$. We have

$$\Lambda^j \Phi = \sum_{\alpha=1}^n \frac{u_{\alpha,k_\alpha+1}^j \Phi_{k_\alpha+1} - u_{\alpha,k_\alpha-1}^j \Phi_{k_\alpha-1}}{2\Delta x_\alpha}$$

$$- \sum_{\alpha=1}^n \frac{\Phi_{k_\alpha}}{2} \cdot \frac{u_{\alpha,k_\alpha+1}^j - u_{\alpha,k_\alpha-1}^j}{2\Delta x_\alpha} - \frac{1}{4}\sum_{\alpha=1}^n \Delta x_\alpha^2 R_\alpha^j. \tag{4.68}$$

Assume that this difference version of the continuity equation is such that

$$\sum_{\alpha=1}^n \frac{u_{\alpha,k_\alpha+1} - u_{\alpha,k_\alpha-1}}{2\Delta x_\alpha} = O(h^2). \tag{4.69}$$

In this case the second sum in (4.68) becomes zero, and we obtain

$$\Lambda^j \Phi = \sum_{\alpha=1}^n \frac{u_{\alpha,k_\alpha+1}^j \Phi_{k_\alpha+1} - u_{\alpha,k_\alpha-1}^j \Phi_{k_\alpha-1}}{2\Delta x_\alpha} + O(h^2), \tag{4.70}$$

where $h = \max_\alpha \{\Delta x_\alpha\}$. Equation (4.70) thus implies that the full operator Λ^j approximates A with an accuracy up to the second order in all spatial variables.

Let us now turn to the splitting of (4.54). For this consider the following two-cycle scheme:

$$\left(E + \frac{\tau}{2}\Lambda_\alpha^j\right)\Phi^{j-[(n-\alpha)/n]} = \left(E - \frac{\tau}{2}\Lambda_\alpha^j\right)\Phi^{j-[(n-\alpha+1)/n]}$$

$$(\alpha = 1, 2, \ldots, n),$$

$$\left(E + \frac{\tau}{2}\Lambda_\alpha^j\right)\Phi^{j+[(n-\alpha+1)/n]} = \left(E - \frac{\tau}{2}\Lambda_\alpha^j\right)\Phi^{j+[(n-\alpha)/n]}$$

$$(\alpha = n, n-1, \ldots, 1).$$

$$\tag{4.71}$$

Here it is necessary that the coefficients in the operator Λ_α^j be chosen according to the equations $u^j = v(x, t_j)$. This will guarantee the second-order approximation of the system on every interval

$$t_{j-1} \leq t \leq t_{j+1}. \tag{4.72}$$

The splitting-up method of (4.71) is in addition absolutely stable. Indeed, since the approximation of the operators Λ_α^j preserves the condition

$$(\Lambda_\alpha^j \Phi, \Phi) = 0, \tag{4.73}$$

it is not difficult to see that the Kellogg lemma implies

$$\|T_\alpha^j\| = 1 \quad (\alpha = 1, 2, \ldots, n)$$

and consequently

$$\|\Phi^{j-[(n-\alpha)/n]}\| = \|\Phi^{j-[(n-\alpha+1)/n]}\| \quad (\alpha = 1, 2, \ldots, n),$$
$$\|\Phi^{j+[(n-\alpha+1)/n]}\| = \|\Phi^{j+[(n-\alpha)/n]}\| \quad (\alpha = n, n-1, \ldots, 1).$$

Eliminating the intermediate values of $\|\Phi\|$, we obtain

$$\|\Phi^{j+1}\| = \|\Phi^{j-1}\|. \tag{4.74}$$

The above discussed method of solving problems of hydrodynamical motion can be easily generalized to the case of quasilinear equations of hydrodynamics. The only thing to be done is to supplement in addition a good scheme for extrapolating the coefficients u_α at time t_j from their preceding history.

In our development we have chosen a second-order approximation relative the spatial coordinates. The extension of the algorithm to higher-order approximations is also possible. To see this, let

$$\Lambda_\alpha \Phi = \sum_{m=1}^p \beta_m \frac{\frac{1}{2}(u_{k+m} + u_k)\Phi_{k+m} - \frac{1}{2}(u_k + u_{k-m})\Phi_{k-m}}{2(m\Delta x_\alpha)} \tag{4.75}$$

where β_m satisfy the following system of equations:

$$\sum_{m=1}^p \beta_m = 1; \quad \sum_{m=1}^p m^2 \beta_m = 0, \ldots; \quad \sum_{m=1}^p m^{2p-2} \beta_m = 0. \tag{4.76}$$

Then, if $x_\alpha = x_{k_\alpha}$, we have

$$\sum_{\alpha=1}^n \Lambda_\alpha \Phi = \left(\sum_{\alpha=1}^n \frac{\partial u_\alpha \Phi}{\partial x_\alpha}\right)_{x_\alpha = x_{k_\alpha}} + O(h^{2p}).$$

At the same time we assume that the coefficients satisfy the following relation:

$$\sum_{\alpha=1}^n \sum_{m=1}^p \beta_m \frac{u_{\alpha, k_\alpha + m} - u_{\alpha, k_\alpha - m}}{2m\Delta x_\alpha} = O(h^{2p}). \tag{4.77}$$

Using further the splitting up algorithm of (4.71), the solution (4.54) can be obtained with an accuracy up to $O(h^{2p} + \tau^2)$.

The given algorithm can be very easily generalized to the case where the hydrodynamical fluid is compressible. In this case the fundamental equations are as follows:

$$\frac{\partial \rho \Phi}{\partial t} + \sum_{\alpha=1}^{n} \frac{\partial \rho v_\alpha \Phi}{\partial x_\alpha} = 0,$$

$$\frac{\partial \rho}{\partial t} + \sum_{\alpha=1}^{n} \frac{\partial \rho v_\alpha}{\partial x_\alpha} = 0.$$

(4.78)

Approximate the above by

$$\frac{\rho^{j+1/2}\Phi^{j+1} - \rho^{j-1/2}\Phi^{j-1}}{2\tau} + \sum_{\alpha=1}^{n} \frac{(\rho u_\alpha)^j_{k_\alpha+1/2}\Phi^j_{k_\alpha+1} - (\rho u_\alpha)^j_{k_\alpha-1/2}\Phi^j_{k_\alpha-1}}{2\Delta x_\alpha} = 0,$$

$$\frac{\rho^{j+1} - \rho^{j-1}}{2\tau} + \sum_{\alpha=1}^{n} \frac{(\rho u_\alpha)^j_{k_\alpha+1} - (\rho u_\alpha)^j_{k_\alpha-1}}{2\Delta x_\alpha} = 0.$$

(4.79)

Here we use the notation

$$\rho^{j+1/2} = \rho^{j+1} - \tfrac{1}{2}(\rho^{j+1} - \rho^j), \qquad \rho^{j-1/2} = \rho^{j-1} + \tfrac{1}{2}(\rho^j - \rho^{j-1})$$

and also the expressions as in (4.65). Rewrite the first equation in (4.79) as follows:

$$\frac{\rho^{j+1}\Phi^{j+1} - \rho^{j-1}\Phi^{j-1}}{2\tau} + \sum_{\alpha=1}^{n} \frac{(\rho u_\alpha)^j_{k_\alpha+1}\Phi^j_{k_\alpha+1} - (\rho u_\alpha)^j_{k_\alpha-1}\Phi^j_{k_\alpha-1}}{2\Delta x_\alpha} = O(\tau^2 + h^2).$$

(4.80)

Here we have made a considerable use of the second equation from (4.79).

In order to solve (4.78), we may use a method similar to (4.71).

The next section is devoted to an efficient and highly accurate method for solving difference equations; the method is based on some special splitting schemes.

6.5. On Increasing the Order of Approximation of Difference Schemes

A lot of attention has recently been paid to constructions of high-accuracy difference approximation schemes. Out of the various approaches to such schemes we will consider only one; it goes back to Richardson, and has been applied and developed by Volkov [4] in the case of the Laplace operator. In order to get the approximate solution of a given order of accuracy, the method makes use of a sequence of nets and the corresponding approximations. The method requires for its application only the standard first- or second-order difference approximations.

Kuznetsov and Shaidurov [4] discussed some other types of equations which allow a similar construction; these include the Helmholtz equation and the equation of hydrodynamical motion. The basic idea is as follows:

Assume that we are to solve the standard problem of mathematical physics

$$A\phi = f \quad \text{in} \quad D,$$
$$l\phi = g \quad \text{on} \quad \partial D, \tag{5.1}$$

where A and l are some linear differential operators. We assume, that A, l, and the boundary of the region ∂D are such that for any sufficiently smooth f and g there exists a unique solution ϕ. Since we are primarily concerned with the idea and not with the actual realizations, we will not worry about the order of smoothness, and we will consider it to be sufficient for the purposes of further development.

Let us construct the following sequence of nets and net functions. Let D_h be the basic net (for simplicity we take it as being uniform with a mesh-size h) and denote by ∂D_h the net points on the boundary. Construct next a new net $D_{h/2}$ with the mesh-size $h/2$ such that it refines D_h. Then construct further refinements $D_{h/3}$ and etc. On each of the net regions $D_h, D_{h/2}, \ldots, D_{h/n}$ we define the difference approximations of the differential problem (5.1) as based on simple schemes (as a rule, no more than second-order accuracy). We thus obtain the sequence of problems

$$A_{h/p}\phi^{h/p} = f^{h/p} \quad \text{in} \quad D_{h/p},$$
$$l_{h/p}\phi^{h/p} = g^{h/p} \quad \text{on} \quad \partial D_{h/p}. \tag{5.2}$$
$$(p = 1, 2, \ldots, n).$$

Suppose, that all the problems from (5.2) have been solved. Define on D_h the vector

$$\bar{\phi} = \sum_{p=1}^{n} \gamma_p \phi^{h/p} \tag{5.3}$$

where the constants γ_p are chosen so as to satisfy the system†

$$\sum_{p=1}^{n} \gamma_p = 1, \quad \sum_{p=1}^{n} \frac{\gamma_p}{p^m} = 0 \quad (m = 1, 2, \ldots, n-1). \tag{5.4}$$

System (5.4) can be solved explicitly in the form

$$\gamma_p = \frac{p^n}{p!(n-p)!}(-1)^{n-p}.$$

We will show that under certain assumptions Form (5.3) of the solution of Problem (5.2) guarantees no less than an $O(h^n)$ approximation of the solution ϕ on D_h.

---

† This system of equations has been presented by the author in his paper at SYNSPADE-1971.

On Increasing the Order of Approximation of Difference Schemes   253

We need the following requirement regarding the difference operators $A_{h/p}$ and $l_{h/p}$:

$$A_{h/p}\psi = A\psi + \sum_{m=1}^{n-1} \frac{h^m}{p^m} \pi_m + \pi_{n,p} \quad \text{in} \quad D_{h/p},$$

$$l_{h/p}\psi = l\psi + \sum_{m=1}^{n-1} \frac{h^m}{p^m} \chi_m + \chi_{n,p} \quad \text{on} \quad \partial D_{h/p}$$

(5.5)

for an arbitrary, sufficiently smooth function ψ, where π_m, χ_m $(m = 1, \ldots, n-1)$ are some smooth functions independent of h and p; $\|\pi_{n,p}\|_{D_{h/p}} = O(h^n)$; $\|\chi_{n,p}\|_{D_{h/p}} = O(h^n)$.

At the next stage of the proof we need to obtain a similar decomposition for the difference solution, namely

$$\phi^{h/p} = \phi + \sum_{m=1}^{n-1} \frac{h^m}{p^m} \Phi_m + \Phi_{n,p} \quad \text{in} \quad D_{h/p},$$

(5.6)

where Φ_m $(m = 1, \ldots, n-1)$ are some smooth functions independent of h and p; $\|\Phi_{n,p}\|_{D_h} = O(h^n)$.

In order to see that the representation of (5.6) is possible, let us substitute (5.6) in place of $\phi^{h/p}$ into the corresponding Equation (5.2). There results

$$A_{h/p}\phi + \sum_{m=1}^{n-1} \frac{h^m}{p^m} A_{h/p}\Phi_m + A_{h/p}\Phi_{n,p} = f^{h/p} \quad \text{in} \quad D_{h/p},$$

$$l_{h/p}\phi + \sum_{m=1}^{n-1} \frac{h^m}{p^m} l_{h/p}\Phi_m + l_{h/p}\Phi_{n,p} = g^{h/p} \quad \text{on} \quad \partial D_{h/p}.$$

(5.7)

By (5.5) we can write

$$A_{h/p}\phi = f + \sum_{r=1}^{n-1} \frac{h^r}{p^r} F_{r,0} + F_{n,0} \quad \text{in} \quad D_{h/p},$$

$$l_{h/p}\phi = g + \sum_{r=1}^{n-1} \frac{h^r}{p^r} G_{r,0} + G_{n,0} \quad \text{on} \quad \partial D_{h/p}$$

(5.8)

and

$$A_{h/p}\Phi_m = A\Phi_m + \sum_{r=1}^{n-m-1} \frac{h^r}{p^r} F_{r,m} + F_{n-m,m} \quad \text{in} \quad D_{h/p},$$

$$l_{h/p}\Phi_m = l\Phi_m + \sum_{r=1}^{n-m-1} \frac{h^r}{p^r} G_{r,m} + G_{n-m,m} \quad \text{on} \quad \partial D_{h/p}.$$

(5.9)

From (5.8) we already know that $F_{m,0}$ and $G_{m,0}$ $(m = 1, \ldots, n-1)$ are independent of h and p, $\|F_{n,0}\|_{D_{h/q}} = O(h^n)$, $\|G_{n,0}\|_{\partial D_{h/q}} = O(h^n)$. This expression is rather formal, except for the independence of Φ_m on h.

Substituting (5.8) and (5.9) into (5.7), and collecting terms with the same powers of h/p, we obtain

$$\sum_{m=1}^{n-1}\frac{h^m}{p^m}\left[A\Phi_m+\sum_{r=0}^{m-1}F_{m-r,r}\right]+A_{h/p}\Phi_{n,p}+\sum_{r=0}^{n-1}\frac{h^r}{p^r}F_{n-r,r}=0\quad\text{in}\quad D_{h/p},$$

$$\sum_{m=1}^{n-1}\frac{h^m}{p^m}\left[l\Phi_m+\sum_{r=0}^{m-1}G_{m-r,r}\right]+l_{h/p}\Phi_{n,p}+\sum_{r=0}^{n-1}\frac{h^r}{p^r}G_{n-r,r}=0\quad\text{on}\quad \partial D_{h/p}.$$

(5.10)

Now it is enough to choose Φ_1 from the solution of the auxiliary problem

$$A\Phi_1+F_{1,0}=0\quad\text{in}\quad D,$$
$$l\Phi_1+G_{1,0}=0\quad\text{on}\quad \partial D.$$

In this case the coefficient standing with the first power of (h/p) turns out to be zero. Note that in this way we are choosing a function which does not depend on h; hence we have the decomposition (5.9) holding true for $m=1$, and for the functions $F_{r,1}$ ($r=1,\ldots,n-2$) independent of h,

$$\|F_{n-1,1}\|_{D_{h/p}}=O(h^{n-1}),\qquad \|G_{n-1,1}\|_{\partial D_{h/p}}=O(h^{n-1}).$$

Next let us find Φ_2 from the solution of

$$A\Phi_2+F_{2,0}+F_{1,1}=0\quad\text{in}\quad D,\quad l\Phi_2+G_{2,0}+G_{1,1}=0\quad\text{on}\quad \partial D,$$

etc. Every time (until the $(n-1)$st step) we obtain smooth solutions of the auxiliary problems, independent of h.

Write now the auxiliary problem for the last, nth step:

$$A_{h/p}\Phi_{n,p}+\sum_{r=0}^{n-1}\frac{h^r}{p^r}F_{n-r,r}=0\quad\text{in}\quad D_{h/p},$$

$$l_{h/p}\Phi_{n,p}+\sum_{r=0}^{n-1}\frac{h^r}{p^r}G_{n-r,r}=0\quad\text{on}\quad \partial D_{h/p}.$$

(5.11)

In order to prove (5.6), assume that all the difference schemes for Problem (5.2) are stable in the net norm D_h, in particular

$$\|\phi^{h/p}\|_{D_h}\leqslant C(\|f^{h/p}\|_{D_{h/p}}+\|g^{h/p}\|_{\partial D_{h/p}}).$$

From (5.11) one can immediately show that

$$\|\Phi_{n,p}\|_{D_h}=O(h^n).$$

This estimate concludes the proof of existence of Decomposition (5.6).

With Decomposition (5.6) at hand, let us now form the following weighted sum of Representations (5.6):

$$\sum_{p=1}^{n}\gamma_p\phi^{h/p}=\phi\sum_{p=1}^{n}\gamma_p+\sum_{p=1}^{n}\gamma_p\sum_{m=1}^{n-1}\frac{h^m}{p^m}\Phi_m+\sum_{p=1}^{n}\gamma_p\Phi_{n,p}.$$

From this it follows that, if the weights γ_p are chosen according to (5.4), then the last expression becomes

$$\sum_{p=1}^{n} \gamma_p \phi^{h/p} = \phi + \sum_{p=1}^{n} \gamma_p \Phi_{n,p}.$$

Taking now into account the immediate inequality

$$\left\| \sum_{p=1}^{n} \gamma_p \Phi_{n,p} \right\|_{D_h} \leq \sum_{p=1}^{n} |\gamma_p| \|\Phi_{n,p}\|_{D_h} = O(h^n),$$

we finally obtain

$$\|\bar{\phi} - \phi\|_{D_h} = O(h^n).$$

Thus the problem of constructing the solution with high-order accuracy has been reduced to solving the problems with first- or second-order accuracy on the sequence of nets. This algorithm is feasible for use on a computer. It is to be noted, that the above approach can be extended to the case of non-uniform nets. In this latter case, however, we have to repeat at every step the analysis of accuracy and convergence similar to that we have just shown.

In the case of evolution problems with smooth solutions, one can construct higher order difference approximations with respect to the spatial variables. In order to see this, consider the evolution equation

$$\frac{\partial \phi}{\partial t} + \Lambda \phi = f. \tag{5.12}$$

Let us approximate this equation on $t_j \leq t \leq t_{j+1}$ by a difference expression of first or second order in h and consider the following sequence of evolution problems:

$$\frac{\partial \phi^h}{\partial t} + \Lambda_h \phi^h = f^h,$$

$$\frac{\partial \phi^{h/2}}{\partial t} + \Lambda_{h/2} \phi^{h/2} = f^{h/2}, \tag{5.13}$$

$$\cdots\cdots\cdots\cdots\cdots\cdots$$

$$\frac{\partial \phi^{h/n}}{\partial t} + \Lambda_{h/n} \phi^{h/n} = f^{h/n}$$

with the corresponding initial data at $t = t_j$.

Let us next construct the difference analogs of (5.13) in the time variable, with the goal of obtaining a highly accurate solution of (5.12). We will point our attention to the homogeneous evolution problems.

Consider any of the homogeneous problems of System (5.13), and write it in the form

$$\frac{\partial \phi}{\partial t} + A\phi = 0, \quad \phi = g \quad \text{for} \quad t = 0. \tag{5.14}$$

Here ϕ and g are vector functions, and the matrix $A \geq 0$ is time invariant.

Equation (5.14) will be solved on a sequence of nets in t, assuming the solution of (5.14) is smooth in t. To this end consider the basic net t_j ($j = 1, 2, \ldots$) with the constant step size $\tau = t_{j+1} - t_j$. Denote this net by D_τ. We introduce next the net $D_{\tau/2}$ obtained from D_τ by adding the net points $t_{j+1/2} = (t_{j+1} - t_j)/2$. By further refinement we get similarly the nets $D_{\tau/3}, D_{\tau/4}$, etc.

Let us now define on this sequence of nets implicit difference schemes which are first-order approximations. Suppose that we know the function value of the solution at the point t_j and that it is required to determine the solution for $t = t_{j+1}$. Then we have the problem

$$\frac{\phi_1^{j+1} - \phi^j}{\tau} + A\phi_1^{j+1} = 0. \tag{5.15}$$

Solving for ϕ_1^{j+1} we obtain

$$\phi_1^{j+1} = (E + \tau A)^{-1} \phi^j. \tag{5.16}$$

Consider next the net $D_{\tau/2}$. The implicit difference scheme on this net will involve two equations:

$$\frac{\phi_2^{j+1/2} - \phi^j}{\tau/2} + A\phi_2^{j+1/2} = 0;$$

$$\frac{\phi_2^{j+1} - \phi^{j+1/2}}{\tau/2} + A\phi_2^{j+1} = 0. \tag{5.17}$$

Solving for ϕ_2^{j+1}, we have

$$\phi_2^{j+1} = \left(E + \frac{\tau}{2} A\right)^{-2} \phi^j \tag{5.18}$$

etc. Thus

$$\phi_p^{j+1} = \left(E + \frac{\tau}{p} A\right)^{-p} \phi^j \quad (p = 1, 2, \ldots).$$

In order to keep the exposition simple, let us confine ourselves to the nets D, $D_{\tau/2}$ and $D_{\tau/3}$, and let us try to construct a third-order approximation of the solution of (5.14). We have the following equations on our sequence of nets:

$$\phi_1^{j+1} = (E + \tau A)^{-1} \phi^j,$$

$$\phi_2^{j+1} = \left(E + \frac{\tau}{2} A\right)^{-2} \phi^j, \tag{5.19}$$

$$\phi_3^{j+1} = \left(E + \frac{\tau}{3} A\right)^{-3} \phi^j.$$

We will expand these relations as follows:

$$\phi_1^{j+1} = \phi^j - \tau A\phi^j + \tau^2 A^2 \phi^j - \tau^3 A^3 \phi^j + \cdots,$$
$$\phi_2^{j+1} = \phi^j - \tau A\phi^j + \tfrac{3}{4}\tau^2 A^2 \phi^j - \tfrac{1}{2}\tau^3 A^3 \phi^j + \cdots, \quad (5.20)$$
$$\phi_3^{j+1} = \phi^j - \tau A\phi^j + \tfrac{2}{3}\tau^2 A^2 \phi^j - \tfrac{10}{27}\tau^3 A^3 \phi^j + \cdots,$$

Multiplying the obtained identities correspondingly by γ_1, γ_2, and γ_3, we obtain

$$\bar{\phi}^{j+1} = (\gamma_1 + \gamma_2 + \gamma_3)(\phi^j - \tau A\phi^j) + (\gamma_1 + \tfrac{3}{4}\gamma_2 + \tfrac{2}{3}\gamma_3)$$
$$\times \tau^2 A^2 \phi^j - (\gamma_1 + \tfrac{1}{2}\gamma_2 + \tfrac{10}{27}\gamma_3)\tau^3 A^3 \phi^j + \cdots. \quad (5.21)$$

where

$$\bar{\phi}^{j+1} = \gamma_1 \phi_1^{j+1} + \gamma_2 \phi_2^{j+1} + \gamma_3 \phi_3^{j+1}.$$

Consider the original evolution Equation (5.14) under the assumption that its formal solution can be written as

$$\phi^{j+1} = e^{-\tau A}\phi^j, \quad (5.22)$$

or, expanding with respect to the small parameter τ,

$$\phi^{j+1} = \phi^j - \tau A\phi^j + \frac{\tau^2}{2!} A^2 \phi^j - \frac{\tau^3}{3!} A^3 \phi^j + \cdots. \quad (5.23)$$

Let us require, that $\bar{\phi}^{j+1}$ approximates the exact solution ϕ^{j+1} up to the third power of τ. Then we obtain the system

$$\gamma_1 + \gamma_2 + \gamma_3 = 1,$$
$$\gamma_1 + \tfrac{3}{4}\gamma_2 + \tfrac{2}{3}\gamma_3 = \tfrac{1}{2}, \quad (5.24)$$
$$\gamma_1 + \tfrac{1}{2}\gamma_2 + \tfrac{10}{27}\gamma_3 = \tfrac{1}{6},$$

the solution of which is given by

$$\gamma_1 = \tfrac{1}{2}, \qquad \gamma_2 = -4, \qquad \gamma_3 = \tfrac{9}{2}. \quad (5.25)$$

As a result we thus have

$$\bar{\phi}^{j+1} = \tfrac{1}{2}\phi_1^{j+1} - 4\phi_2^{j+1} + \tfrac{9}{2}\phi_3^{j+1}.$$

Thus, if the "initial" solution at $t = t_j$ is taken to be the exact solution of the problem, then on each interval $t_j \leq t \leq t_{j+1}$ the approximate solution obtained differs from the exact one by a term of order no less than $O(\tau^4)$.

Assuming that (5.19) are to be solved on the whole interval $0 \leq t \leq T$, we obtain the three approximate solutions defined at exactly the same time instants belonging to the net D_τ. The linear combination of the solutions $\bar{\phi}$ will then differ from the exact solution ϕ by $O(\tau^3)$ at any instant of time, i.e.

$$\|\bar{\phi}^j - \phi^j\| = O(\tau^3).$$

With the help of the implicit difference scheme one can obtain similarly more accurate approximations. Such constructions have been given by Shaidurov for the case of splitting up schemes.

Let

$$A = \sum_{\alpha=1}^{n} A_\alpha, \qquad A_\alpha \geq 0.$$

By the above, the operator does not depend on time, and the problem is taken as homogeneous. In place of Schemes (5.19) consider the implicit splitting up schemes of first-order accuracy. More specifically, consider these schemes on the net $D_{\tau/p}$. They turn out to be p-cyclic on the interval $t_j \leq t \leq t_{j+1}$. The first cycle has the following form:

$$\frac{\phi_p^{j+(1/pn)} - \phi_p^j}{\tau/p} + A_1 \phi_p^{j+(1/pn)} = 0,$$

$$\cdots\cdots\cdots\cdots\cdots\cdots\cdots\cdots\cdots\cdots\cdots\cdots\cdots\cdots \qquad (5.26)$$

$$\frac{\phi_p^{j+(1/p)} - \phi_p^{j+[(n-1)/pn]}}{\tau/p} + A_n \phi_p^{j+(1/p)} = 0.$$

Each cycle is to be repeated p times in the sequence. As a result, we obtain the solution for ϕ_p^{j+1}.

Take now the p-cycle splitting-up Scheme (5.26), and eliminate all the auxiliary variables. There results

$$\phi_p^{j+1} = \left[\left(E + \frac{\tau}{p} A_1 \right)^{-1} \cdots \left(E + \frac{\tau}{p} A_n \right)^{-1} \right]^p \phi_p^j \quad (p = 1, 2, 3, \ldots).$$

If we restrict ourselves to the indices $p = 1, 2, 3$, we then can show similarly as before, that the corrector

$$\bar\phi^{j+1} = \gamma_1 \phi_1^{j+1} + \gamma_2 \phi_2^{j+1} + \gamma_3 \phi_3^{j+1}$$

yields the approximate solution of the problem with an accuracy up to the terms of order τ^3 (relative to the exact solution), i.e.

$$\|\bar\phi^j - \phi^j\| = O(\tau^3).$$

Here γ_1, γ_2, and γ_3 are the same as in (5.25).

The methods we have exhibited permit one to solve problems of mathematical physics by an extensive application of the most simple difference schemes (obtained, for instance, by the finite element method).

Chapter 7

Numerical Methods in the Theory of Radiative Transfer

This chapter deals with the application of the splitting-up method to one of the modern branches of mathematical physics, namely the theory of radiative transfer. Reduction methods for replacing complicated problems with ones easily realizable on a computer will be demonstrated with a particular problem.

These days it would be difficult to name a branch of science or technology which could do without the ideas and methods from the theory of radiative transfer. This theory has been part of theoretical astrophysics from as early as the beginning of our century (Schwartzchild, 1906). Later it began infiltrating physcis of the atmosphere, the atmospheric optics, optics of the sea, and also geophysics. Hand-in-hand with this development came applications in technology. Engineering optics in the twenties and, somewhat later, thermo-engineering were among the first.

We would like especially to underline that, starting with the forties, the leading role in the development of mathematical methods for problems in radiative propagation was undertaken by atomic reseach, where transfer theory not only confronted fundamentally new problems but also new mathematical and physical challenges. It should be noted at this point that the development of powerful mathematical methods for transfer problems took place in the framework of atomic physics.

The basic results of this chapter are devoted to the following objectives: to develop a method for constructing monotonic second-order approximation schemes; to develop a numerical method for solving problems in the theory of radiative transfer with a nonspherical scattering function; and finally, to use the splitting-up method as a constructive apparatus for solving these problems.

7.1. Problem Statement

Consider the Euclidean space R_3 of points $x \equiv (x_1, x_2, x_3)$, along with the vectors r connecting the origin and the points x. Denote by Ω the unit sphere of R_3 centered at the origin, and let us use the same letter Ω for vectors $(\Omega_1, \Omega_2, \Omega_3)$ on this sphere (i.e. $\Omega_1^2 + \Omega_2^2 + \Omega_3^2 = 1$). Using the spherical coordinates ϑ and ψ ($0 \leq \vartheta \leq \pi$, $0 \leq \psi \leq 2\pi$) we can write

$$\Omega_1 = \sin\vartheta \cos\psi, \qquad \Omega_2 = \sin\vartheta \sin\psi, \qquad \Omega_3 = \cos\vartheta.$$

The surface element of the sphere Ω can be expressed by the formula

$$d\Omega = \sin\vartheta \, d\vartheta \, d\psi.$$

Let further $R_3 \times \Omega$ be the Cartesian product of R_3 and Ω. Let $\phi(x, \Omega, t)$ be a nonnegative function, defined on the set $D \times \Omega \times T$, where D is a region in R_3 in which the transport process is being considered, and T is a corresponding interval of time ($0 \leq t \leq T_0$). The boundary of D will be denoted by ∂D.

Introduce in D and Ω the Lebesgue measure and the corresponding integrals, so that in particular

$$\phi_0 = \frac{1}{4\pi} \int_\Omega \phi(x, \Omega', t) \, d\Omega',$$

where

$$\int_\Omega \phi(x, \Omega', t) \, d\Omega' = \int_0^{2\pi} \int_0^\pi \phi(x, \psi', \vartheta', t) \sin \vartheta' \, d\psi' \, d\vartheta'.$$

We introduce further the expressions

$$\Omega \nabla \phi = \sum_{i=1}^3 \Omega_i \frac{\partial \phi}{\partial x_i}$$

and denote

$$\mu_0 = \Omega \Omega' = \sum_{i=1}^3 \Omega_i \Omega_i',$$

where $\Omega, \Omega' \in \Omega$.

Using the considerations regarding the balance of particles in a region of the phase space, the equations describing radiative transfer can be written in the following form:

$$\frac{1}{c} \cdot \frac{\partial \phi}{\partial t} + \Omega \nabla \phi + \sigma \phi = \sigma_s \int_\Omega \phi \gamma(x, \mu_0) \, d\Omega' + f. \tag{1.1}$$

Here $\phi = cn = cn(x, \Omega, t)$ is the density of the particles with a velocity c in the direction Ω at the point x; $\gamma(x, \mu_0)$ is the scattering function. $f(x, \Omega, t)$ are the radiation sources specified in D; $\sigma(x) > 0$, $\sigma_s(x) \geq 0$, $\sigma - \sigma_s \geq 0$. By its very definition, the scattering function satisfies

$$\int_\Omega \gamma(x, \mu_0) \, d\Omega' = 1.$$

Since we are interested in the fundamentals of the numerical algorithms for problems of transfer theory, we will take the scattering function to be isotropic and set

$$\gamma(x, \mu_0) = \frac{1}{4\pi}.$$

Suppose that the function f belongs to a certain subspace, the structure of which will become fully clear later.

Consider the boundary conditions

$$\phi(x, \Omega, t) = 0 \quad \text{on} \quad \partial D \quad \text{for} \quad n\Omega < 0, \tag{1.2}$$

where n is the outer normal to the convex surface ∂D.

Condition (1.2) indicates the absence of a radiation source in a vacuum. This means that we restrict our considerations to internal radiation sources—a simplification. Let us take

$$\phi = \phi^0 \quad \text{for} \quad t = 0 \quad \text{in} \quad D \times \Omega \tag{1.3}$$

as the initial data.

The problems related to existence, uniqueness, and smoothness of solutions of kinetic equations (thus including the transfer equations) have been studied by Vladimirov [17], Jorgens [17], Shikhlov [17], Germogenova [17], Bardos [17], Sultangazin [17], Marchuk [17], and others.

7.2. The Transport Equation in Various Geometries

Consider the equation

$$\frac{1}{c} \cdot \frac{\partial \phi}{\partial t} + \Omega \nabla \phi + \sigma \phi = \frac{\sigma_s}{4\pi} \int d\Omega' \phi(r, \Omega', t) + f(r, \Omega, t). \tag{2.1}$$

The following relation holds along the ray ξ coinciding with the vector Ω (Vladimirov [17], Marchuk and Lebedev [17]):

$$\frac{1}{c} \cdot \frac{d\phi}{d\xi} = \frac{1}{c} \cdot \frac{\partial \phi}{\partial t} + \Omega \nabla \phi,$$

(ξ represents the coordinate of a point on the ray).

In coordinate form we have

$$\frac{1}{c} \cdot \frac{d\phi}{d\xi} = \frac{1}{c} \cdot \frac{\partial \phi}{\partial t} + \sum_{i=1}^{3} \frac{\partial \phi}{\partial x_i} \cdot \frac{dx_i}{d\xi}, \tag{2.2}$$

where

$$\frac{dx_1}{d\xi} = \Omega_1 = \sin \vartheta \cos \psi; \quad \frac{dx_2}{d\xi} = \Omega_2 = \sin \vartheta \sin \psi; \quad \frac{dx_3}{d\xi} = \Omega_3 = \cos \vartheta \tag{2.3}$$

Using Equations (2.2) and (2.3), Equation (2.1) can be written in the form

$$\frac{1}{c} \cdot \frac{\partial \phi}{\partial t} + \sin \vartheta \cos \psi \frac{\partial \phi}{\partial x_1} + \sin \vartheta \sin \psi \frac{\partial \phi}{\partial x_2} + \cos \vartheta \frac{\partial \phi}{\partial x_3} + \sigma \phi$$

$$= \frac{\sigma_s}{4\pi} \int_0^{2\pi} d\psi' \int_0^{\pi} \phi(x, \psi', \vartheta', t) \sin \vartheta' d\vartheta' + f(x, \psi, \vartheta, t). \tag{2.4}$$

Suppose the input data of the problem σ, σ_s, and f are independent of the coordinates x_1 and x_2 and are functions of $x_3 = z$, ϑ, ψ, and t. In addition assume that the region of the three-dimensional space in which we seek the solution coincides with the strip $0 \le z \le H$. Then the solution of Equation (2.1) will clearly depend only on z, ϑ, ψ, and t. As a result we obtain the equation in the parallel-plane geometry

$$\frac{1}{c} \cdot \frac{\partial \phi}{\partial t} + \cos \vartheta \frac{\partial \phi}{\partial z} + \sigma \phi = \frac{\sigma_s}{4\pi} \int_0^{2\pi} d\psi' \int_0^{\pi} \phi(z, \psi', \vartheta', t) \sin \vartheta' d\vartheta' + f(z, \phi, \vartheta, t). \tag{2.5}$$

If the solution ϕ and the source f do not depend on the azimuth ψ, then (2.5) can be somewhat simplified by writing

$$\frac{1}{c} \cdot \frac{\partial \phi}{\partial t} + \mu \frac{\partial \phi}{\partial z} + \sigma \phi = \frac{\sigma_s}{2} \int_{-1}^{1} d\mu' \phi(z, \mu', t) + f(z, \mu, t). \tag{2.6}$$

On the boundaries of the region, i.e. for $z = 0$ and $z = H$, we set

$$\begin{aligned} \phi(0, \mu, t) = 0 & \quad \text{for} \quad \mu > 0, \\ \phi(H, \mu, t) = 0 & \quad \text{for} \quad \mu < 0. \end{aligned} \tag{2.7}$$

Consider the spherically symmetric problem. The coordinates in this case can be taken as the distance r from the center of the spherically-symmetrical system to the point P under consideration, and the angle between the radius vector of the point P and the z-axis. We will thus have

$$\frac{1}{c} \cdot \frac{d\phi}{d\xi} = \frac{1}{c} \cdot \frac{\partial \phi}{\partial t} + \frac{\partial \phi}{\partial r} \cdot \frac{dr}{d\xi} + \frac{\partial \phi}{\partial \vartheta} \cdot \frac{d\vartheta}{d\xi}.$$

From geometric considerations we obtain easily that

$$\frac{dr}{d\xi} = \Omega_r = \cos \vartheta, \qquad \frac{\partial \vartheta}{d\xi} = \Omega_v = -\frac{\sin \vartheta}{r}.$$

Similarly as before, we obtain the following equation for problems with spherical symmetry and with sources and solutions independent of the azimuth ψ:

$$\frac{1}{c} \cdot \frac{\partial \phi}{\partial t} + \mu \frac{\partial \phi}{\partial r} + \frac{1-\mu^2}{r} \cdot \frac{\partial \phi}{\partial \mu} + \sigma \phi = \frac{\sigma_s}{2} \int_{-1}^{1} d\mu' \phi(r, \mu', t) + f(r, \mu, t), \tag{2.8}$$

where $\mu = \cos \vartheta$.

Equation (2.8) can be rewritten in the divergence form:

$$\frac{1}{c} \cdot \frac{\partial \phi}{\partial t} + \frac{1}{r^2} \cdot \frac{\partial}{\partial r}(r^2 \mu \phi) + \frac{\partial}{\partial \mu}\left(\frac{1-\mu^2}{r} \phi\right) + \sigma \phi$$

$$= \frac{\sigma_s}{2} \int_{-1}^{1} d\mu' \phi(r, \mu, t) + f(r, \mu, t). \tag{2.9}$$

If there is no flow of the particles from the vacuum, then (2.9) can be supplemented by the condition

$$\phi(R, \mu, t) = 0 \quad \text{for} \quad \mu < 0, \tag{2.10}$$

where R is the radius of the outer boundary of the region D.

The transfer equation in cylindrical geometry can be treated similarly.

7.3. Numerical Solution of the Transport Equation in the Parallel-Plane Geometry

Let us turn to numerical algorithms for problems in transfer theory. Since our goals are primarily methodic in nature, we point our attention to the simplest and at the same time practically interesting models. Although we have chosen the splitting-up method as the basic mathematical tool for solving nonstationary problems, the fundamental ideas in constructing the difference analogs of transfer problems can also be used in connection with other approaches.

Consider the following simple problem in the theory of radiative transfer in parallel-plane geometry:

$$\frac{1}{c} \cdot \frac{\partial \phi}{\partial t} + \mu \frac{\partial \phi}{\partial z} + \sigma \phi = \frac{\sigma_s}{2} \int_{-1}^{1} \phi \, d\mu + f, \tag{3.1}$$

$$\phi = 0 \quad \text{for} \quad z = 0, \mu > 0, \tag{3.2}$$

$$\phi = 0 \quad \text{for} \quad z = H, \mu < 0,$$

$$\phi = \phi^0 \quad \text{for} \quad t = 0. \tag{3.3}$$

Let us bring in some convenient transformations of the problem introduced by Kuznetsov [17]. Multidimensional analogs have been introduced by Vladimirov [17]. Denote by ϕ^+ and ϕ^- the solution of (3.1) corresponding to $\mu > 0$ and $\mu < 0$. The equation of the transfer can then be represented by the following two equations:

$$\frac{1}{c} \cdot \frac{\partial \phi^+}{\partial t} + \mu \frac{\partial \phi^+}{\partial z} + \sigma \phi^+ = \frac{\sigma_s}{2} \int_0^1 (\phi^+ + \phi^-) \, d\mu' + f^+,$$

$$\frac{1}{c} \cdot \frac{\partial \phi^-}{\partial t} - \mu \frac{\partial \phi^-}{\partial z} + \sigma \phi^- = \frac{\sigma_s}{2} \int_0^1 (\phi^+ + \phi^-) \, d\mu' + f^-. \tag{3.4}$$

The boundary conditions for ϕ^+ and ϕ^- are as follows:

$$\phi^+(z, \mu) = 0 \quad \text{for} \quad z = 0.$$

$$\phi^-(z, \mu) = 0 \quad \text{for} \quad z = H. \tag{3.5}$$

Now adding up these two equations, and then subtracting one from another, we obtain two new equations

$$\frac{1}{c} \cdot \frac{\partial u}{\partial t} + \mu \frac{\partial v}{\partial z} + \sigma u = \sigma_s \int_0^1 u d\mu' + g,$$

$$\frac{1}{c} \cdot \frac{\partial v}{\partial t} + \mu \frac{\partial u}{\partial z} + \sigma v = r,$$
(3.6)

where

$$u = \tfrac{1}{2}(\phi^+ + \phi^-), \qquad v = \tfrac{1}{2}(\phi^+ - \phi^-),$$
$$g = \tfrac{1}{2}(f^+ + f^-), \qquad r = \tfrac{1}{2}(f^+ - f^-).$$

It is not difficult to verify that the boundary conditions (3.5) are then changed as follows:

$$u + v = 0 \quad \text{for} \quad z = 0,$$
$$u - v = 0 \quad \text{for} \quad z = H;$$
(3.7)

the initial data become

$$u = u^0, \, v = v^0 \quad \text{for} \quad t = 0.$$
(3.8)

In order to write (3.6)–(3.8) in operator form, let us introduce the vector functions w, w^0, F and the operator A as follows:

$$w = \begin{vmatrix} u \\ v \end{vmatrix}, \qquad w^0 = \begin{vmatrix} u^0 \\ v^0 \end{vmatrix}, \qquad F = \begin{vmatrix} g \\ r \end{vmatrix},$$

$$A = \begin{Vmatrix} \sigma - \sigma_s \int_0^1 d\mu' & \mu \frac{\partial}{\partial z} \\ \mu \frac{\partial}{\partial z} & \sigma \end{Vmatrix}.$$
(3.9)

Introduce further the Hilbert space $L^2(D)$ with the inner product

$$(a, b) = \sum_{i=1}^{2} \int_0^1 d\mu \int_0^H a^i b^i dx,$$
(3.10)

where a^i, b^i are the components of the vector functions a and b.

Define in this space the subspace Φ, the elements of which satisfy the condition

$$(Aw, w) < +\infty.$$
(3.11)

Note that the inner product above depends on time. We will not repeat this fact in the sequel. The components of the vector functions w are required to be continuous in D and to have absolutely continuous derivatives $\mu \partial w / \partial z$. Note that the smoothness of u and v is the direct consequence of ϕ^+ and ϕ^-

Numerical Solution of the Transport Equation

having the same property. Finally we define the subspace Φ^0 of Φ to be the set of those vector functions from Φ which satisfy (3.7) and which in addition possess absolutely continuous derivatives with respect to time. Clearly, Φ^0 is the domain of the operator

$$L \equiv \frac{1}{c} \cdot \frac{\partial}{\partial t} + A.$$

Thus, we obtain the following problem:

$$\frac{1}{c} \cdot \frac{\partial w}{\partial t} + Aw = F \quad \text{in} \quad D \times D_t, \tag{3.12}$$

$$w = w^0 \quad \text{for} \quad t = 0 \quad \text{in} \quad D,$$

while

$$F(t) \in L_2(D \times \Omega); \quad w^0 \in \Phi; \quad w(t) \in \Phi^0.$$

It is not difficult to verify that Φ^0 is also the domain of A and that on the elements from this domain

$$(Aw, w) > 0. \tag{3.13}$$

Moreover, as shown by Vladimirov [17], the operator A is positive definite, i.e.

$$(Aw, w) \geq \gamma(w, w), \tag{3.14}$$

where γ is a positive constant related to the characteristic geometrical dimension of the region.

Let us consider next the difference approximation of Problems (3.6)–(3.8) in the spatial variable z. Thus, integrate the first of Equations (3.6) with respect to z in the limits $(z_{k-1/2}, z_{k+1/2})$, and the second in the limits (z_k, z_{k+1}), where $\{z_k\}$ ($\{z_{k+1/2}\}$) is the system of basic (auxiliary) net points. Let

$$z_{k+1/2} = \tfrac{1}{2}(z_k + z_{k+1}), \qquad \Delta z_k = z_{k+1/2} - z_{k-1/2}, \qquad \Delta z_{k+1/2} = z_{k+1} - z_k.$$

Then Equation (3.6) and boundary Conditions (3.7) become

$$\frac{1}{c} \cdot \frac{\partial}{\partial t} \int_{z_{k-1/2}}^{z_{k+1/2}} u \, dz + \mu \int_{z_{k-1/2}}^{z_{k+1/2}} \frac{\partial v}{\partial z} dz + \int_{z_{k-1/2}}^{z_{k+1/2}} \sigma u \, dz$$

$$= \int_{z_{k-1/2}}^{z_{k+1/2}} \sigma_s \int_0^1 u d\mu dz + \int_{z_{k-1/2}}^{z_{k+1/2}} g \, dz; \tag{3.15}$$

$$\frac{1}{c} \cdot \frac{\partial}{\partial t} \int_{z_k}^{z_{k+1}} v \, dz + \mu \int_{z_k}^{z_{k+1}} \frac{\partial u}{\partial z} dz + \int_{z_k}^{z_{k+1}} \sigma v \, dz = \int_{z_k}^{z_{k+1}} r \, dz.$$

Let us introduce the following notation:

$$u_k = u(z_k, \mu, t); \qquad v_{k+1/2} = v(z_{k+1/2}, \mu, t),$$

$$\sigma_k = \frac{1}{\Delta z_k} \int_{z_{k-1/2}}^{z_{k+1/2}} \sigma\, dz; \qquad \sigma_{sk} = \frac{1}{\Delta z_k} \int_{z_{k-1/2}}^{z_{k+1/2}} \sigma_s\, dz; \tag{3.16}$$

$$\sigma_{k+1/2} = \frac{1}{\Delta z_{k+1/2}} \int_{z_k}^{z_{k+1}} \sigma\, dz;$$

$$g_k = \frac{1}{\Delta z_k} \int_{z_{k-1/2}}^{z_{k+1/2}} g\, dz; \qquad r_{k+1/2} = \frac{1}{\Delta z_{k+1/2}} \int_{z_k}^{z_{k+1}} r\, dz. \tag{3.17}$$

Under the assumptions of continuity of u and v almost everywhere in z and μ from D, and piece-wise continuity of σ, σ_s, g, and r with possible discontinuities of the first kind at the points z_k, we then obtain the following difference approximations of Equations (3.6), using the methods from Section 2.3:

$$\frac{1}{c} \cdot \frac{\partial u_k}{\partial t} + \mu \frac{v_{k+1/2} - v_{k-1/2}}{\Delta z_k} + \sigma_k u_k = \sigma_{sk} \int_0^1 u_k\, d\mu + g_k,$$

$$\frac{1}{c} \cdot \frac{\partial v_{k+1/2}}{\partial t} + \mu \frac{u_{k+1/2} - u_k}{\Delta z_{k+1/2}} + \sigma_{k+1/2} v_{k+1/2} = r_{k+1/2}. \tag{3.18}$$

The boundary Conditions (3.7) have for their difference analogs

$$u_1 + v_{1/2} = 0, \qquad u_N - v_{N-1/2} = 0 \tag{3.19}$$

and the initial data from (3.8) become

$$u_k = u_k^0,\ v_{k+1/2} = v_{k+1/2}^0 \quad \text{for} \quad t = 0. \tag{3.20}$$

In the same manner as (3.9), let us introduce the vector functions w_k, w_k^0, F_k and the matrix A^h as follows:

$$w_k = \begin{Vmatrix} u_k \\ v_{k+1/2} \end{Vmatrix}, \qquad w_k^0 = \begin{Vmatrix} u_k^0 \\ v_{k+1/2}^0 \end{Vmatrix}, \qquad F_k = \begin{Vmatrix} g_k \\ r_{k+1/2} \end{Vmatrix},$$

$$A_k^h = \begin{Vmatrix} \sigma_k - \sigma_{sk} \int_0^1 d\mu & \mu \nabla_k \\ \mu \nabla_{k+1/2} & \sigma_{k+1/2} \end{Vmatrix}.$$

Then (3.18), (3.19) can be formally written as

$$\frac{1}{c} \cdot \frac{\partial w_k}{\partial t} + A_k^h w_k = F_k,$$

$$w_k - w_k^0 \quad \text{for} \quad t = 0, \tag{3.21}$$

where ∇_k and $\nabla_{k+1/2}$ are defined by the expressions

$$\nabla_k a_k = \frac{a_{k+1/2} - a_{k-1/2}}{\Delta z_k}, \qquad \nabla_{k+1/2} a_k = \frac{a_{k+1} - a_k}{\Delta z_{k+1/2}}.$$

Numerical Solution of the Transport Equation

Introduce further the Hilbert space of functions w with the components $\{w_k\}$ and endowed with the inner product

$$(ab) = \sum_{k=1}^{N-1} \int_0^1 d\mu (a_k^{(1)} b_k^{(1)} \Delta z_k + a_{k+1/2}^{(2)} b_{k+1/2}^{(2)} \Delta z_{k+1/2}).$$

Similarly as before, introduce the subspaces Φ_h^0 and Φ_h, which differ from Φ^0 and Φ only in that the differential conditions of smoothness of the solutions in z are replaced by weaker conditions, related to the features of the finite-dimensional (with respect to z) Euclidean space, and accounting for the boundary conditions in the form of (3.19). It is not difficult to show that for $w \in \Phi_k$ (the domain of A^h) we have the conditional positive definiteness:

$$(A^h w, w) \geq \gamma(w, w). \tag{3.22}$$

Here w is a vector function with components $\{w_k\}$ and A^h is a block-diagonal matrix with elements A_k^h. If we also denote by F the vector function with components F_k, then (3.21) can be rewritten in the form

$$\frac{1}{c} \cdot \frac{\partial w}{\partial t} + A^h w = F, \tag{3.23}$$

$$w = w^0 \quad \text{for} \quad t = 0.$$

Note that Relation (3.22) holds true only when the boundary conditions are approximated in the form of (3.19). Let us show that A^h is positive definite. For that we use the inner product introduced above. We have

$$(A^h \omega, \omega) = \sum_{k=1}^{N-1} \int_0^1 \left[\left(\mu \nabla_k v_k + \sigma_k u_k - \sigma_{sk} \int_0^1 u_k d\mu \right) u_k \Delta z_k \right.$$
$$\left. + (\mu \nabla_{k+1/2} u_k + \sigma_{k+1/2} v_{k+1/2}) v_{k+1/2} \Delta z_{k+1/2} \right] d\mu$$

$$= \sum_{k=1}^{N-1} \int_0^1 \left[\left(\mu \frac{v_{k+1/2} - v_{k-1/2}}{\Delta z_k} + \sigma_k u_k - \sigma_{sk} \int_0^1 u_k d\mu \right) u_k \Delta z_k \right.$$
$$\left. + \left(\mu \frac{u_{k+1} - u_k}{\Delta z_{k+1/2}} + \sigma_{k+1/2} v_{k+1/2} \right) v_{k+1/2} \Delta z_{k+1/2} \right] d\mu$$

$$= \sum_{k=1}^{N-1} \int_0^1 \mu(v_{k+1/2} u_{k+1} - v_{k-1/2} u_k) d\mu + \sum_{k=1}^{N-1} \left[\sigma_k \int_0^1 u_k^2 d\mu \right.$$
$$\left. - \sigma_{sk} \left(\int_0^1 u_k d\mu \right)^2 \right] \Delta z_k + \sum_{k=1}^{N-1} \sigma_{k+1/2} \left(\int_0^1 v_{k+1/2}^2 d\mu \right) \Delta z_{k+1/2}$$

$$= -\int_0^1 \mu v_{1/2} u_1 d\mu + \int_0^1 \mu v_{N-1/2} u_N d\mu + \sum_{k=1}^{N-1} \left[\sigma_k \int_0^1 u_k^2 d\mu \right.$$
$$\left. - \sigma_{sk} \left(\int_0^1 u_k d\mu \right)^2 \right] \Delta z_k + \sum_{k=1}^{N-1} \sigma_{k+1/2} \left(\int_0^1 v_{k+1/2}^2 d\mu \right) \Delta z_{k+1/2}.$$

Taking into account that

$$\sum_{k=1}^{N-1}\left[\sigma_k\int_0^1 u_k^2 d\mu - \sigma_{sk}\left(\int_0^1 u_k d\mu\right)^2\right]\Delta z_k \geqslant \sum_{k=1}^{N-1}\sigma_{ck}\left(\int_0^1 u_k^2 d\mu\right)\Delta z_k$$

$$(\sigma_{ck} = \sigma_k - \sigma_{sk})$$

and the boundary Condition (3.19), we finally obtain

$$(A^h\omega, \omega) \geqslant \int_0^1 \mu u_1^2 d\mu + \int_0^1 \mu u_N^2 d\mu + \sum_{k=1}^{N-1}\sigma_{ck}\left(\int_0^1 \mu u_k^2 d\mu\right)\Delta_k$$

$$+ \sum_{k=1}^{N-1}\sigma_{k+1/2}\left(\int_0^1 v_{k+1/2}^2 d\mu\right)\Delta z_{k+1/2} > 0 \quad \text{for all} \quad \|\omega\| \neq 0.$$

Considering the relations between the functions u, v and ϕ^+, ϕ^-, it can be seen that boundary Conditions (3.19) transform exact Relations (3.5)

$$\phi^+ = 0, \qquad \phi^- = 0$$

into the approximate ones*

$$\phi_{1/2}^+ = O(\Delta x), \qquad \phi_{N-1/2}^- = O(\Delta x). \tag{3.24}$$

Turning now to the formulation of the splitting-up method, we introduce the following two matrices:

$$A_{1k}^h = \left\|\begin{matrix} \sigma_{sk} - \sigma_{sk}\int_0^1 d\mu & 0 \\ 0 & 0 \end{matrix}\right\|,$$

$$A_{2k}^h = \left\|\begin{matrix} \sigma_{ck} & \mu\nabla_k \\ \mu\nabla_{k+1/2} & \sigma_{k+1/2} \end{matrix}\right\|. \tag{3.25}$$

It is easy to see that

$$A_k^h = A_{1k}^h + A_{2k}^h.$$

Further,

$$(A_1 w, w) \geqslant 0; \qquad (A_2 w, w) > \gamma(w, w). \tag{3.26}$$

The second condition is based on the immediate inequality

$$\sum_{k=1}^{N-1}\Delta z_k \sigma_{ck}\left[\int_0^1 u_k^2 d\mu - \left(\int_0^1 u_k d\mu\right)^2\right] \geqslant 0.$$

The positivity Conditions (3.26) allow the formulation of the algorithm on the basis of the two-cycle component-by-component splitting-up method (see Section 4.3). To this end let us formulate the following problems.

---

* Agoshkov V. I. (Computing Center of the Siberian Branch of the USSR Academy of Sciences) has shown that in the case of a uniform net with the mesh size h the constructed approximation at $\sigma(z)$, $\sigma_s(z) \in C^{(1)}[0, H]$ on smooth solutions ensures the accuracy $O(h^2|\ln h|^{\frac{1}{2}})$ of the approximate solution in the L_2-metric.

Numerical Solution of the Transport Equation

For $t_{j-1} \leqq t \leqq t_j$ let

$$\left(E + \frac{\tau}{2} A_1^h\right) w^{j-2/3} = \left(E - \frac{\tau}{2} A_1^h\right) w^{j-1},$$

$$\left(E + \frac{\tau}{2} A_2^h\right) w^{j-1/3} = \left(E - \frac{\tau}{2} A_2^h\right) w^{j-2/3},$$

(3.27)

for $t_{j-1} \leqq t \leqq t_{j+1}$ let

$$w^{j+1/3} = w^{j-1/3} + 2\tau F^j \tag{3.28}$$

and on the interval $t_j \leqq t \leqq t_{j+1}$ let

$$\left(E + \frac{\tau}{2} A_2^h\right) w^{j+2/3} = \left(E - \frac{\tau}{2} A_2^h\right) w^{j+1/3},$$

$$\left(E + \frac{\tau}{2} A_1^h\right) w^{j+1} = \left(E - \frac{\tau}{2} A_1^h\right) w^{j+2/3},$$

(3.29)

where $\tau = c\Delta t$.

Rewrite now the first of Equation (3.27) in scalar (component) form. We have

$$u_k^{j-2/3} + \frac{\tau \sigma_{sk}}{2} \left(u_k^{j-2/3} - \int_0^1 u_k^{j-2/3} d\mu \right) = u_k^{j-1}$$

$$- \frac{\tau \sigma_{sk}}{2} \left(u_k^{j-1} - \int_0^1 u_k^{j-1} d\mu \right); \quad v_{k+1/2}^{j-2/3} = v_{k+1/2}^{j-1}. \tag{3.30}$$

Integrating further the first of Equations (3.30) with respect to μ in the limits $0 \leqq \mu \leqq 1$, we obtain

$$\int_0^1 u_k^{j-2/3} d\mu = \int_0^1 u_k^{j-1} d\mu.$$

Using this relation, the solution of System (3.30) can be obtained explicitly:

$$u_h^{j-2/3} = \frac{1 - \frac{\tau \sigma_{sk}}{2}}{1 + \frac{\tau \sigma_{sk}}{2}} u_k^{j-1} + \frac{\tau \sigma_{sk} \int_0^1 u_k^{j-1} d\mu}{1 + \frac{\tau \sigma_{sk}}{2}}, \quad v_{k+1/2}^{j-2/3} = v_{k+1/2}^{j-1}. \tag{3.31}$$

Similarly one can find the solution of the second of Equations (3.29):

$$u_k^{j+1} = \frac{1 - \frac{\tau \sigma_{sk}}{2}}{1 + \frac{\tau \sigma_{sk}}{2}} u_k^{j+2/3} + \frac{\tau \sigma_{sk} \int_0^1 \bar{u}_k^{j+2/3} d\mu}{1 + \frac{\tau \sigma_{sk}}{2}}, \quad v_{k+1/2}^{j+1} = v_{k+1/2}^{j+2/3}. \tag{3.32}$$

We rewrite now the second equation in (3.27) in the scalar component-by-component form:

$$u_k^{j-1/3} + \frac{\tau}{2}\left(\mu \frac{v_{k+1/2}^{j-1/3} - v_{k-1/2}^{j-1/3}}{\Delta z_k} + \sigma_{ck} u_k^{j-1/3}\right) = p_k,$$

$$v_{k+1/2}^{j-1/3} + \frac{\tau}{2}\left(\mu \frac{u_{k+1}^{j-1/3} - u_k^{j-1/3}}{\Delta z_{k+1/2}} + \sigma_{k+1/2} v_{k+1/2}^{j-1/3}\right) = q_{k+1/2},$$

(3.33)

where p_k and $q_{k+1/2}$ are known from the previous computational stage.

Let us supplement this system by boundary conditions in the form (3.19), i.e.

$$u_1^{j-1/3} + v_{1/2}^{j-1/3} = 0, \qquad u_N^{j-1/3} - v_{N-1/2}^{j-1/3} = 0. \tag{3.34}$$

Solving for $v_{k+1/2}^{j-1/3}$ in the second equation from System (3.33), we obtain

$$v_{k+1/2}^{j-1/3} = \frac{-\tau\mu}{1 + \frac{\tau\sigma_{k+1/2}}{2}} \cdot \frac{u_{k+1}^{j-1/3} - u_k^{j-1/3}}{2\Delta z_{k+1/2}} + \frac{q_{k+1/2}}{1 + \frac{\tau\sigma_{k+1/2}}{2}}. \tag{3.35}$$

Using this expression in the first of Equations (3.33), we obtain

$$a_k u_{k-1}^{j-1/3} - b_k u_k^{j-1/3} + c_k u_{k+1}^{j-1/3} = -\rho_k \qquad (k = 1, 2, \ldots, N-1), \tag{3.36}$$

where

$$a_k = \frac{\tau^2}{4\Delta z_k \Delta z_{k-1/2}} \cdot \frac{\mu^2}{\left(1 + \frac{\tau\sigma_{ck}}{2}\right)\left(1 + \frac{\tau\sigma_{k-1/2}}{2}\right)};$$

$$b_k = 1 + a_k + c_k;$$

$$c_k = \frac{\tau^2}{4\Delta z_k \Delta z_{k+1/2}} \cdot \frac{\mu^2}{\left(1 + \frac{\tau\sigma_{ck}}{2}\right)\left(1 + \frac{\tau\sigma_{k+1/2}}{2}\right)};$$

$$\rho_k = \frac{1}{1 + \frac{\tau\sigma_{ck}}{2}}\left(p_k + \frac{\tau}{2\Delta z_k} \cdot \frac{\mu q_{k-1/2}}{1 + \frac{\tau\sigma_{k-1/2}}{2}} - \frac{\tau}{2\Delta z_k} \cdot \frac{\mu q_{k+1/2}}{1 + \frac{\tau\sigma_{k+1/2}}{2}}\right).$$

The system of Equations (3.36) is to be supplemented by boundary conditions in the form

$$u_1^{j-1/3} + v_{1/2}^{j-1/3} = 0, \qquad u_N^{j-1/3} - v_{N-1/2}^{j-1/3} = 0. \tag{3.37}$$

Eliminate next $v_{1/2}$ and $v_{N-1/2}$ from boundary Conditions (3.37), using Relation (3.35). Then we obtain the following two relations for $u_k^{j-1/2}$:

$$(1 - \alpha_{1/2})u_1^{j-1/3} + \alpha_{1/2} u_0^{j-1/3} = -\beta_{1/2} q_{1/2};$$

$$-\alpha_{N-1} u_{N-1}^{j-1/3} + (1 + \alpha_{N-1/2}) u_N^{j-1/3} = \beta_{N-1/2} q_{N-1/2},$$

(3.38)

Numerical Solution of the Transport Equation

where

$$\alpha_{k-1/2} = \frac{\tau}{2\Delta z_{k-1/2}} \cdot \frac{\mu}{1 + \frac{\tau\sigma_{k-1/2}}{2}}; \quad \beta_{k-1/2} = \frac{1}{1 + \frac{\tau\sigma_{k-1/2}}{2}}.$$

System (3.36) along with (3.38) turn out to be full rank and allow us to solve the problem for any μ with the help of the factorization method. Similarly we find the solution of the first equation in (3.29). The solution of (3.28) is obtained in an explicit form. Thus the numerical algorithm for the system of the split Equations (3.27)–(3.29) has been found. Note that so far we have not considered the difference approximation of the equations with respect to μ; we will do that now. Divide the interval $0 \leq \mu \leq 1$ into the subintervals $\Delta\mu_l$ by means of the net points μ_l so as to guarantee the best approximations of the intervals in (3.30) and (3.31) on the given class of solutions.

Let

$$\int_0^1 u\,d\mu = \sum_{l=1}^m s_l u_l,$$

where s_l are the coefficients of the quadrature formula with the net points $\{\mu_l\}$.

From (3.30) and (3.31) we obtain the expressions for $u_{k,l}^{j-2/3}$, $v_{k+1/2,l}^{j-2/3}$ and $u_{k,l}^{j+1}$, $v_{k+1/2,l}^{j+1}$:

$$u_{k,l}^{j-2/3} = \frac{1 - \frac{\tau\sigma_{sk}}{2}}{1 + \frac{\tau\sigma_{sk}}{2}} u_{k,l}^{j-1} + \frac{\tau\sigma_{sk}\sum_{l=1}^m s_l u_{k,l}^{j-1}}{1 + \frac{\tau\sigma_{sk}}{2}}; \quad v_{k+1/2,l}^{j-2/3} = v_{k+1/2,l}^{j-1}. \tag{3.39}$$

Such a solution can also be written for the second equation of System (3.29). Problem (3.36) becomes

$$a_{k,l} u_{k-1,l}^{j-1/3} - b_{k,l} u_{k,l}^{j-1/3} + c_{k,l} u_{k+1,l}^{j-1/3} = -p_{k,l}$$
$$(k = 1, 2, \ldots, N-1; l = 1, 2, \ldots, m) \tag{3.40}$$

under the condition

$$(1 - \alpha_{1/2,l}) u_{1,l}^{j-1/3} + \alpha_{1/2,l} u_{0,l}^{j-1/3} = -\beta_{1/2,l} q_{1/2,l},$$
$$-\alpha_{N-1/2,l} u_{N-1,l}^{j-1/3} + (1 + \alpha_{N-1/2,l}) u_{N,l}^{j-1/3} = \beta_{N-1/2,l} q_{N-1/2,l}. \tag{3.41}$$

All the quantities indexed by l are taken for $\mu = \mu_l$.

Thus, the numerical algorithm for a nonstationary equation of radiation transfer has been fully determined. The resulting scheme is absolutely stable. The accuracy of the algorithm requires a careful treatment, involving the special features of the input data of the problem and its solution.

In conclusion let us note, that a good approximation of the solution of the problem is guaranteed by taking

$$(\tau\sigma/2) \ll 1,$$

where $\sigma = \max_k \{\sigma_k, \sigma_{k+1/2}\}$. This condition restricts the choice of the time-step τ.

7.4. The Stationary Transport Problem

The algorithm we have given can be also used to solve stationary problems by means of the steady-state method. This means that the solution of the stationary problem is taken to be the limit as $t \to \infty$ of the corresponding nonstationary solution.

Finally let us note that in order to solve the stationary problem

$$\mu \frac{\partial \phi}{\partial z} + \sigma \phi = \frac{\sigma_s}{2} \int_{-1}^{1} \phi \, d\mu + f, \qquad (4.1)$$

$$\phi = 0 \quad \text{for} \quad z = 0, \qquad \mu > 0,$$
$$\phi = 0 \quad \text{for} \quad z = H, \qquad \mu < 0,$$

it is economical to use difference approximations in the form of (3.18) and (3.19), where the time derivatives are taken to be zero. Then we have

$$\mu \frac{v_{k+1/2} - v_{k-1/2}}{\Delta z_k} + \sigma_k u_k = \sigma_{sk} \int_0^1 u_k \, d\mu + g_k,$$

$$\mu \frac{u_{k+1} - u_k}{\Delta z_{k+1/2}} + \sigma_{k+1/2} v_{k+1/2} = r_{k+1/2} \qquad (4.2)$$

under the condition

$$u_1 + v_{1/2} = 0, \qquad u_N - v_{N-1/2} = 0. \qquad (4.3)$$

Eliminating from System (4.2) and boundary Conditions (4.3) the unknowns $v_{k+1/2}$, we obtain the difference equations for u_k:

$$\frac{\mu^2}{\Delta z_k} \left(\frac{u_k - u_{k-1}}{\sigma_{k-1/2} \Delta z_{k-1/2}} - \frac{u_{k+1} - u_k}{\sigma_{k+1/2} \Delta z_{k+1/2}} \right) + \sigma_k u_k - \sigma_{sk} \int_0^1 u_k \, d\mu = f_k, \qquad (4.4)$$

where

$$f_k = g_k + \frac{\mu}{\Delta z_k} \left(\frac{r_{k-1/2}}{\sigma_{k-1/2}} - \frac{r_{k+1/2}}{\sigma_{k+1/2}} \right).$$

We have the following boundary conditions for (4.4):

$$\mu \frac{u_1 - u_0}{\sigma_{1/2} \Delta z_{1/2}} = u_1 + \frac{r_{1/2}}{\sigma_{1/2}},$$

$$-\mu \frac{u_N - u_{N-1}}{\sigma_{N-1/2} \Delta z_{N-1/2}} = u_N - \frac{r_{N-1/2}}{\sigma_{N-1/2}}. \qquad (4.5)$$

The Stationary Transport Problem

Let us note at this point that the finite-difference Expression (4.4), approximating the differential operator, fully coincides with the scheme in Section 2.1 for the diffusion equation with discontinuous coefficients.

Let us approximate Problems (4.4) and (4.5) on the net $\{\mu_l\}$ corresponding to the parameter μ. There results

$$\frac{\mu_l^2}{\Delta z_k}\left(\frac{u_{k,l} - u_{k-1,l}}{\sigma_{k-1/2}\Delta z_{k-1/2}} - \frac{u_{k+1,l} - u_{k,l}}{\sigma_{k+1/2}\Delta z_{k+1/2}}\right) + \sigma_k u_{k,l} - \sigma_{sk}\sum_{l=1}^{m} s_l u_{k,l} = f_{k,l}, \quad (4.6)$$

where

$$f_{k,l} = g_{k,l} + \frac{\mu_l}{\Delta z_k}\left(\frac{r_{k-1/2,l}}{\sigma_{k-1/2}} - \frac{r_{k+1/2,l}}{\sigma_{k+1/2}}\right)$$

under the assumptions

$$\mu_l \frac{u_{1,l} - u_{0,l}}{\sigma_{1/2}\Delta z_{1/2}} = u_{1,l} + \frac{r_{1/2,l}}{\sigma_{1/2}},$$

$$-\mu_l \frac{u_{N,l} - u_{N-1,l}}{\sigma_{N-1/2}\Delta z_{N-1/2}} = u_{N,l} - \frac{r_{N-1/2,l}}{\sigma_{N-1/2}}. \quad (4.7)$$

Let us introduce the vectors u_k, f_k, and $r_{k+1/2}$ in correspondence with the components $\{u_{k,l}\}, \{f_{k,l}\}$, and $\{r_{k+1/2,l}\}$, the diagonal matrix M^2 with nonzero diagonal elements $\{\mu_l^2\}$ and the matrix S defined by the relation

$$Su_k = \sum_{l=1}^{m} s_l u_{k,l}.$$

We have then the following discrete (in μ) analog of Problem (4.4) and (4.5) written in the vector-matrix form:

$$\frac{1}{\Delta z_k}M^2\left(\frac{u_k - u_{k-1}}{\sigma_{k-1/2}\Delta z_{k-1/2}} - \frac{u_{k+1} - u_k}{\sigma_{k+1/2}\Delta z_{k+1/2}}\right) + \sigma_k E u_k - \sigma_{sk} S u_k = f_k \quad (4.8)$$

under the conditions

$$M\frac{u_1 - u_0}{\sigma_{1/2}\Delta z_{1/2}} = u_1 + \frac{r_{1/2}}{\sigma_{1/2}},$$

$$-M\frac{u_N - u_{N-1}}{\sigma_{N-1/2}\Delta z_{N-1/2}} = u_N - \frac{r_{N-1/2}}{\sigma_{N-1/2}}. \quad (4.9)$$

Note that the diagonal matrices M and M^2 are related by the following simple equation:

$$M^2 = (M)^2.$$

Introduce further the vector-function w with the components $\{u_k\}$ and the three-diagonal matrix

$$A = \begin{Vmatrix} b_1 & -c_1 & 0 & 0 & \cdots & 0 & 0 \\ -a_2 & b_2 & -c_2 & 0 & \cdots & 0 & 0 \\ 0 & -a_3 & b_3 & -c_3 & \cdots & 0 & 0 \\ \multicolumn{7}{c}{\dotfill} \\ 0 & 0 & 0 & 0 & \cdots & -a_{N-1} & b_{N-1} \end{Vmatrix},$$

where

$$a_k = \frac{1}{\sigma_{k-1/2}\Delta z_{k-1/2}\Delta z_k} M^2 \quad (k = 2, 3, \ldots, N-1),$$

$$c_k = \frac{1}{\sigma_{k+1/2}\Delta z_{k+1/2}\Delta z_k} M^2 \quad (k = 1, 2, \ldots, N-2),$$

$$b_k = \begin{cases} \dfrac{1}{\Delta z_1} M + c_1 + \sigma_1 E - \sigma_{s1} S & (k = 1), \\ a_k + c_k + \sigma_k E - \sigma_{sk} S & (k = 2, 3, \ldots, N-2), \\ \dfrac{M}{\Delta z_{N-1}}(M + \sigma_{N-1/2}\Delta z_{N-1/2} E)^{-1} M + a_{N-1} + \sigma_{N-1} E \\ \quad - \sigma_{sN-1} S & (k = N-1). \end{cases} \quad (4.10)$$

Since

$$\Delta z_{k+1} a_{k+1} = \Delta z_k c_k,$$

we can show that A is symmetric. Indeed, consider the vectors a and b with the components $a_{k,l}$ and $b_{k,l}$, and define the inner product

$$(a, b) = \sum_{k=1}^{N-1} \sum_{l=1}^{m} s_l a_{k,l} b_{k,l} \Delta z_k.$$

Then it is not difficult to verify that

$$(Aw, w^*) = (Aw^*, w),$$

which says that A is symmetric on the corresponding subspace of vector functions.

The fact that A is symmetric implies that the problem

$$Aw = \lambda w \tag{4.11}$$

defines a real spectrum $\lambda(A)$.

Since the operator A is positive-definite, the spectrum is positive. Finally, since the original problem has been reduced to a problem in linear algebra, the spectrum is bounded. Thus the spectral Problem (4.11) defines the spectrum $\lambda(A)$ which is contained in the interval

$$\alpha(A) \leq \lambda \leq \beta(A).$$

The Stationary Transport Problem

Our next problem is to find the largest and smallest eigenvalues of the matrix A. For this we use the method of Section 1.1.4. Thus we find first $\beta(A)$ and then use the shift of origin method to find $\alpha(A)$ and its corresponding eigenvector w_1.

Having found the spectral boundaries, Problems (4.6) and (4.7) can be solved by an iterative method. Let us write this problem in the operator form

$$Aw = F, \tag{4.12}$$

where the vector F has been obtained in the course of reducing the problem to its form (4.12). We may choose the following iterations:

$$w^{j+1} = w^j - \tau_j(Aw^j - F), \tag{4.13}$$

where τ_j depends on the particular optimization method (see Chapter 3).

Problem (4.11) can be also solved by means of the splitting-up method described in Section 3.3. Let us introduce for this purpose the matrices

$$A_1 = \begin{Vmatrix} \sigma_{s1}(E-S) & 0 & 0 & \cdots & 0 \\ 0 & \sigma_{s2}(E-S) & 0 & \cdots & 0 \\ 0 & 0 & \sigma_{s3}(E-S) & \cdots & 0 \\ \vdots & & & & \vdots \\ 0 & 0 & 0 & \cdots & \sigma_{s,N-1}(E-S) \end{Vmatrix},$$

$$A_2 = \begin{Vmatrix} \hat{b}_1 & -c_1 & 0 & 0 & \cdots & 0 & 0 \\ -a_2 & \hat{b}_2 & -c_2 & 0 & \cdots & 0 & 0 \\ 0 & -a_3 & \hat{b}_3 & -c_3 & \cdots & 0 & 0 \\ \vdots & & & & & & \vdots \\ 0 & 0 & 0 & 0 & \cdots & -a_{N-1} & \hat{b}_{N-1} \end{Vmatrix},$$

where E is the identity block matrix; a_k and c_k are defined by (4.10);

$$\hat{b}_k = \begin{cases} \dfrac{1}{\Delta z_1} M + c_1 + \sigma_{c,1} E & (k=1), \\ a_k + c_k + \sigma_{c,k} E & (k = 2, 3, \ldots, N-2), \\ \dfrac{M}{\Delta z_{N-1}} (M + \sigma_{N-1/2} \Delta z_{N-1/2} E)^{-1} M + a_{N-1} \\ \quad + \sigma_{c,N-1} E & (k = N-1); \end{cases}$$

The matrices A_1 and A_2 satisfy the conditions

$$(A_1 w, w) \geq 0, \qquad (A_2 w, w) > 0 \tag{4.14}$$

and hence we can apply the splitting-up method, since (4.14) meets the applicability assumptions of the method, namely that one of the operators A_j must be positive.

7.5. Nonisotropic Particle Scattering

Above we have considered the transport problems with isotropic scattering. Now we are going to show that the numerical scheme, as derived, can be retained also for problems with a nonisotropic scattering function. To see this, consider the following problem:

$$\frac{1}{c} \cdot \frac{\partial \phi}{\partial t} + \mu \frac{\partial \phi}{\partial z} + \sigma \phi = \frac{\sigma_s}{2} \int_{-1}^{1} d\mu' \int_{1}^{2\pi} d\alpha' \phi \gamma(\mu_0) + f,$$

$$\phi = 0 \quad \text{for} \quad z = 0, \quad \mu > 0, \tag{5.1}$$

$$\phi = 0 \quad \text{for} \quad z = H, \quad \mu < 0,$$

where $\mu_0 = \mu\mu' - \sqrt{1-\mu^2}\sqrt{1-\mu'^2}\cos(\alpha - \alpha')$ is the cosine function of the scattering angle.

Suppose that the input data of problem (5.1) do not depend on the azimuth α. In this case the scattering function can be integrated with respect to α' (see Marchuk and Lebedev [17]). As a result we obtain the following problem:

$$\frac{1}{c} \cdot \frac{\partial \phi}{\partial t} + \mu \frac{\partial \phi}{\partial z} + \sigma \phi = \frac{\sigma_s}{2} \int_{-1}^{1} \phi \gamma(z; \mu, \mu') \, d\mu' + f,$$

$$\phi = 0 \quad \text{for} \quad z = 0, \quad \mu > 0, \tag{5.2}$$

$$\phi = 0 \quad \text{for} \quad z = H, \quad \mu < 0.$$

The function $\gamma(z; \mu, \mu_1)$ can be represented as a Legendre polynomial series

$$\gamma(z; \mu, \mu') = \sum_n \gamma_n(z) P_n(\mu) P_n(\mu').$$

Rewrite it in the following form:

$$\gamma(z; \mu, \mu') = \gamma^+(z; \mu, \mu') + \gamma^-(z; \mu, \mu'),$$

where

$$\gamma^+(z; \mu, \mu') = \sum_n \gamma_{2n}(z) P_{2n}(\mu) P_{2n}(\mu'),$$

$$\gamma^-(z; \mu, \mu') = \sum_n \gamma_{2n+1}(z) P_{2n+1}(\mu) P_{2n+1}(\mu').$$

Let us perform a transformation of (5.2) similar to that in the isotropic case.

Nonisotropic Particle Scattering

For this purpose we rewrite the transport equation into two separate equations corresponding to $\mu > 0$ and $\mu < 0$ respectively. Thus

$$\frac{1}{c} \cdot \frac{\partial \phi^+}{\partial t} + \mu \frac{\partial \phi^+}{\partial z} + \sigma \phi^+ = \frac{\sigma_s}{2} \int_0^1 [(\phi^+ + \phi^-)\gamma^+$$
$$+ (\phi^+ - \phi^-)\gamma^-] \, d\mu' + f^+, \quad (5.3)$$

$$\frac{1}{c} \cdot \frac{\partial \phi^-}{\partial t} - \mu \frac{\partial \phi^-}{\partial z} + \sigma \phi^- = \frac{\sigma_s}{2} \int_0^1 [(\phi^+ + \phi^-)\gamma^+$$
$$- (\phi^+ - \phi^-)\gamma^-] \, d\mu' + f^-,$$

$$\phi^+ = 0 \quad \text{for} \quad z = 0,$$
$$\phi^- = 0 \quad \text{for} \quad z = H. \quad (5.4)$$

We introduce now the new functions u and v by the relations

$$u = \tfrac{1}{2}(\phi^+ + \phi^-), \quad v = \tfrac{1}{2}(\phi^+ - \phi^-).$$

Adding and subtracting the obtained equations, we obtain [similarly to (3.6)]

$$\frac{1}{c} \cdot \frac{\partial u}{\partial t} + \mu \frac{\partial v}{\partial z} + \sigma u = \sigma_s \int_0^1 u \gamma^+(\mu, \mu') \, d\mu' + q,$$
$$\frac{1}{c} \cdot \frac{\partial v}{\partial t} + \mu \frac{\partial u}{\partial z} + \sigma v = \sigma_s \int_0^1 v \gamma^-(\mu, \mu') \, d\mu' + r \quad (5.5)$$

under the assumption

$$u + v = 0 \quad \text{for} \quad z = 0,$$
$$u - v = 0 \quad \text{for} \quad z = H. \quad (5.6)$$

Note that by the normalization condition on the coefficient γ, and by the fact that $\gamma^+(\gamma^-)$ is an even (odd) function, we have

$$\int_0^1 \gamma^+(\mu, \mu') \, d\mu = 1. \quad (5.7)$$

In this fashion Problems (5.5) and (5.6) have been brought to a form convenient for the realization of the numerical algorithm which we have considered above. In the present case it is only necessary to replace the operator A_k^h from (3.21) by

$$A_k^h = \left\| \begin{array}{cc} \sigma_k - \sigma_{sk} \int_0^1 \gamma_k^+(\mu, \mu') \, d\mu' & \mu \nabla_k \\ \mu \nabla_{k+1/2} & \sigma_{k+1/2} - \sigma_{sk+1/2} \int_0^1 \gamma_k^-(\mu, \mu') \, d\mu' \end{array} \right\|$$

and similarly in (3.25) to take

$$A_{1k}^k = \left\| \begin{matrix} \sigma_{sk} - \sigma_{sk} \int_0^1 \gamma_k^+(\mu, \mu') d\mu' & 0 \\ 0 & \bar{\mu}_{k+1/2} \sigma_{sk+1/2} - \sigma_{sk+1/2} \int_0^1 \gamma_k^-(\mu, \mu') d\mu' \end{matrix} \right\|,$$

$$A_{2k}^h = \left\| \begin{matrix} \sigma_{ck} & \mu \nabla_k \\ \mu \nabla_{k+1/2} & \sigma_{k+1/2}^{tr} \end{matrix} \right\|,$$

rather than A_1^h, A_2^h; here $\sigma^{tr} = \sigma_s(1 - \bar{\mu}) + \sigma_c$; $\bar{\mu}$ is a constant determined from the condition $(A^h w, w) \geq 0$.

In conclusion let us note that the above-discussed principles of constructing the difference equations of transport theory can serve as a basis for building algorithms for one-dimensional problems with spherical or cylindrical symmetry and, more importantly, for multidimensional problems of transport theory.

Chapter 8

A Review of the Methods of Numerical Mathematics

The development of large-scale computers have formed a basis for algorithmic constructions and extensive mathematical experiments in many areas of science and technology, thereby attracting a new generation of scientists to problems of numerical mathematics. The valuable experience accumulated by solving applied problems has been later used for constructing effective methods and algorithms for numerical mathematics.

In this chapter we will briefly review the fundamental directions in numerical mathematics as of the present time and indicate the trends of their development.

8.1. The Theory of Approximation, Stability, and Convergence of Difference Schemes

Wide applications of the finite-difference method in differential equations of mathematical physics have made it necessary to study in detail those properties of difference equations which have direct bearing on the performance of difference schemes, particularly the stability and convergence properties.

The development of the theory of stability and convergence started in response to the discovery that the difference scheme for a well-posed differential problem can become unstable (or ill-posed). An unstable scheme is sensitive to the round-off errors arising in the numerical process and can lead to a solution which differs considerably from the one corresponding to the differential problem.

This particular feature of difference equations triggered intensified theoretical investigations regarding the relation between the convergence on one side and the stability on the other side.

In the midfifties Lax [6, 7], Richtmyer [3, 6, 7], Ryabenkii and Filippov [6], and Meiman [4] formulated—almost simultaneously and from different viewpoints—the following fundamental result which became known as the equivalence theorem: Consider a well-posed linear homogeneous differential problem and an approximating difference scheme; in order for the solution of the difference problem to converge to the solution of the original differential problem it is necessary and sufficient that the difference scheme be stable. The final formulation of this theorem and its proof for an abstract evolution equation in a Banach space were given by Ryabenkii and Filippov [6]. Richtmyer [3] generalized the equivalence theorem to the case of nonhomogeneous linear differential equations. The equivalence theorem is formulated in terms of one single norm. Convergence with respect to other norms can be established on the bases of the imbedding theorems of Sobolev [1]. With the initial data having stronger smoothness assumptions, the requirement for stability of the scheme can be weakened, as was originally pointed out by Ryabenkii and Filippov [6]. This idea has been later reflected in the equivalence theorem due to Strang [7] in connection with the concept of weak stability.

Regarding the choice of efficient stability criteria one must primarily refer to Neumann and Richtmyer [7]. They formulated the so-called local criterion of stability. This criterion however holds true only for equations with constant coefficients describing selfadjoint problems. This circumstance prompted intensive search for the limits of applicability of the local criterion.

Lax and Nirenberg [6, 7] have developed a stability theory for hyperbolic difference schemes in terms of the so-called symbol of the difference scheme. In the case of explicit difference equations the symbol coincides with the usual transition matrix obtained by the Fourier method; here the local stability criterion holds true if the coefficients have bounded second derivatives with respect to x.

Strang [7] has formulated a convergence theorem for systems of quasi-linear hyperbolic equations under the condition of local stability of the difference equations corresponding to the first variation of the differential system and also assuming a sufficient smoothness of the solution.

The development of difference schemes with variable coefficients is associated with the concept of dissipativity. Here we note first of all the work by Kreiss [6]. His theorems relate the order of dissipativity of the difference equations approximating systems of hyperbolic equations to the order of their accuracy. It is assumed that the matrix coefficients of the difference equations are Hermitian and Lipschitz continuous in x.

An interesting approach to the problem of stability has been given by Yanenko and Shokin [7]: instead of the difference equation we consider some accompanying differential equation, the so-called first-differential approximation, which, being well-posed, implies the stability of the difference scheme.

A very important class of difference schemes consists of those with positive coefficients; it was considered by Friedrichs [2]. He introduced a general concept of positive schemes and established an L^2-stability criterion.

Godunov and Ryabenkii [6, 7] have introduced the notion of a spectrum of a family of difference operators. This concept allows them to form necessary conditions for stability of difference schemes which nicely expose the roots of instability. A new concept, the kernel of the spectrum of the family of difference operators, was introduced.

The above authors gave estimates of the norms of powers of the operators from the family in terms of the radii of the spectral kernels. These estimates turn out to be uniform on the family and can be conveniently used for stability investigations.

All the stability criteria we have considered so far can be classified as spectral criteria, since they are based on the spectral properties of difference operators. These criteria can be used to establish L^2-convergence. It is preferable, however, to have the convergence in norm of the C-space. The related results can be found in Serdyukova [6], Thomee [6], Samarskii [7], and others.

Among the spectral approaches to the stability problem regarding the difference analogues of parabolic or hyperbolic equations we point out a fairly general theory due to Samarskii [3, 6, 7], based on energy inequalities and *a priori* estimates. This theory contains necessary and sufficient conditions of stability for a wide class of two-layer and three-layer schemes. The conditions have the form of inequalities relating the coefficients operators of the difference schemes. They are very constructive, and in addition to the stability analysis they permit one to devise new stable schemes.

The energy method is based on the concept of strong stability. The idea of the method is to choose a norm, in which the vector solution grows from step to step no faster than $1 + O(\Delta t)$, which means stability in that norm. Later on, the proof is reduced to showing the equivalence of this norm to L^2-norm.

The energy method goes back to Courant, Friedrichs, and Lewy [7]. It was successfully developed by Ladyzhenskaya [7], Lees [7], Lax [7], Kreiss [6], Samarskii [3], Konovalov [15], and others.

The theory of boundary-value difference problems has not reached the completeness of the Cauchy problem. The reason is that, as a rule, Fourier analysis is not applicable in this case.

The development of the theory of approximation and convergence in the general framework of functional analysis has been provided by Kantorovich and Akilov [1] who considered wide classes of operator equations with a special emphasis on the problem of numerical solution of integral equations.

Of importance for the theory of convergence is the closure theory due to Sobolev. It is widely used for the purposes of theoretical justification of approximation methods in mathematical physics.

8.2. Numerical Methods for Problems of Mathematical Physics

The concepts of approximation, stability, and convergence have provided the necessary basis for wide research of efficient difference schemes for problems of mathematical physics. The algorithms of the finite-difference methods combine, as a rule, constructions of the difference analogues and methods for their solutions. Therefore the advances in the constructive theory of finite difference methods depend on a mutually coordinated development of the two fields.

In trying to summarize the vast growth in the development of finite difference methods in the recent years, we will provisionally distinguish the following important trends.

8.2.1. Constructions of Difference Schemes

A particular trend is related to the development of methods for constructing conservative difference schemes based on the laws of conservation, characteristic of the majority of physical processes. The starting point for devising conservative difference schemes is to write the balance equations for a single cell of the net region and to apply subsequently the quadrature and interpolation formulas. Performing then the necessary transformations and summing over the net region, we obtain equations satisfying the integral laws of conservation.

Such approaches have been considered by Ladyzhenskaya [7]. She constructed difference operators with discontinuous coefficients which have an identical form for any internal point of the region. In order to get a solid theoretical base for the algorithms, she uses the concept of generalized solutions and proves that the solution of the difference problem is represented by a certain functional converging to the functional of the differential problem as $h \to 0$.

A class of conservative difference schemes in hydrodynamics, based on explicit difference approximations, have been considered by Godunov [4] and Lax and Wendroff [6]. Of considerable importance for hydrodynamical problems has been the method of integral relations (suggested by Dorodnitsyn [3] and developed by Belotserkovskii, Chushkin, and others), which uses a partial difference approximation of the equations in a divergence form, based on the method of straight lines. These methods played a fundamental role in forming the general approach to difference schemes for quasilinear equations. Interesting general approaches to problems of hydrodynamics have been also suggested by Babenko and Rusanova [12], Fromm [4], Crowley [4], Kuropatenko [4].

The above methods as applied to elliptic and parabolic equations with discontinuous coefficients have been studied by Tikhonov and Samarskii [4] and others, by means of the integrointerpolation method.

8.2.2. Variational Methods

In recent years there has been a marked interest in variational methods for problems of mathematical physics. The variational methods of Ritz, Galerkin, Treftz, and others have long occupied an important place in numerical mathematics. These methods are especially effective for computing functionals of the solutions. It has turned out that functionals can be obtained highly accurately even for comparatively poor approximations. The most complete justification of these methods has been given by Mikhlin [1], who has established the necessary and sufficient conditions of stability of the variational methods for problems with energy norms. The active development of variational methods has also displayed some of their deficiencies related to the difficulties in constructing the test functions which would reflect special features of the solution of the problem, and which would give satisfactory approximations of the solution when taken in small numbers.

A new stage in the development of variational methods was marked by the active deployment of computers, when it became possible to obtain test functions for a variational problem by solving simpler problems using difference methods. It turned out that in a number of cases one could effectively exploit successive approximations of the solution for constructing test functions in the very course of the iteration process and thus combine the successive approximations method with the variational approach, thereby accelerating considerably the process of the solution.

A new twist in methods of constructing difference equations of mathematical physics has been brought about by combining variational methods with a special design of test functions, taken identically zero outside some rather small regions belonging to the domain of definition of the solution. The first results of Courant [5], Oganesyan and Rukhovets [5], Lions and Temam [5], Aubin [5], Birkhoff, Schultz and Varga [5], Bramble and Schatz [5], and others stimulated a development of the ideas which have fundamentally enlarged the applicability of variational methods to problems with selfadjoint and nonselfadjoint operators, and increased interest in schemes of high-order accuracy based on finite element methods, spline approximations, etc. The development of these methods is due primarily to Babuška [5], Strang and Fix [5], Zlámal [5], Douglas and Dupont [5], Rivkind [4], and others.

8.2.3. Multidimensional Stationary Problems

An intensive development of methods for solving linear algebraic equations with Jacobi and tridiagonal block matrices resulted in a number of excellent numerical algorithms for stationary processes based on the factorization of the corresponding difference operator. We mention in particular various methods of matrix factorization studied by Keldysh, Gel'fand and Lokutsievskii [12], Babenko and Rusanov [12], Chentsov and Godunov [12], Abramov [12], and others.

At the same time that the exact factorization methods were being developed, there has been a development of approximate factorization, in which the factorization of the operator is combined with the successive approximation method. A need for such algorithms emerged as early as the problems of mathematical physics were being reduced to large algebraic systems. The first results of Buleev [15] and Baker and Oliphant [15] triggered the development of new methods for the multidimensional problems based on the fast-converging processes.

The early sixties were marked by a major contribution in computational mathematics associated with the names of Douglas, Peaceman, and Rachford, who suggested the so-called alternating-direction method. The success of the method was ensured by the use of a simple reduction of a multidimensional problem to a sequence of one-dimensional problems with Jacobi matrices easily manageable by a computer. Essentially the method can be viewed as an iteration method in which the computations are optimized by a special choice of the contraction operator consisting of a product of simpler operators and a number of free relaxation parameters. The succession of inversions of the simple operators is implemented by the linear factorization method. Such iteration schemes are very economical and effective. An increase in the amount of numerical operations is insignificant in comparison with the explicit method of Richardson. The alternating-direction method has considerably influenced constructions of algorithms in various areas of applied mathematics and the investigations of nonlocal and block-iterative processes. The theoretical investigations devoted to this method can be found in Douglas [15], Birkhoff, Varga and, Young [15], Wachspress [15], Kellogg [15], Vorobyev [15], and others.

New methods are being developed by considering homogeneous and nonhomogeneous approximations. In the case of nonhomogeneous approximations, any of the auxiliary problems may not possibly approximate the original problem; but as a whole, and using special norms, such approximations hold good. These methods have become known as the splitting-up methods. They have been studied by the Soviet mathematicians Yanenko [3, 15], Dyakonov [3, 15], Samarskii [3, 15], Saulyev [3], Marchuk [15], and others.

A large volume of research regarding the splitting-up methods has been devoted to choosing the parameters to be optimized by means of spectral or variational techniques.

Various aspects of the alternating direction method and the splitting-up method have been considered in papers by Andreev [15], Widlund [15], Fairweather and Mitchell [15], and others.

8.2.4. Multidimensional Nonstationary Problems

Experience with the one-dimensional problems has formed a basis for devising algorithms for more complex problems of mathematical physics. An important stage in the development of the methods for nonstationary two-dimensional problems is the alternating-direction method based on the homogeneous approximations. Originally the method was applied to multidimensional parabolic equations, and it since has been widely used in many problems of mathematical physics.

Further advances in multidimensional nonstationary problems are related to the splitting-up techniques, based as a rule on the nonhomogeneous difference approximations of the original problem. Essentially, the splitting-up method consists in a reduction of a complicated operator to simpler operators. Thus the problem of solving a given equation is replaced by a sequence of problems with a simpler structure. Only at the end must the difference schemes satisfy at the same time the conditions of approximation and stability. This allows flexibility in constructing the schemes for virtually all fundamental problems of mathematical physics. In the case of explicit schemes the splitting-up method was suggested by Godunov and Bagrinovskii [15]. In the implicit case splitting-up schemes have been given by Yanenko [15], Dyakonov [15], Samarskii [15], and others. These methods have found wide applications in diverse problems and have stimulated a more general approach to problems of mathematical physics on the basis of the so-called weak approximation, introduced by Yanenko [3, 15] and Samarskii [3, 15]. It has turned out that the splitting-up method can be viewed as the weak approximation of the original equation by a simpler one. Convergence criteria for the weak approximation method are given by the Yanenko–Demidov theorem [15] and can also be found in Lebedev [15] and Dyakonov [3, 15]. The method has found natural applications in problems of hydrodynamics, meteorology, oceanology, radiative transfer theory, etc. (See Marchuk [3, 17] and Yanenko [3].)

Another original scheme, widely applied in hydrodynamics, meteorology, and oceanology is the predictor-corrector method of Lax and Wendroff, in which the corrector is represented by an explicit difference scheme. This scheme is conditionally stable, has a very simple realization, and is second-order accurate with respect to all the variables. The detailed exposition is given in the book by Richtmyer and Morton [3].

Various variants of the predictor-corrector method, as based on implicit difference approximations, have been introduced by Bryan [4], Douglas [15], Sofronov [12], Marchuk and Yanenko [15], and others. It turns out that all these schemes are in a sense equivalent, differing only by their realizations. In the last of the sources just quoted the predictor is taken as an implicit splitting-up scheme with first-order accuracy and with the factorized operator. In problems of hydrodynamics the predictor is usually taken as an implicit dominated scheme.

Of particular interest is the method of decomposition and decentralization formulated by Lions and Temam [5]. It is similar to the splitting-up methods and to weak approximations.

The Particles-in-the-cell method. Recently there has been an intensive development of a new method for solving multidimensional problems of mathematical physics heralded by Harlow [19]. This method has become known as the method of large particles. It is widely used for computing the multidimensional hydrodynamical flows of highly compressible liquids with large relative displacements and colliding separation surfaces. The essence of the method is as follows: Using a weak approximation at every small time interval, the equations of hydrodynamics are reduced to two simpler systems. The first system describes the mutual adaptation of the hydrodynamical fields with no account of the advection terms and is integrated by usual methods on a fixed Euler net. The second system describes the transport of the fluid in a Lagrangian coordinate system. In order to solve the second system we use a phenomenological simplification of the fluid model, replacing it by a system of particles in every cell of the Euler system so that the overall balance of the mass, impulse, and energy of the particles in the cell coincides with the corresponding characteristics of the fluid. As soon as the trajectory (computed individually) of a particle, "bearing" certain mass, crosses the borders of a cell, the mass, impulse, and energy corresponding to this particle are subtracted from the abandoned cell and added to the new cell in which the particle is now located. The Harlow scheme is based on explicit solution methods at the first and second stages. It is globally conditionally stable and is especially fruitful at the first stages of the implicit schemes. In this case the stability criterion of the overall system coincides with the well-known Courant condition. Absolutely stable schemes of the particle method have not been obtained so far. The next few years should bring progress in this regard.

Recently Dyachenko [19], Belotserkovskii and Davydov [19], and Yanenko, Anuchina, Petrenko, and Shokina [19] have given various modifications of this method, which considerably reduce the density and the damping fluctuations characteristic of the method with improved stability. The above authors have also considered various realization schemes.

It is hoped that absolutely stable methods in the absence of the fluctuations will permit one to extend the applications of this technique to low-compressible liquids and to multidimensional problems in general.

The Monte Carlo method. This method, suggested by Neumann and Ulam, has been in development for more than two decades. The initial optimism for this method has eventually yielded to an equally unjustified pessimism. The heart of the matter is that it became clear at the early stages of the development that the efficiency of the Monte Carlo method would depend on computer speed. Millions of operations per second are required to implement the statistical tests for reducing the mean-squares error of the result.

Nevertheless, in spite of the difficulties in implementing this method on an average-size computer, or possibly thanks to them, the theory of the method has been significantly improved and its efficiency increased. The most important improvements have been achieved by bringing in the conditional probabilities of the processes and the statistical weights, defined by means of the information about the solutions of the adjoint equations related to the key functionals of the problem. Such methods reduce the dispersion error by an order of one or even two, and they consequently cut down the required solution time by an order of one or two as compared to methods of direct statistical modelling.

The third-generation computers have now brought in the necessary basis for the reactivation of this method. The Monte Carlo method has now a solid position in the theory of radiative transfer, public-service systems, quadrature and interpolation processes, integral equations, and algebraic equations. Lately the method has been tried on the nonlinear Boltzmann equation, for problems of linear programming etc.

In connection with the Monte Carlo method important contributions were made by Vladimirov and Sobol [18], Chentsov [18], Fano, Spencer and Berger [18], Ermakova, Zolotukhina [18], Mikhailova [18], Buslenko and Golenko [18], and others. Simple and universal as it stands, the Monte Carlo method will become undoubtedly an important tool of computational mathematics.

8.3. Conditionally Well-Posed Problems

An important role in the numerical treatment of problems of mathematical physics is played by the concept of well-posed problems, a concept first introduced by Hadamard at the beginning of the century. While there are a variety of classical problems of mathematical physics which are well posed in the sense of Hadamard, a deeper understanding of various problems in natural sciences and technology requires solving the so-called conditionally well-posed problems. Tikhonov [16] has formulated natural requirements on the ill-conditioned (in the sense of Hadamard) problems. Essentially, these conditions require that the solution of the problem be *a priori* assumed to exist in a given compact. In order to establish well posedness, it is necessary to prove uniqueness.

Well-posed problems have been extensively studied by Lavrentiev [16]. Various aspects in the theory of the conditionally well-posed problems of mathematical physics have been considered by John [16], Mergelyan [16], Douglas [16], Krein [16], and others.

Tikhonov [16] has introduced the concept of regularization; that is, in place of the unbounded operator which gives the exact formula for the solution of the ill-posed problem, we consider a sequence (the regularization family) of continuous operators such that, on every element belonging to the existence domain, the corresponding sequence converges to the solution.

An interesting approach to ill-conditioned problems is provided by the theory of probability. Such investigations in their most complete form can be found in Lavrentiev and Vasyl'ev [16]. In the framework of probability one defines the concept of stability and constructs the optimal (in a sense) algorithms for various classes of problems, under some assumptions on the probabilistic properties of the input data errors and of the required solutions.

Lattes and Lions [16] have formulated a numerical method for inverse evolution problems using quasi-inversion. In their method, the evolution equation is supplemented by an additional regularization operator with a small parameter. This operator coincides with the product of the original operator and its adjoint. The small parameter is chosen by considering certain optimal estimates. The method is very simple as far as its realization is concerned.

In a joint paper with Atanbayev [16], the present author has developed a method for solving conditionally well-posed problems of the evolution type by means of the residual method on the whole space-time domain of definition of the solution. In this method one regularizes the problem by choosing the optimal number of steps of the iterative process on the basis of an *a priori* error estimate for the input data.

As evidenced by the tendencies in solving conditionally well-posed problems, the techniques used are closely related to those of optimization of numerical processes.

8.4. Numerical Methods in Linear Algebra

There is an ever increasing interest in solving large systems of linear algebraic equations with both sparse and dense matrices, in solving ill-conditioned systems, and in spectral problems for matrices of arbitrary structure. At the same time considerable attention has been paid to the possibility of exploiting the *a priori* as well as the *a posteriori* information about the problem in the course of the solution process. Computer technology has greatly influenced the process of reviewing the old linear algebraic methods and has stimulated interest in new computer-oriented algorithms.

8.4.1. Direct Methods of Linear Algebra

By direct methods of linear algebra we usually mean methods which solve a given problem in a finite number of arithmetic operations. Direct methods play a prominent role in solving systems of linear equations, inversions of matrices, and computing determinants. Direct methods can be used to decompose the original matrix into a product of two matrices each of which is easily invertible.

A classical example is the Gauss elimination method. Other distinguished examples fall under the heading of conjugate gradients methods: the method of conjugate gradients of Hestenes and Stiefel [11] and the residual method of Lanczos [3]. We can also trace back to the methods based on orthogonalization.

Lately there has been a considerable development of direct methods by Faddeev, Faddeeva and Kublanovskaya [8], Bauer [8], Householder [3], Wilkinson [8], Henrici [4], Forsythe and Moler [8], Golub [9], Voevodin [8], and others.

Ill-conditioned systems still remain a problem. They are closely connected with the conditionally well-posed problems of mathematical physics. The difficulty here is the strong sensitivity of the solution to the errors in the matrix and in the right side of the system. Although there is already a number of significant results, we are still at the start of an extensive scientific development aimed at a general theory.

8.4.2. Iterative Methods

Iterative methods represent an important tool for solving problems of linear algebra. Their development has led to a number of good, computer-oriented algorithms. This progress was spurred first of all by the necessity to deal with problems of mathematical physics, economics, and control theory, which reduce eventually to high-order systems with matrices of special kinds. In such cases the direct methods are more or less powerless, although the progress in hardware increases their chances.

At present we can distinguish certain directions in the construction of iterative methods. We restrict our attention to only two of them. The first relies on the spectral characteristics of the operators involved. Methods of this type can be described as follows: One constructs an iterative process with the transition matrix depending on a set of parameters. These parameters can be taken either identical for all steps and chosen by minimizing the spectral radius of the transition matrix; or they are chosen at each step in such a way that the error vector approaches zero as fast as possible, uniformly in the initial approximations. In both cases we use the *a priori* information about the spectra of the corresponding matrices. The choice of such parameters is an integral part of the optimization of the numerical algorithm. As a rule, the main problem is in finding the corresponding spectral bounds.

Advances in spectral optimization of numerical algorithms have stimulated the formulations of a number of problems as exemplified below. It is to be kept in mind that the spectral optimization methods are especially convenient when dealing with a whole family of problems having the same operator and different input data. Increasingly greater interest has been recently paid to the Lanczos transformation [3] of matrices of arbitrary structure; it takes the original problem into one with a symmetric matrix, the spectrum of which consists of two intervals symmetrically placed about the origin. Such a symmetrization (along with some other techniques) can be used to accelerate convergence by exploiting the polynomials with the smallest deviation from zero on the above intervals. Another problem arises in connection with optimization of processes involving matrices with the spectra consisting of many intervals. Here we need first to develop methods for determining these intervals.

The second direction makes use of variational principles. Methods of this class are used to implement a sequential minimization of certain functionals (quadratic as a rule), the minimum of which is achieved at the required solution. The foundation of the variational approach have been laid by Kantorovich [11], Lanczos [3], Hestenes and Stiefel [11],

Krasnoselskii and Krein [11], and others. Among the recent contributions we note Petryshin [9, 10], Forsythe [11], Danial [11], Marchuk and Kuznetsov [3, 11], Godunov and Prokopov [11], Lebedev [9], and others.

The advantages of the variational methods of the steepest descent type and the iterative processes of minimal residuals stem from the fact that the relaxation parameters can be chosen by using the *a posteriori* information generated at each step. The rate of convergence of these methods does not drop below the one corresponding to Chebyshev methods. It is also very important that such methods converge, disregarding whether the matrix is symmetric or not (assuming it is positive). A number of efficient residual-type methods have been proposed recently.

An important circumstance which is still delaying development of nonstationary variational methods is the necessity to store a larger amount of transient information, much larger than that for the Chebyshev optimization methods.

Recently there is a trend to combine the spectral and variational techniques. Lebedev has formulated the conditions under which the iterative process has a nonimprovable estimate of the number of arithmetic operations required. There is also a probabilistic technique of choosing the optimal number of iterations. A number of interesting results in this area have been obtained by Vorob'ev [9]. The overrelaxation method of Young and Frankel [10] did not loose its great appeal, despite its becoming a classic. This method has been generalized by Wasov and Forsythe [3], Varga [3], Isaacson and Keller [3], and others. A systematic review of the iteration methods can be found in the book by Marchuk and Lebedev [17].

Let us next turn to the iterative processes for the full eigenvalue problem with general matrices. We consider only the power methods since this trend has been recently enriched by some significant results due to Wilkinson [8], Bauer [8], Collatz [3], Voevodin [8], Frencis [8], Kublanovskaya [8], Eberlein [8], and many others.

The power methods are based on sequential transformations of the original matrix to one with easily obtainable eigenvalues, such as diagonal or triangular matrices, or a block-triangular matrix (the diagonal blocks being at most second order). For this purpose one can use either unitary similarity transformations (Jacobi method, QR-algorithm) or similarity transformations with triangular matrices (LR-algorithm).

Until recently the existing algorithms for the eigenvalue problem, such as the Jacobi algorithm (see Faddeev and Faddeeva [8] and Rutishauser [10]) required symmetric matrices. The discovery of the QR-algorithms (see Kublanovskaya [8] and Francis [8]) and the generalized method of rotations (see Vocvodin [8]) permit one to consider problems with arbitrary matrices. At present the emphasis is put on the development of various modifications of the QR-algorithm.

Advances in the eigenvalue problem are also due in part to the research related to nuclear reactor design. From this comes the interest in iterative methods for solving partial eigenvalue problems with non-negative matrices. The foundations of these methods were built by Perron and Frobenius, and considerably developed by Varga [3], Traub [8] and others.

8.4.3. Round-Off Error Analysis

Until recently computational methods were judged by the number of arithmetic operations and the memory size required for their realization. Now we have to consider one more factor, namely the accuracy. In other words, the analysis of round-off errors has become an integral part of the algorithm.

The development in this area has been initiated by J. von Neumann. A systematic study regarding round-off errors has been done by Wilkinson [8]. The basis of his mathematical apparatus is the method of equivalent perturbations, which is used for obtaining the norm-estimates of all linear algebraic transformations. He has also obtained the norm-estimates of the equivalent perturbations for many methods.

The development of the equivalent perturbations method has been paralleled by that of statistical round-off error analysis. The origins of the latter are traced to Bakhvalov [8], Voevodin [8], Kim [8], and others. Statistical methods will play unquestionably an important role in the analysis of round-off errors in numerical schemes.

8.4.4. Complexes of Standard Programs

Achievements in numerical linear algebra have prompted the development of high-performance standard programs for solving systems of linear equations and finding eigenvalues. The journal *Numerische Mathematik*, for instance, has already come up with a large number of various procedures which are widely used for solving general problems of linear algebra, as well as special problems of mathematical physics, economics, and procedures related to matrices of special kinds.

This problem will surely attract the attention of researchers; the goal should be to develop a universal numerical system for solving problems of linear algebra. There are at least two trends already in progress in the pursuit of this goal. One is a thorough elaboration of complexes of algorithms and programs for the general problems; the other consists in developing universal methods adaptable to the special features of the given families of problems. Both trends are extremely interesting since they pave the way to a universal, problem-oriented software for the computers of fourth and later generations.

8.5. Optimization Problems in Numerical Methods

An important goal of computational mathematics is to find the fastest and most economical algorithms, in other words, optimization of numerical algorithms. In dealing with optimization problems under some given constraints, one usually has to rely on general mathematical theorems and estimate the lowest possible cost of solving a specific problem from a given class. Solution of a single isolated optimization problem usually does not give any practical answer. Nevertheless, if one knows how to find the conditionally extremal solution, i.e., the best technique of solving the local problem given the numerical possibilities and means, then we actually have the general problem in our hands. This concept of numerical optimization has been formulated by Babuška and Sobolev [20], and it sufficiently reflects the essence of the present problem.

Yet in many cases the optimal algorithm cannot be constructed, although it may be possible to construct a nearly optimal one. This situation is typical, for instance, of asymptotically optimal algorithms. We note that at the present time the theory of asymptotic estimates is in fact an effective tool for optimizing algorithms for various classes of problems.

The most developed subject from the computational point of view is unfortunately the theory of quadrature formulas due to Sobolev [1, 20] and Babuška [20]. Here the problem of estimating the quadrature formulas has been reduced to the minimization of a linear functional of errors. The above authors have obtained the quadrature formulas estimates on classes of periodic functions and infinitely differentiable functions. The methods are based on the asymptotic approximation estimates. The theory of quadrature formulas is also the subject of the contributions by Bakhvalov [20] and Sobol; they consider the optimal convergence estimates for the quadrature processes, the integration methods of the Monte Carlo type, and also the problem of designing the best techniques of numerical integration.

A different, number theoretical approach to quadrature formulas has been suggested by Vinogradov [2] and further pursued by Korobov [20]. The formulas obtained are exact on finite trigonometric polynomials. The loss estimates are derived on the class of periodic functions.

Kolmogorov [20] has introduced a number of general set-theoretical notions which can be used to estimate the bounds for the amount of numerical operations needed for solving a given problem. These estimates are of special importance for finding algorithms in the case where the upper and lower asymptotics diverge. He gave an estimate for problems involving linear differential operators with compact inverses. His estimate permits one to find algorithms which are asymptotically close to the optimal algorithms (in the sense of the number of arithmetic operations).

Bakhvalov [20] has studied a complex of algorithms for problems of mathematical physics by means of finite difference methods. In particular he has given lower estimates on the amount of operations needed for the solution of the Dirichlet problem for the Laplace equation. By a special choice of the net one can get also an upper estimate. The above kinds of estimates have been also obtained for the Peaceman–Rachford method [15] and the Douglas–Rachford method [15]. The method of Young and Frankel [10] and some other have also been considered.

An interesting trend in optimization has been developed by Lebedev

[9, 15], by considering various difference methods. For fundamental minimization functionals he takes the so-called value of the algorithm and the entropy. The method has been used in transport theory.

In optimization problems we often have to abstract from many factors such as rounding off of numbers in the realization process, various details in the implementation of arithmetic operations in the registers of a particular computer, etc. On the other hand, these factors sometimes determine the efficiency of the algorithm at hand. But this takes us to optimization problems regarding the computational process itself.

The theory of computational processes and their optimization have been dealt with in many contributions by Babuška [20], Dahlquist [20], Henrici [3], and others. Babuška, Vitásek, and Práger [3] have introduced the concept of α_k-sequences of numerical processes, reflecting thereby the fact that the accuracy of the computations should increase as a power of the increasing length of the computational process. In connection with the theory of α_k-sequences the concepts of local and global stability of numerical processes have been introduced. These concepts allow one to analyze large classes of actual algorithms of numerical mathematics.

Moor [20], Nickel [20], and others have introduced recently a new trend in estimating the accuracy of an actual algorithm; it has become known as interval arithmetic. The main purpose of interval arithmetic is to obtain *a posteriori* round-off-error estimates which can be analyzed by a double procedure on a single computer.

8.6. Some Trends in Numerical Mathematics

Advances in computer technology in recent years have had a significant effect on many areas of numerical sciences which show merging tendencies. The relationship between the hardware, methods of numerical and applied mathematics, the theory of automatic programming, and the theory of languages has become so close that choosing a strategy for solving a specific problem is presently of paramount significance. Although the optimization of the individual components of the numerical process is the core of the theory, as before, the attention becomes ever increasingly focused on the optimization of the process as a whole.

The optimization of numerical processes is presently, without any doubt, one of the central problems of computational sciences; it stimulates explorations of new numerical algorithms and methods of their realizations.

The next trend is related to the shift from solving particular problems to solving classes of problems and the standardization of algorithms. There is a need for systematization and ordering of the large flow of the computer-processed information. The valuable experience accumulated by solving problems of sciences and technology makes it possible in many cases to aim at universal methods suitable for handling wide classes of problems of the same type. A rational strategy in this direction depends on what class of problems we deal with. For diverse and rarely repeated problems we need to construct universal numerical algorithms which would adapt to the optimal regime by using the *a posteriori* information. In the case of frequently repeated problems, a reasonable strategy is a careful elaboration of special algorithms. These two approaches complement one another and form a basis for an economical exploitation of resources in creating an effective software system. The first steps in the development of the theory of universal optimal algorithms, selfadjusting to the optimal (in some sense) regime, have already been made, as well as the projections regarding the further development.

The computer systems of new generations, featuring high speeds and large memories, are becoming an effective storage place of the valuable and immediately accessible information. In addition, the multi-access systems permit one to materialize new dialog forms of the man-computer interaction. Therefore, the standardization of software in general and the numerical algorithms in particular is becoming an urgent problem.

The development in software confronted the numerical mathematics with a number of new problems, such as the problem of constructing a net for a complicated region, which would in some sense uniformly cover the domain of definition of the solution of the given problem. While this problem is nearly solved in the case of two-dimensional regions, it is still in a beginning stage in three and more dimensions. The problem of information inputs and outputs on a computer has also created a number of problems, namely with regard to the graphical representations of the information. This triggered, for instance, the development of new interpolation methods on various classes of functions.

Computer achievements in the area of analytic mappings have facilitated practical possibilities for solving problems of mathematical physics using well-developed methods of the theory of continuous argument. These methods will penetrate more and more extensively into software as the necessary hardware for implementing the analytic operations grows. Advances in the computer realizations of analytic transformations will bring new possibilities in the development of effective methods for problem solving.

Finally let us note that the rate of growth in the development of numerical mathematics is determined by the level of research in the fundamental areas of mathematics. The importance of basic research and the rate with which it develops have increased significantly in the era of technological progress. Only harmonious research in all branches of mathematics will create necessary and favorable conditions for the spontaneous development of mathematics and its applications.

References

1. Functional Analysis and Numerical Mathematics

Collatz, L.: *Functional Analysis and Numerical Mathematics*. New York: Academic Press, 1966.
Kantorovich, L. V.: Functional analysis and applied mathematics. *Usp. Math. Nauk*, **3**, 6 (1948) [Russian].
Kantorovich, L. V. and Akilov, G. P.: *Functional Analysis in Normed Spaces*. New York: Macmillan, 1964.
Keldysh, M. V. and Lidskii, V. B.: On the spectral theory of non-selfadjoint operators. In: *Trudy IV Vsesoyuz. Mat. S'ezda*, Vol. 1, Moscow, AN SSSR, 1963 [Russian].
Kolmogorov, A. N., Fomin, S. V.: *Introductory Real Analysis*. Englewood Cliffs, N.Y.: Prentice-Hall, Inc., 1970.
Krasnosel'skii, M. A., Vainikko, G. M., Zabreiko, P. P., Rutickii, Ya. B., and Stetsenko, V. Ya.: *Approximate Solutions of Operator Equations*. Moscow: Nauka, 1969 [Russian].
Krein, S. G.: *Linear Differential Equations in Banach Space*. Providence, R.I.: American Mathematical Society, 1971.
Lavrentiev, M. A. and Shabat, B. V.: *Methods of Functions of a Complex Variable*. Moscow: Fizmatgiz, 1965 [Russian].
Lions, J. L.: *Equations Différentielles Opérationelles*. Berlin–Göttingen–Heidelberg: Springer-Verlag, 1961.
Liusternik, L. A. and Sobolev, V. I.: *Elements of Functional Analysis*. New York: Ungar, 1964.
Mikhlin, S. G.: *Variational Methods in Mathematical Physics*. New York: Macmillan, 1964.
Natanson, I. P.: *Constructive Function Theory*. New York: Ungar, 1964.
Nikol'skii, S. M.: *Approximation of Functions of Many Variables and Imbedding Theorems*. Moscow: Nauka, 1969 [Russian].
Sobolev, S. L.: *Applications of Functional Analysis in Mathematical Physics*. Providence, R.I.: American Mathematical Society, 1963.
Sobolev, S. L.: *Lectures on Two-Dimensional Numerical Integration Formulas*. Novosibirsk: Izd. Novosib. Univ., 1964, 1965 [Russian].
Varga, R.: *Functional Analysis and Approximation Theory in Numerical Analysis*. Philadelphia: The Society for Industrial and Applied Mathematics, 1971.

2. Partial Differential Equations and Mathematical Physics

Bitsadze, A. V.: *Boundary Value Problems for Elliptic Equations of the Second Order*. Moscow: Nauka, 1966 [Russian].
Courant, R.: *Partial Differential Equations*.
Courant, A., Hilbert, D.: *Methods of Mathematical Physics*. New York: Interscience, 1953.
Friedrichs, K.: Non-linear hyperbolic differential equations for functions of two independent variables. *Amer. J. Math.*, **70** (1948).
Godunov, S. K.: *Equations of Mathematical Physics*. Moscow: Nauka, 1971 [Russian].

Kantorovich, L. V. and Krylov, V. I.: *Approximate Methods of Higher Analysis.* Moscow–Leningrad: Fizmatgiz, 1962 [Russian].
Lavrentiev, M. A.: *Variational Methods for Boundary Value Problems for Equations of Elliptic Type.* Moscow: Izd. AN SSSR, 1952 [Russian].
Lavrentiev, M. M.: *On Some Ill-Posed Problems of Mathematical Physics.* Novosibirsk: Izd. Sib. Otd. AN SSSR, 1962 [Russian].
Ladyzhenskaya, O. A.: *Mixed Problems for Hyperbolic Equations.* Moscow: Gostekhizdat, 1953 [Russian].
Lions, J. L.: *Quelques Méthodes de Résolution des Problèmes aux Limites non Linéaires.* Paris: Dunod, 1969.
Lions, J. L. and Magenes, E.: *Non-Homogeneous Boundary Value Problems and Applications.* Berlin–New York: Springer-Verlag, 1972, 1973.
Petrovskii, I. G.: *Lectures on Partial Differential Equations.* New York: Interscience, 1954.
Rozhdestvenskii, B. L. and Yanenko, N. N.: *Systems of Quasi-Linear Equations.* Moscow: Nauka, 1968 [Russian].
Smirnov, V. I.: *Lectures in Higher Mathematics,* Vols. 1–5. Moscow: Gostekhizdat, 1948 [Russian].
Sobolev, S. L.: *Equations of Mathematical Physics.* Moscow: Nauka, 1966 [Russian].
Tikhonov, A. N. and Samarskii, A. A.: *Equations of Mathematical Physics.* Moscow: Nauka, 1966 [Russian].
Vekua, I. N.: *New Methods for Elliptic Equations.* Moscow: Gostekhizdat, 1948 [Russian].
Vladimirov, V. S.: *Equations of Mathematical Physics.* Moscow: Nauka, 1967 [Russian].

3. Numerical Methods for Solving Differential Equations

Collatz, L.: *The Numerical Treatment of Differential Equations.* Berlin: Springer-Verlag, 1960.
Babuška, I., Vitásek, E., and Práger, M.: *Numerical Processes for Solving Differential Equations.* New York: Interscience, 1966.
Balakrishnan, A. V. and Neustadt, L. W.: *Computing Methods in Optimization Problems.* New York: Academic Press, 1964.
Bellman, R., Kalaba, R., and Lockett, J.: *Numerical Inversion of the Laplace Transform.* New York: American Elsevier, 1966.
Berezin, I. S. and Zhidkov, N. P.: *Numerical Methods,* Vols. 1–2. Moscow: Fizmatgiz, 1962, 1966 [Russian].
Cea, J.: *Optimization Théorie et Algorithmes.* Paris: Dunod, 1971.
Dorodnitsyn, A. A.: On a numerical method for certain non-linear problems of aerodynamics. In: *Trudy III Vsesoyuznovo Mat. S'ezda,* Vol. 2. Moscow: Izd. AN SSSR, 1969 [Russian].
Dorodnitsyn, A. A.: *Lectures on Numerical Methods for Equations of Viscous Liquids.* Moscow: Izd. VC AN SSSR, 1969 [Russian].
Dyakonov, E. G.: *Iterative Methods for Solving Difference Analogs of Boundary Value Problems for the Equations of Elliptic Type.* Kiev: Izd. Inst. Kibernet. AN USSR, 1970 [Russian].
Gel'fond, A. O.: *The Method of Finite Differences.* Moscow: Nauka, 1967 [Russian].
Godunov, S. K.: *Difference Equations for Equations of Aerodynamics.* Novosibirsk: Izd. Novosib. Univ., 1962 [Russian].
Godunov, S. K. and Ryabenkii, V. S.: *Introduction to the Theory of Difference Schemes.* Moscow: Fizmatgiz, 1962 [Russian].
Henrici, P.: *Error Propagation for Difference Methods.* New York: Wiley, 1963.
Householder, A. S.: *Principles of Numerical Analysis.* New York: McGraw-Hill, 1953.
Isaacson, E. and Keller, H. B.: *Analysis of Numerical Methods.* New York: Wiley, 1966.

Konovalov, A. N.: *Numerical Solution of Problems of the Theory of Elasticity*. Novosibirsk: Izd. Novosib. Univ., 1968 [Russian].
Lanczos, C.: *Applied Analysis*. Englewood Cliffs, N.J.: Prentice Hall, 1956.
Marchuk, G. I.: *Design of Nuclear Reactors*. Moscow: Atomizdat, 1961 [Russian].
Marchuk, G. I.: *Numerical Methods in Weather Prediction*. Leningrad: Gidrometizdat, 1967 [Russian].
Marchuk, G. I.: Methods and problems of computational mathematics. In: *Proceedings of the International Congress of Mathematicians*. Nice, 1970.
Marchuk, G. I. and Kuznetsov, Yu. A.: *Iteration Methods and Quadratic Functionals*. Novosibirsk: Nauka, 1972 [Russian].
Miller, J., Strang, G.: Matrix theorems for partial differential and difference equations. *Math. Scand.*, **18**, 2 (1966).
Mitchell, A. R.: *Computational Methods in Partial Differential Equations*. London: Wiley, 1970.
Mysovskikh, I. P.: *Lectures on Numerical Methods*. Moscow: Fizmatgiz, 1962 [Russian].
Polozhii, G. N.: *Numerical Solution of Two-Dimensional and Three-Dimensional Problems of Mathematical Physics and Functions with the Discrete Argument*. Kiev: Izd. Kiev. Univ., 1962 [Russian].
Pontryagin, L. S., Boltianskii, V. G., Gamkrelidze, R. V., and Mischenko, E. F.: *Mathematical Theory of Optimal Processes*. New York: Interscience, 1962.
Richtmyer, R. D.: *Difference Methods for Initial Value Problems*. New York: Interscience, 1957.
Richtmyer, R. and Morton, K. W.: *Difference Methods for Initial Value Problems*. New York: Wiley, 1967.
Samarskii, A. A.: *Introduction to the Theory of Difference Schemes*. Moscow: Nauka, 1971 [Russian].
Saulyev, V. K.: *Integration of the Equations of Parabolic Type by the Method of Nets*. Moscow: Fizmatgiz, 1960 [Russian].
Varga, R. S.: *Matrix Iterative Analysis*. New York, 1963.
Vorob'ev, Yu. V.: *Method of Moments in Applied Mathematics*. Moscow: Fizmatgiz, 1958 [Russian].
Wasov, W. R. and Forsythe, G. E.: *Finite Difference Methods for Partial Differential Equations*. New York: Wiley, 1960.
Yanenko, N. N.: *The Method of Fractional Steps for Solving Multi-Dimensional Problems of Mathematical Physics*. Novosibirsk: Nauka, 1967 [Russian].
Yanenko, N. N.: *Introduction to Difference Schemes of Mathematical Physics*. Novosibirsk: Izd. Novosib. Univ., 1968 [Russian].

4. The Method of Nets

Bryan, K.: A Scheme for numerical integration of the equations of motion on an irregular grid free of non-linear instability. *Monthly Weather Review*, **94**, 1 (1966).
Chudov, L. A., Kudryavtsev, V. P.: On round off errors of difference methods for elliptic equations and systems with initial conditions. In: *Numerical Methods in Gas Dynamics*. Moscow: Izd. Mosk. Univ., 1963 [Russian].
Crowley, W.: Second order numerical advection. *J. Comp. Phys.*, **1**, 4 (1967).
Demyanovich, Yu. K.: Method of nets for certain problems of mathematical physics. *Doklady AN SSSR*, **159**, 2 (1964) [Russian].
Fichera, G.: Further development in the approximation theory of eigen-values. In: *Numerical Solutions of Partial Differential Equations II, SYNSPADE 1970*. New York–London: Academic Press, 1971.
Fox, L., Henrici, P., and Moler, C.: Approximations and bounds for eigen-values of elliptic operators. *SIAM J. Numer. Anal.*, **4**, 1 (1967).

Fromm, J. E.: Numerical methods for computing non-linear time-dependent buoyancy of air in rooms. *J. Res. Dev. IBM*, **15**, 5 (1971).

Godunov, S. K.: Difference methods for numerical solution of problems in gas dynamics. *J. Vych. Matem. Matem. Fiz.*, **2**, 7 (1962) [Russian].

Godunov, S. K. and Zaborodin, A. V.: On difference scheme of second order accuracy for multi-dimensional problems. *Zh. Vych. Matem. Matem. Fiz.*, **2**, 4 (1962) [Russian].

Il'in, V. P.: *Difference Methods for Elliptic Equations.* Novosibirsk: Izd. Novosib. Univ., 1970 [Russian].

Keller, H.: A new difference scheme for parabolic problems. In: *Numerical Solutions of Partial Differential Equations, II, SYNSPADE 1970.* New York–London: Academic Press, 1971.

Kellogg, R.: Singularities in interface problems. In: *Numerical Solutions of Partial Differential Equations II, SYNSPADE 1970.* New York–London: Academic Press, 1971.

Kuznetsov, Yu. A. and Shaydurov, V. V.: On uniform convergence of difference schemes. In: *Numerical Methods of Linear Algebra.* Novosibirsk: Izd. VC Sib. otd. AN SSSR, 1972 [Russian].

Kurihara, Y. and Holloway, I.: Numerical integration of a nine-level global primitive equations model formulated by the box method. *Monthly Weather Review*, **95**, 8 (1967).

Kuropatenko, V. F.: A Method of constructing difference schemes for numerical solution of the equations of gas dynamics. *Izv. Vuzov. Matematika*, **3**(28) (1962) [Russian].

Landau, L. D., Meiman, N. N., and Khalatnikov, I. M.: Numerical methods for solving partial differential equations based on the method of nets. *Trudy III Vsesoyuznovo Mat. S'ezda, Vol. II,* Moscow: Izd. AN SSSR, 1956 [Russian].

Lebedev, V. I.: The method of nets for the equations of Sobolev type. *Doklady AN SSSR*, **114**, 6 (1957) [Russian].

Lebedev, V. I.: On the method of nets for a system of partial differential equations. *Izv. AN SSSR, Ser. Mat.*, **22**, 5 (1958) [Russian].

Lebedev, V. I.: On the Dirichlet and Neumann problems on triangular and hexagonal nets. *Doklady AN SSSR*, **138**, 1 (1961) [Russian].

Lyusternik, L. A.: On difference equations of the Laplace operator. *Usp. Mat. Nauk*, **IX**, 2 (1954) [Russian].

Raviart, P. A.: Sur l'approximation de certaines équations d'évolution linéaires et non linéaires. *J. Mathem. Pures Appliq.*, **46**, 1 (1967).

Rivkind, V. Ya.: An approximate method for the Dirichlet problem, and on rate estimates for convergence of solutions of difference equations to solutions of elliptic equations with discontinuous coefficients. *Vestnik Leningradskogo Univ., Ser. Mat.*, **3** (1964) [Russian].

Rivkind, V. Ya.: On rate convergence estimates for homogeneous difference schemes corresponding to elliptic and parabolic equations with discontinuous coefficients. In: *Problems of Mathematical Analysis.* Leningrad: Izd. Leningr. Univ., 1966 [Russian].

Samarskii, A. A.: On monotone difference schemes for elliptic and parabolic equations in the case of non-selfadjoint elliptic operator. *Zh. Vych. Matem. Matem. Fiz.*, **5**, 3 (1965) [Russian].

Samarskii, A. A.: On accuracy of the method of nets for the Dirichlet problem in an arbitrary region. *Apl. Mat.* **10**, 3 (1965) [Russian].

Samarskii, A. A.: Some problems in the theory of difference schemes. *Zh. Vych. Matem. Matem. Fiz.*, **6**, 4 (1966) [Russian].

Tikhonov, A. N. and Samarskii, A. A.: On difference schemes for equations with discontinuous coefficients. *Doklady AN SSSR*, **108**, 3 (1956) [Russian].

Tikhonov, A. N. and Samarskii, A. A.: On homogeneous difference schemes. *Zh. Vych. Matem. Matem. Fiz.*, **1**, 1 (1961) [Russian].

Tikhonov, A. N., Samarskii, A. A.: Homogeneous difference schemes on non-uniform nets. *Zh. Vych. Matem. Matem. Fiz.*, **2**, 6 (1962) [Russian].

Valitskii, Yu. N.: On the convergence of difference approximations of eigen-values and eigen-vectors of a two-dimensional elliptic operator. *Doklady AN SSSR*, **198**, 2 (1971) [Russian].

Volkov, E. A.: Solving the Dirichlet problem by the method of increased accuracy, I, II. *Differential Equations*, **1**, 7 and 8 (1965) [Russian].

Volkov, E. A.: The method of nets for the Laplace equation on finite and infinite regions with piece-wise continuous boundaries. Dissertation, Moscow, 1967.

Wachspress, E. L.: The numerical solution of boundary value problems. In: *Mathematical Methods for Digital Computers*, New York, 1960.

Yanenko, N. N., Suchkov, V. A., and Pogodin, Yu. Ya.: On a difference solution of the heat equation in the curvilinear coordinates. *Doklady AN SSSR*, **128**, 5 (1959) [Russian].

5. Variational-Difference Methods

Aubin, J. P.: Approximation des espaces des distributions et des opérateurs différentiels. *Bull. Soc. Math. France, Memoire*, **12** (1967).

Aubin, J. P.: Behavior of the error of the approximate solutions of boundary value problems for linear elliptic equations by Galerkin's and finite difference methods. *Ann. Scuola Norm. Super., Pisa*, **21**, 4 (1967).

Aubin, J. P.: Best approximations of linear operators in Hilbert space. *SIAM J. Numer. Anal.*, **5**, 3 (1968).

Aubin, J. P., Burchard, H. G.: Some aspects of the method of hypercircle applied to elliptic variational problems. In: *Numerical Solutions of Partial Differential Equations II, SYNSPADE 1970.* New York–London: Academic Press, 1971.

Babuška, I.: The finite element method for elliptic differential equations. In: *Numerical Solutions of Partial Differential Equations II, SYNSPADE 1970.* New York–London: Academic Press, 1971.

Babuška, I.: The rate of convergence for finite element method. *SIAM J. Numer. Anal.*, **8**, 2 (1971).

Birkhoff, G., Schultz, M. H., and Varga, R. S.: Hermite interpolation in one and two variables with applications to partial differential equations. *Numer. Math.*, **11**, 3 (1968).

Bramble, J.: A second order finite difference analog of the first biharmonic boundary value problem. *Numer. Math.*, **9**, 3 (1966).

Bramble, J. and Hubbard, B.: On the formulation of finite difference analogues of the Dirichlet problem for Poisson's equation. *Numer. Math.*, **4**, 4 (1962).

Bramble, J. and Schatz, A.: On the numerical solution of elliptic boundary value problems by least squares approximations of the data. In: *Numerical Solutions of Partial Differential Equations II, SYNSPADE 1970.* New York–London: Academic Press, 1971.

Cea, J.: Approximation opérationelle des problèmes aux limites. *Ann. Inst. Fourier, Grenoble*, **14**, 2 (1964).

Courant, R.: Variational methods for the solutions of problems of equilibrium and variations. *Bull. Amer. Math. Soc.*, **49** (1943).

Douglas, J. and Dupont, T.: Alternating-direction Galerkin methods on rectangles. In: *Numerical Solutions of Partial Differential Equations II, SYNSPADE 1970.* New York–London: Academic Press, 1971.

Hubbard, B.: Remarks on the convergence in the discrete Dirichlet problem. In: *Numerical Solutions of Partial Differential Equations* (Edited by J. H. Bramble). New York–London: Academic Press, 1965.

Keldysh, M. V.: On the Galerkin method for boundary value problems. *Izv. AN SSSR, Ser. Matam.*, 6 (1942) [Russian].

Lebedev, V. I.: Difference analogues for orthogonal expansions of the fundamental differential operators and certain boundary value problems of mathematical physics. *Zh. Vych. Matem. Matem. Fiz.*, **4**, 3, 4 (1964) [Russian].

Lions, J. L., Teman, R.: Une méthode d'éclament des opérateurs et des contraintes en calcul des variations. *C. R. Acad. Sci., Paris*, **263** (1966).

Oganesyan, L. A.: A variational-difference scheme on a regular net for the Dirichlet problem. *Zh. Vych. Matem. Matem. Fiz.*, **11**, 6 (1971) [Russian].

Oganesyan, L. A. and Rukhoviets, L. A.: On variational difference schemes for second-order linear elliptic equations in a two-dimensional region with the piece-wise continuous boundary. *Zh. Vych. Matem. Matem. Fiz.*, **8**, 1 (1968) [Russian].

Oganesyan, L. A. and Rukhoviets, L. A.: A study of the rate of convergence of variational-difference schemes for the second-order elliptic equations in a two-dimensional region with smooth boundary. *Zh. Vych. Matem. Matem. Fiz.*, **9**, 5 (1969) [Russian].

Rukhovets, L. A.: A study of the rate of convergence of variational-difference schemes for elliptic equations of second order. Dissertation, Leningrad, 1970 [Russian].

Strang, G.: The finite element method and approximation theory. In: *Numerical Solutions of Partial Differential Equations II, SYNSPADE 1970*. New York–London: Academic Press, 1971.

Strang, G., Fix, G.: A Fourier analysis of the finite element variational method. Preprint (1970).

Zienkiewicz, O.: *The Finite Element Method in Structural and Continuum Mechanics*. London: McGraw-Hill, 1967.

Zlámal, M.: On some finite element procedures for solving second order boundary value problems. *Numer. Math.*, **14**, 1 (1969).

Zlámal, M.: On the finite element method. *Numer. Math.*, **12**, 5 (1968).

6. The Theory of Stability of Difference Schemes

Fedorov, M. V.: On c-stability of the Cauchy problem for difference equations and partial differential equations. *Zh. Vych. Matem. Matem. Fiz.*, **7**, 3 (1967) [Russian].

Filippov, A. F.: On stability of difference equations. *Doklady AN SSSR*, **100**, 6 (1955) [Russian].

Il'in, A. M.: Stability of difference schemes for the Cauchy problem for systems of partial differential equations. *Doklady AN SSSR*, **164**, 3 (1965) [Russian].

Keller, H. B. and Thomee, V.: Unconditionally stable difference methods for mixed problems for quasi-linear hyperbolic systems in two dimensions. *Comm. Pure Appl. Math.*, **15**, 1 (1962).

Kreiss, H. O.: On difference approximations of the dissipative type for hyperbolic differential equations. *Comm. Pure Appl. Math.*, **17**, 3 (1964).

Kreiss, H. O.: Initial boundary value problems for partial differential and difference equations in one space dimension. In: *Numerical Solutions of Partial Differential Equations II, SYNSPADE 1970*. New York–London: Academic Press, 1971.

Lax, P. D.: On stability of finite difference approximations for hyperbolic equations with variable coefficients. *Matematika* (Selected translations), **6**, 3 (1962) [Russian].

Lax, P. D. and Wendroff, B.: On the stability of difference schemes with variable coefficients. *Comm. Pure Appl. Math.*, **15**, 4 (1962).

Lax, P. D. and Nirenberg, L.: On stability of difference schemes: Exact form of Garding's inequality. *Matamatika* (Selected Translations), **11**, 6 (1967) [Russian].

Richtmyer, R.: On the non-linear instability of difference schemes. In: *Some Problems of Numerical and Applied Mathematics*. Novosibirsk: Nauka, 1966 [Russian].

Ryabenkii, V. S. and Filippov, A. F.: *On Stability of Difference Equations*. Moscow: Gostekhizdat, 1956 [Russian].

Samarskii, A. A.: Necessary and sufficient conditions of stability of two-layer difference schemes. *Doklady AN SSSR*, **181**, 4 (1968) [Russian].

Serdyukova, S. I.: C-stability of explicit difference schemes with constant real l_2-stable coefficients. *Zh. Vych. Matem. Matem. Fiz.*, **3**, 2 (1963) [Russian].

Strang, G.: Difference methods for mixed boundary value problems. *Duke Math. J.*, **27**, 2 (1960).

Thomee, V.: Generally unconditionally stable difference operators. *SIAM J. Numer. Anal.*, **4**, 1 (1967).

7. Stability and Convergence

Andreev, B. B.: On the convergence of difference schemes approximating the second and the third boundary value problems of elliptic equations. *Zh. Vych. Matem. Matem. Fiz.*, **8**, 6 (1968) [Russian].

O'Brien, G. G., Hyman, M. A., and Kaplan, S.: A study of the numerical solution of partial differential equations. *J. Math. Phys.*, **29**, 4 (1951).

Courant, R., Friedrichs, K., and Lewy, H.: Über die partiellen Differenzengleichungen der mathematischen Physik. *Math. Ann.*, **100**, 32 (1928).

Du Fort, E. C. and Frankel, S. P.: Stability conditions in the numerical treatment of parabolic differential equations. *Math. Tables Other Aids Comput.*, **7**, 43 (1953).

Godunov, S. K. and Ryabenkii, V. S.: Canonical forms of systems of linear ordinary difference equations with constant coefficients. *Zh. Vych. Matem. Matem. Fiz.*, **3**, 2 (1963) [Russian].

Godunov, S. K. and Ryabenkii, V. S.: Spectral stability criteria of boundary value problems for non-selfadjoint difference equations. *Usp. Matem. Nauk*, XVIII, 3 (111) (1963) [Russian].

John, F.: On the integration of parabolic equations by difference methods. I. Linear and quasi-linear equations for the infinite interval. *Comm. Pure Appl. Math.*, **5**, 2 (1952).

Ladyzhenskaya, O. A.: The method of finite differences in the theory of partial differential equations. *Usp. Matem. Nauk*, XII, 5 (1957) [Russian].

Lax, P. D., Wendroff, B.: System of Conservation laws. *Comm. Pure Appl. Math.*, **13**, 2 (1960).

Lax, P. D. and Richtmyer, R. D.: Survey of the stability of linear finite difference equations. *Comm. Pure. Appl. Math.*, **9**, 2 (1956).

Lees, M.: *A priori* estimate for the solution of difference approximations to parabolic partial differential equations. *Duke Math. J.*, **27**, 3 (1960).

Lees, M.: Energy inequalities for the solution of differential equations. *Trans. Amer. Math. Soc.*, **94**, 1 (1960).

Lions, J. P.: Equations différentielles opérationnelles dans les espaces de Hilbert. Centro Int. Mat. Estivo, Varenna (1963). (Equazioni Differenziali Astratte, Cremonese, Roma (1963).)

Neumann, J. and Richtmyer, R. D.: A method for the numerical calculation of hydrodynamic shocks. *J. Appl. Physics.* **21**, 3 (1950).

Phillips, N. A.: *The Atmosphere and the Sea in Motion. Scientific Contributions to the Rossby Memorial Volume.* The Rockefeller Institute, 1959.

Ryabenkii, V. S.: Structure of spectra of families of non-selfadjoint difference operators. In: *Materials of Joint Soviet–American Symposium on Partial Differential Equations.* Novosibirsk, 1963. [Russian].

Ryabenkii, V. S.: The spectrum of a family of difference operators on functions on the net graph. *Zh. Vych. Matem. Matem. Fiz.*, **7**, 6 (1967) [Russian].

Samarskii, A. A.: Some problems in the general theory of difference schemes. In: *Partial Differential Equations* (Papers of the symposium honoring the sixtieth birthday of academician S. L. Sobolev). Moscow: Nauka, 1970. [Russian].

Sobolev, S. L.: Some remarks on numerical solutions of integral equations. *Izv. AN SSSR, Ser. Matem.*, **20**, 4 (1956) [Russian].
Strang, G.: Accurate partial difference methods. I. Linear Cauchy problem. *Arch. Rational Mech. Anal.*, **12**, 5 (1963).
Thomee, V.: On the rate of convergence of difference schemes for hyperbolic equations. In: *Numerical Solutions of Partial Differential Equations II, SYNSPADE 1970.* New York-London: Academic Press, 1971.
Yanenko, N. N. and Boyarintsev, Yu. E.: On the convergence of difference schemes for the heat equation with variable coefficients. *Doklady AN SSSR*, **139**, 6 (1961) [Russian].
Yanenko, N. N. and Shokin, Yu. I.: On the relation between well-posed first differential approximations and the stability of difference schemes for hyperbolic systems. *Matematicheskie zametki*, **4**, 5 (1968) [Russian].
Yanenko, N. N., Shokin, Yu. I.: On well-posed first differential approximations of difference schemes. *Doklady AN SSSR*, **182**, 4 (1968) [Russian].

8. Numerical Methods of Linear Algebra

Abramov, A. A.: Ideas of perturbation theory in some algorithms of linear algebra. In: *Numerical Methods of Linear Algebra, Vol. 1.* Moscow: VC AN SSSR, 1968 [Russian].
Bakhvalov, N. S.: Foundations of Numerical Mathematics. Moscow: *Izd. Moskov. Univ.*, 1970 [Russian].
Bauer, F. L., Fike, C. T.: Norms and Exclusion Theorems. *Numer. Math.*, **2**, 3 (1960).
Bellman, R.: *Introduction to Matrix Analysis.* New York: McGraw-Hill 1970.
Dorodnitsyn, A. A.: On the problem of computing eigenvalues and eigenvectors of matrices. *Doklady AN SSSR*, **126**, 6 (1959) [Russian].
D'yakonov, E. G.: On the solution of some elliptic difference equations. *J. Inst. Math. Applics.*, **7** (1971).
Eberlein, P.: A Jacobi-like method for the automatic computation of eigenvalues and eigenvectors of an arbitrary matrix. *J. Soc. Ind. Appl. Math.*, **10** (1962).
Faddeev, D. K. and Faddeeva, V. N.: *Computational Methods of Linear Algebra.* San Francisco: H. W. Freeman, 1963.
Faddeev, D. K., Faddeeva, V. N., and Kublanovskaya, V. N.: Linear algebraic systems with rectangle matrices. In: *Numerical Methods of Linear Algebra.* Moscow: Nauka, 1968 [Russian].
Forsythe, G. E. and Moler, C. B.: *Computer Solution of Linear Algebraic Systems.* Englewood Cliffs, N.J.: Prentice Hall, 1967.
Frencis, J.: The QR-transformation, Parts I, II. *Computer J.*, **4** (1961, 1962).
Gantmakher, F. R.: *The Theory of Matrices.* New York: Chelsea, 1959.
Ikramov, Kh. D.: *Matrix Norms and Jacobi-Like Methods.* Moscow: Izv. Mosc. Univ., 1969 [Russian].
Kellogg, R., Noderer, L.: Sealed iterations and linear equations. *SIAM J.*, **8**, 4 (1960).
Kim, G.: *On the Distribution of the Round-Off Errors for Iteration Methods for Linear Algebraic Systems.* Moscow: Izd. Mosc. Univ., 1969 [Russian].
Kublanovskaya, V. N.: Orthogonal transformations in algebraic problems. Dissertation, Leningrad, 1972 [Russian].
Marek, I.: Iterations for linear bounded operators and Kellogg's process. Dissertation, Prague, 1962 [Czech].
Marek, I.: On iteration of linear bounded operators and the convergence of Kellogg's iteration process. *Czech Math. J.*, **12** (1962).
Nemchinov, S. V. and Libov, S. L.: A direct method of increasing accuracy of solutions of boundary value problems for the Helmholtz equation on a rectangular net region. *Zh. Vych. Matem. Matem. Fiz.*, **4**, 4 (1964) [Russian].

Stiefel, E.: Kernel polynomials in linear algebra and their numerical applications. *NBS, Appl. Math., Ser. 49*, 1 (1958).
Traub, J.: *Iterative Methods for the Solution of Equations.* Englewood Cliffs, N.J.: Prentice Hall, 1964.
Voevodin, V. V.: *Numerical Methods of Algebra. Theory and Algorithms.* Moscow: Nauka, 1966. [Russian].
Wilkinson, J.: *The Algebraic Eigenvalue Problem.* Oxford: Clarendon Press, 1965.

9. Optimization of Iterative Processes by Spectral Methods

Abramov, A. A.: On an acceleration method for iterative processes. *Doklady AN SSSR*, **74**, 6 (1950) [Russian].
Bakhvalov, N. S.: On the convergence of a relaxation method under the natural constraints on the elliptic operator. *Zh. Vych. Matem. Matem. Fiz.*, **6**, 5 (1966) [Russian].
Collatz, L.: Fahlerabschatzung für das Iterationsverfahren zur Auflösung linearer Gleichungssysteme. *Z. Angew. Math. Mech.*, 22 (1942).
Dyakonov, E. G.: Constructions of iteration methods using spectrum-equivalent operators. *Zh. Vych. Matem. Matem. Fiz.*, **6**, 1, 4 (1966) [Russian].
Fedorenko, R. P.: Solution of difference elliptic equations by the relaxation method. *Zh. Vych. Matem. Matem. Fiz.*, **1**, 5 (1961) [Russian].
Fedorenko, R. P.: On the rate of convergence of an iterative process. *Zh. Vych. Matem. Matem. Fiz.*, **4**, 3 (1964) [Russian].
Gavurin, M. K.: Nonlinear functional equations and continuous analogs of iterative methods. *Izv. Vuzov. Matematika*, **5**, 6 (1958) [Russian].
Gavurin, M. K.: An application of the best approximation polynomials for improving convergence of iterative processes. *Usp. Matem. Nauk*, V, 3 (1950) [Russian].
Golub, G. H. and Varga, R. S.: Chebyshev semi-iterative methods, successive overrelaxation iterative methods and second-order Richardson iterative methods. Parts I, II. *Numer. Math.*, **3**, 2 (1961).
Juncosa, M. L. and Milliken, T. M.: On the increase of convergence rates of relaxation procedures for elliptic partial difference equations. *J. Assos. Comp. Math.*, **7**, 1 (1960).
Ivanov, V. K.: On the convergence of iterative processes for linear algebraic systems. *Izv. AN SSSR, Ser. Matem.*, 4 (1939) [Russian].
Lanczos, C.: An iteration method for the solution for the eigenvalue problem for linear differential and integral operators. *J. Res. Nat. Bur. Stand.*, **45**, 1 (1950).
Lanczos, C.: Chebyshev polynomials in the solution of large-scale linear systems. *Proc. Assoc. Comput. Math.*, Toronto, September, 1952.
Lebedev, V. I.: On iteration methods of solving operator equations with the spectrum formed by several segments, *Zh. Vych. Matem. Matem. Fiz.*, **9**, 6 (1969) [Russian].
Lebedev, V. I.: On a construction of the p-operation in the kp-method. *Zh. Vych. Matem. Matem. Fiz.*, **9**, 4 (1969) [Russian].
Lebedev, V. I. and Finogenov, S. A.: On the order of choosing the iteration parameters in the Chebyshev cyclic iteration method. *Zh. Vych. Matem. Matem. Fiz.*, **11**, 2 (1971) [Russian].
Marchuk, G. I. and Sarbasov, K. E.: On a method of solving stationary problems. *Doklady AN SSSR*, **182**, 1 (1968) [Russian].
Ostrowski, A. M.: On the linear iteration procedures for symmetric matrices. *Univ. Roma, Inst. Naz. Alta Math. Rend. Mat. e Appl.*, 14, 1–2 (1954).
Petryshin, W.: On a general iterative method for the approximate solution of linear operator equations. *Math. Comput.*, 17, 1 (1963).
Reich, I.: On the convergence of the classical iterative method of solving linear simultaneous equations. *Ann. Math. Statist.*, 20, 3 (1949).
Vorob'ev, Yu. B.: A random iterative process. *Zh. Vycisl. Matem. Matem. Fiz.*, **4**, 6 (1964) and **5**, 5 (1965) [Russian].

10. The Over-Relaxation Method

Broyden, C. G.: Some generalizations of the theory of successive over-relaxation. *Numer. Math.*, **6**, 4 (1964).

Broyden, C. G.: On convergence criteria for the method of successive over-relaxation. *Math. Comut.*, **18**, 85 (1964).

Evans, D. J.: Note on the linear over-relaxation factor for small mesh size. *Comput. J.*, **5**, 1 (1962).

Evans, D. J. and Forington, C. V.: An iterative process for optimizing symmetric successive over-relaxation. *Comput. J.*, **6**, 3 (1963).

Faddeev, D. K.: On the over-relaxation method in systems of linear equations. *Izv. Vuzov. Matematika*, **5** (1958) [Russian].

Garabedian, P.: Estimation of the relaxation factor for small mesh size. *Math. Tables Other Aids Comput.*, **10**, 56 (1956).

Gastinel, N.: Sur le meilleur des paramètres de sur-relaxation (Procédé de Peaceman–Rachford). *Chiffres*, **5**, 2 (1962).

Golub, G. H.: The use of Chebyshev matrix polynomials in the iterative solution of linear equations compared with the method of successive over-relaxation. Doct. Thesis, Univ. of Illinois, **133** (1959).

Hageman, L. A. and Kellogg, R. B.: Estimating optimum over-relaxation parameters. *Math. Comput.*, **22**, 101 (1968).

Linn, M. S.: On the round-off error in the method of successive over-relaxation. *Math. Comput.*, **18**, 85 (1964).

Ostrowski, A. M.: On over- and under-relaxation in the theory of the cyclic single step iteration. *Math. Tables Other Aids Comput.*, **7**, 43 (1953).

Petryshin, W.: On generalized over-relaxation method for the approximate solution of operator equations in Hilbert space. *SIAM J.*, **10**, 4 (1962).

Petryshin, W. V.: On the extrapolated Jacobi or simultaneous displacement method in the solution of matrix and operator equations. *Mat. Comput.*, **19**, 89 (1965).

Rutishauser, H.: The Jacobi Method for Real Symmetric Matrices. *Numer. Math.*, **9**, 1 (1966).

Varga, R. S.: P-Cyclic matrices: The generalization of the Young–Francel successive over-relaxation scheme. *Pacif. J. Math.*, **9** (1959).

Varga, R. S.: Orderings of the successive over-relaxation scheme. *Pacif. J. Math.*, **9** (1959).

Young, D. M.: Iterative methods for solving partial difference equations of elliptic type. *Trans. Amer. Math. Soc.*, **74** (1954).

Young, D. M.: A bound for the optimum relaxation factor for the successive over-relaxation method. *Numer. Math.*, **16**, 5 (1971).

Young, D. M.: Convergence properties of the symmetric and unsymmetric successive over-relaxation methods and related methods. *Math. Comput.*, **24**, 112 (1971).

11. The Gradient Methods

Birman, M. Sh.: Some estimates for the steepest descent method. *Usp. Matem. Nauk*, **V**, 3 (1950) [Russian].

Danial, J. W.: The conjugate gradient method for linear and non-linear operator equations. *SIAM J. Num. Anal.*, **4**, 1 (1967).

Danial, J. W.: Convergence of the conjugate gradient method with computationally convenient modifications. *Numer. Math.*, **10**, 2 (1967).

Forsythe, G. E.: On the asymptotic directions of the s-dimensional optimum gradient method. *Numer. Math.*, **11**, 1 (1968).

Forsythe, G. I. and Motzkin: Asymptotic properties of the optimum gradient method. *Bull. Amer. Math. Soc.*, **57**, 2 (1951).

Forsythe, G. E. and Motzkin: Acceleration of the optimum gradient method. *Bull. Amer. Math. Soc.*, **57**, 4 (1951).

Forsythe, A. I., Forsythe, G. E.: Punchedcard experiments with accelerated gradient methods for linear equations. Contributions to the solution of linear equations and the determination of eigenvalues. *NBS Appl. Math., Ser. 39* (1954).

Fridman, V. M.: New methods for solving linear operator equations. *Doklady AN SSSR*, **128**, 3 (1959) [Russian].

Godunov, S. K. and Prokopov, G. P.: Variational approach to large systems of linear equations arising in strongly elliptic problems. Preprint of the Inst. Appl. Math. AN SSSR, Moscow, 1968 [Russian].

Godunov, S. K. and Prokopov, G. P.: On the difference Laplace equation. *Zh. Vych. Matem. Matem. Fiz.*, **9**, 2 (1969) [Russian].

Hestenes, M. R. and Stiefel, E.: Method of conjugate gradients for solving linear systems. *J. Res. NBS*, **49** (1952).

Kantorovich, L. V.: On the steepest descent method. *Doklady AN SSSR*, **56**, 3 (1947) [Russian].

Krasnosel'skii, M. A. and Krein, S. G.: An iterative process with minimal residuals. *Matem. Sb.*, **31** (1952) [Russian].

Kuznetsov, Yu. A.: On the theory of iterative processes. *Doklady AN SSSR*, **184**, 2 (1969) [Russian].

Kuznetsov, Yu. A.: On the symmetrization of iterative processes. In: *Numerical Methods of Linear Algebra*. Novosibirsk: Izd. VC Sib. Otd. AN SSSR, 1969 [Russian].

Kuznetsov, Yu. A.: Some problems in the theory and applications of iterative methods. Dissertation, Novosibirsk, 1969 [Russian].

Lanczos, C.: Solution of the system of linear equations by minimized iterations. *J. Res. NBS*, **49**, 1 (1952).

Marchuk, G. I. and Kuznetsov, Yu. A.: On optimal iterative processes. *Doklady AN SSSR*, **181**, 6 (1968) [Russian].

Marchuk, G. I. and Kuznetsov, Yu. A.: Some problems in the theory of multi-step processes. In: *Numerical Methods of Linear Algebra*. Novosibirsk: Izd. VC Sib. Otd. AN SSSR, 1969 [Russian].

Marchuk, G. I. and Kuznetsov, Yu. A.: On solution of systems of linear equations by iterative methods. In: *Problems of Accuracy and Efficiency of Numerical Algorithms, V. 1*. Kiev: Izd. Inst. kibernetiki AN USSR, 1969 [Russian].

Samokish, B. A.: A study of the rate of convergence of the steepest descent method. *Usp. Matem. Nauk*, **XII**, 1 (1957) [Russian].

12. The Factorization Method

Abramov, A. A. and Andreev, V. B.: An application of the factorization method for finding periodic solutions of differential and difference equations. *Zh. Vych. Matem. Matem. Fiz.*, **3**, 2 (1963) [Russian].

Ains, E.: *Ordinary Differential Equations*. Moscow: ONTI, 1939.

Degtyarev, L. M. and Favorskii, A. P.: Flow variant of the factorization method. *Zh. Vych. Matem. Matem. Fiz.*, **8**, 3 (1968) [Russian].

Degtyarev, L. M. and Favorovskii, A. P.: A flow variant of the factorization method for difference problems with strongly varying coefficients. *Zh. Vych. Matem. Matem. Fiz.*, **9**, 1 (1969) [Russian].

Fage, M. K.: On the factorization method. *Doklady AN SSSR*, **191**, 2 (1970) [Russian].

Gel'fang, I. M. and Lokutsievskii, O. B.: The factorization method for difference equations. In: *Introduction to the Theory of Difference Schemes* (S. K. Godunov and V. S. Ryabenkii, Eds.), Moscow: Fizmatgiz, 1962 [Russian].

Godunov, S. K.: The orthogonal factorization method for systems of difference equations. *Zh. Vych. Matem. Matem. Fiz.*, **2**, 6 (1962) [Russian].

Ogneva, V. V.: The factorization method for solving difference equations. *Zh. Vych. Matem. Matem. Fiz.*, **7**, 4 (1967) [Russian].

Rusanov, V. V.: Stability of the matrix factorization method. In: *Numerical Mathematics, No. 6*. Moscow: 1960 [Russian].

Sofronov, I. D.: On the factorization method for solving boundary value problems for difference equations. *Zh. Vych. Matem. Matem. Fiz.*, **4**, 2 (1964) [Russian].

Sofronov, I. D.: A Diagonal Matrix Factorization Scheme for the Heat Equation. *Zh. Vych. Matem. Matem. Fiz.*, **5**, 2 (1965) [Russian].

13. The Fast Fourier Transform

Bingham, C., Godfrye, M. D., and Tukey, J.: Modern techniques of power spectrum estimation. *IEEE Trans., Audio and Electroacoustics*, AU-15 (1967).

Buzbee, B., Golub, G., and Nilson, E.: On direct methods of solving Poisson's equations. *SIAM J. Numer. Anal.*, **7**, 4 (1970).

Cooley, G. W., Lewis, P. A., and Welch, P. D.: The Fast Fourier Transform Algorithms and Applications. IBM Research Paper RC-1743, Feb. (1967).

Cooley, J. W. and Tukey, J. W.: An algorithm for machine calculation of complex Fourier series. *Math. Comp.*, **19**, 90 (1965).

Gold, B. and Rader, C. M.: *Digital Processing of Signals*. New York: McGraw-Hill, 1969.

Helms, R. D.: Fast Fourier transform method for computing difference equations and simulating filters. *IEEE Trans.*, AU-15 (1967).

Hockney, R. W.: A fast direct solution of Poisson's equation using Fourier analysis. *J. Assoc. Comp. Mach.*, **12**, 1 (1965).

Klauder, J. R., Price, A. C., Darlington, S., and Albersheim, W. J.: The theory and design of chirp radars. *Bell Syst. Tech. J.*, **39** (1960).

Kuznetsov, Yu. A. and Matsokin, A. M.: A solution of Helmholtz equation by the method of fictive regions. In: *Numerical Methods of Linear Algebra*. Novosibirsk: Izd. VC Sib. Otd. AN SSSR, 1972 [Russian].

Nemchinov, S. V.: An application of the method of nets to boundary value problems for partial differential equations with periodic boundary conditions. In: *Dynamic Meteorology*. Tashkent: Nauka, 1965 [Russian].

Tukey, J. W.: An introduction to the calculations of numerical spectrum analysis. In: *Spectral Analysis in Time Series* (Harris, E.). New York: Wiley, 1967.

14. Interpolation Using Spline Functions

Alberg, J. H., Nilson, E. N., and Walsh, J. L.: Extremal orthogonal lines. *J. Math. Anal. Appl.*, **12**, 1 (1965).

Alberg, J. H., Nilson, E. N., and Walsh, J. L.: *The Theory of Splines and Their Applications*. New York: Academic Press, 1967.

Anselon, P. M. and Laurent, P. J.: A General method for construction of interpolating or smoothing spline-functions. *Numer. Math.*, **12**, 1 (1968).

Atteia, M.: Généralisation de la définition et des propriétés des "spline fonctions". *C. R. Acad. Sci., Paris*, **260** (1965).

Birkhoff, G. and Garabedian, P.: Smooth surface interpolation. *J. Math. Phys.*, **39**, 3 (1960).

de Boor, C.: Bicubic spline interpolation. *J. Math. Phys.*, **41**, 2 (1962).

Holladey, J. C.: Smoothest curve approximation. *Math. Tables Aid Comp.*, **11**, 60 (1957).

Mikhalevich, Yu. I. and Omel'chenko, O. K.: *Procedures of Piece-Wise Polynomial Interpolation of Functions of One and Two Arguments*. Novosibirsk: Izd. VC Sib. Otd. AN SSSR, 1970 [Russian].

Reinsch, C. H.: Smoothing by spline functions. *Numer. Math.*, **10**, 4 (1967).
Schoenberg, I. J.: Contributions to the problem of approximation of equidistant data by analytic functions. *Quart. Appl. Math.*, **4** (1946).
Schumaker, L. L.: *Approximation by Splines: Theory and Application of Spline Functions*. New York–London: Academic Press, 1969.
Walsh, J. L., Alberg, J. H., and Nilson, E. N.: Best approximation properties of the spline fit. *J. Math. Mech.*, **11**, 2 (1962).
Yanenko, N. N. and Kvasov, B. I.: Iterative methods of constructing poly-cubic spline functions. *Numerical Methods in Fluid Mechanics (Informal bull.)*, **1**, 3 (1970) [Russian].
Zavyalov, Yu. S.: Interpolation by cubic multiresponses. In: *Computational Systems, V. 38*. Novosibirsk: Izd. Inst. Math. Sib. Otd. AN SSSR, 1970 [Russian].
Zavyalov, Yu. S.: Interpolation by bi-cubic multiresponses. In: *Numerical Systems, V. 38*. Novosibirsk: Izd. Inst. Math. Sib. Otd. AN SSSR, 1970 [Russian].
Zavyalov, Yu. S.: An extremal property of the cubic multiresponses and the smoothing problem. In: *Numerical Systems, V. 42*, Novosibirsk: Izd. Inst. Math. Sib. Otd. AN SSSR, 1970 [Russian].

15. Splitting-Up Methods

Andreev, V. B.: On the splitting up difference schemes for general p-dimensional parabolic equations of second order with mixed derivatives. *Zh. Vych. Matem. Matem. Fiz.*, **7**, 2 (1967) [Russian].
Bagrinovskii, K. A. and Godunov, S. K.: Difference methods for multi-dimensional problems. *Doklady AN SSSR*, **115**, 3 (1957) [Russian].
Baker, G. A.: An implicit numerical method for the n-dimensional heat equation. *Quart. Appl. Math.*, **17**, 4 (1960).
Baker, G. A. and Oliphant, T. A.: An implicit numerical method for solving the two-dimensional heat equation. *Quart. Appl. Math.*, **17**, 4 (1960).
Bensoussan, A.: Pure decentralization for interrelated payoffs. Symposium on Optimization, Los Angeles, 1971.
Bensoussan, A., Lions, J. L., and Teman, R.: Sur les méthodes de décomposition de décentralisation et de coordination et application. Cahiers IRIA n° a, Tome 2, Paris, 1972.
Birkhoff, G., Varga, R. S.: Implicit alternating direction methods. *Trans. Amer. Math. Soc.*, **92**, 1 (1959).
Birkhoff, G., Varga, R., and Young, D.: *Alternating Direction Implicit Methods*. (Advances in Computing, 3) New York–London: Academic Press, 1962.
Buleev, N. I.: A numerical method for the two- and three-dimensional diffusion equations. *Matem. Sb.*, **51**, 2 (1960) [Russian].
Douglas, J. and Gunn, J. I.: Two high-order correct difference analogues for the equation of multi-dimensional heat flow. *Math. Comput.*, **17**, 81 (1963).
Douglas, J. and Gunn, J. E.: A general formulation of alternating direction methods. Part I. Parabolic and hyperbolic problems. *Numer. Math.*, **6**, 5 (1964).
Douglas, J. and Jones, B. F.: On predictor-corrector methods for non-linear parabolic differential equations. *J. Soc. Ind. Appl. Math.*, **11**, 1 (1963).
Douglas, J., Kellogg, R. B., and Varga, R. S.: Alternating direction methods for n space variables. *Math. Comput.*, **17**, 83 (1963).
Douglas, J. and Pearcy, C. M.: On convergence of alternating directions procedures in the presence of singular operators. *Numer. Math.*, **5**, 2 (1963).
Douglas, J. and Rachford, H.: On the numerical solution of the heat conduction problems in two and three space variables. *Trans. Amer. Math. Soc.*, **82**, 2 (1956).
Dupont, T.: A factorization procedure for the solution of elliptic difference equations. *SIAM J. Numer. Anal.*, **5**, 4 (1968).

Dyakonov, E. G.: The alternating direction method for solving systems of equations in finite differences. *Doklady AN SSSR*, **138**, 2 (1961) [Russian].

Dyakonov, E. G.: On certain difference schemes for boundary value problems. *Zh. Vych. Matem. Matem. Fiz.*, **2**, 1 (1962) [Russian].

Dyakonov, E. G.: Difference schemes with split up operators for multidimensional stationary problems. *Zh. Vych. Matem. Matem. Fiz.*, **2**, 4 (1962) [Russian].

Dyakonov, E. G.: Solution of some multi-dimensional problems of mathematical physics using nets. Dissertation, Moscow, 1962 [Russian].

Dyakonov, E. G. and Lebedev, V. I.: The splitting up method for the third boundary value problem. In: *Numerical Methods and Programming, V. IV.* Moscow: Izd. Mosc. Univ., 1967 [Russian].

Fairweather, G. and Mitchell, A. R.: Some computational results of an improved A. D. I. method for the Dirichlet problem. *Comput. J.*, **9**, 3 (1966).

Fryazinov, I. V.: On difference schemes for Poisson's equation in polar, cylindrical and spherical coordinates. *Zh. Vych. Matem. Matem. Fiz.*, **11**, 5 (1971) [Russian].

Gunn, G. E.: The solution of elliptic difference equations by semi-explicit iterative techniques. *SIAM J. Numer. Anal.*, **2**, 1 (1965).

Hubbard, B. I.: Alternating direction schemes for the heat equation in a general domain. *SIAM J. Numer. Anal.*, **2**, 3 (1966).

Il'in, V. P.: On splitting up of difference equations of the parabolic and elliptic types. *Sib. Math. J.*, **VI**, 1 (1965) [Russian].

Kellogg, R. B.: Another alternating direction implicit method. *J. Soc. Ind. Appl. Math.*, **11**, 4 (1963).

Kellogg, R. B.: An alternating direction method for operator equations. *J. Soc. Ind. Appl. Math.*, **12**, 4 (1964).

Kellogg, R. B. and Spanier, J.: On optimal alternating direction parameters for singular matrices. *Math. Comput.*, **19**, 91 (1965).

Konovalov, A. N.: The method of fractional steps for solving the Cauchy problem for the multi-dimensional wave equation. *Doklady AN SSSR*, **147**, 1 (1962) [Russian].

Konovalov, A. N.: The application of the splitting up method in dynamical problems of the theory of elasticity. *Zh. Vych. Matem. Matem. Fiz.*, **4**, 4 (1964) [Russian].

Konovalov, A. N.: *Filtering Problems of Multi-phase Incompressible Liquids.* Novosibirsk: Izd. Novosib. Univ., 1972 [Russian].

Kuznetsov, B. G.: Numerical methods for solving some problems of viscous liquids. *Fluid Dynamics Trans.*, **4**, 1969.

Lees, M.: Alternating direction methods for hyperbolic differential equations. *J. Soc. Ind. Appl. Math.*, **10**, 4 (1960).

Lees, M.: Alternating direction and semi-explicit difference methods for parabolic partial differential equations. *Numer. Math.*, **3**, 5 (1961).

Marchuk, G. I.: On the theory of the splitting-up method. In: *Numerical Solutions of Partial Differential Equations II, SYNSPADE 1970.* New York–London: Academic Press, 1971.

Marchuk, G. I. and Sultangazin, U. M.: The splitting-up method for the transport equation. *Zh. Vych. Matem. Matem. Fiz.*, **5**, 5 (1965) [Russian].

Marchuk, G. I. and Yanenko, N. N.: The application of the splitting-up method (fractional steps) to problems of mathematical physics. In: *Some Problems of Numerical and Applied Mathematics.* Novosibirsk: Nauka, 1966 [Russian].

Peaceman, D. W. and Rachford, H. H.: The numerical solution of parabolic and elliptic differential equations. *SIAM J.*, **3**, 1 (1955).

Samarskii, A. A.: On an economical difference method for multi-dimensional parabolic equations in arbitrary regions. *Zh. Vych. Matem. Matem. Fiz.*, **2**, 5 (1962) [Russian].

Samarskii, A. A.: On convergence of the fractional steps method for the heat equation. *Zh. Vych. Matem. Matem. Fiz.*, **2**, 6 (1962) [Russian].

Samarskii, A. A.: Locally one-dimensional difference schemes on non-uniform nets. *Zh. Vych. Matem. Matem. Fiz.*, **3**, 3 (1963) [Russian].

Samarskii, A. A.: On an economical algorithm for numerical solutions of differential and algebraic equations. *Zh. Vych. Matem. Matem. Fiz.*, **4**, 3 (1964) [Russian].

Samarskii, A. A.: Economical difference schemes for hyperbolic systems with mixed derivatives and their applications to the equations of the theory of elasticity. *Zh. Vych. Matem. Matem. Fiz.*, **5**, 1 (1965) [Russian].

Samarskii, A. A.: Additive schemes. In: *Papers of the International Symposium of Mathematicians in Moscow*, 1966 [Russian].

Teman, R.: Sur la stabilité et la convergence de la méthode des pas fractionnaires. *Annali. di Mat. Pura Appl.*, IV, 79, 1968.

Teman, R.: Quelques méthodes de décomposition en analyse numérique. *Acte du Congres Intern. Math.*, V. 3, 1970.

Varga, R.: Some results in approximation theory with applications to numerical analysis. In: *Numerical Solutions of Partial Differential Equations II, SYNSPADE 1970*. New York–London: Academic Press, 1971.

Vorob'yev, Yu. V.: A random iterative process in the alternating direction method. *Zh. Vych. Matem. Matem. Fiz.*, **8**, 3 (1968) [Russian].

Wachspress, I. L.: Optimum alternating direction implicit iteration parameters for a model problem. *SIAM J.*, **10**, 2 (1962).

Wachspress, E. L.: Extended application of alternating direction implicit iteration model problem theory. *SIAM J.*, **11**, 3 (1963).

Wachspress, E. L.: *Iterative Solution of Elliptic Systems and Applications to the Neutron Diffusion Equations of Reactor Physics*. Englewood Cliffs, N.Y.: Prentice Hall, 1966.

Wachspress, E. L. and Habetler, G. J.: An alternating direction implicit iteration technique. *SIAM J.*, **8**, 2 (1960).

Widlund, O.: On the rate of an alternating direction implicit method in a non-commutative case. *Math. Comp.*, **20**, 96 (1966).

Widlund, O.: On the effects of scaling of the Peaceman–Rachford method. *Math. Comput.*, **25**, 113 (1971).

Yanenko, N. N.: On a difference method for the multi-dimensional heat equation. *Doklady AN SSSR*, **125**, 6 (1959) [Russian].

Yanenko, N. N.: On economic implicit schemes (the method of fractional steps). *Doklady AN SSSR*, **134**, 5 (1960) [Russian].

Yanenko, N. N.: On implicit difference methods for the multi-dimensional heat equation. *Izv. Vuzov. Matematika*, **4**, 23 (1961) [Russian].

Yanenko, N. N.: On convergence of the splitting-up method for the heat equations with variable coefficients. *Zh. Vych. Matem. Matem. Fiz.*, **2**, 5 (1962) [Russian].

Yanenko, N. N.: On the weak approximation of differential equations. *Sib. Math. J.*, V, 6 (1964) [Russian].

Yanenko, N. N. and Demidov, G. V.: The method of weak approximation as a constructive method for the Cauchy problem. In: *Some Problems of Numerical and Applied Mathematics*. Novosibirsk: Nauka, 1966 [Russian].

16. Conditionally Well-Posed Problems and Some Inverse Problems of Mathematical Physics

Douglas, J.: On the relation between stability and convergence in the numerical solution of linear parabolic and hyperbolic differential equations. *J. Soc. Ind. Appl. Math.*, **4**, 1 (1956).

Frank, L. S. and Chudov, L. A.: Difference methods for ill-posed Cauchy problems. In: *Numerical Methods in Gas Dynamics*. Moscow: Izd. Mosc. Univ., 1965 [Russian].

Fuks, K.: Perturbation theory in neutron multiplication problems. *Proc. Phys. Soc.*, **62**, 791 (1949).

Ivanov, V. K.: On ill-posed problems. *Mat. Sb.*, **61**, 2 (1963) [Russian].

John, F.: Differential equations with approximate and improper data. Lectures New York University, 1955.

Kadomtsev, B. B.: On the effect function in the theory of radiative transfer. *Doklady AN SSSR*, **113**, 3 (1957) [Russian].

Krein, S. G.: On classes of certain well-posed problems. *Doklady AN SSSR*, **114**, 6 (1957) [Russian].

Krein, S. G. and Prozorovskaya, O. I.: On approximate methods for ill-posed problems. *Zh. Vych. Matem. Matem. Fiz.*, **3**, 1 (1963) [Russian].

Lavrentiev, M. M.: On the Cauchy problem for the Laplace equation. *Doklady AN SSSR*, **102**, 2 (1956) [Russian].

Lavrentiev, M. M.: Formulation of certain ill-posed problems of mathematical physics. In: *Some Problems of Numerical and Applied Mathematics*, Novosibirsk: Nauka, 1966 [Russian].

Lavrentiev, M. M.: Numerical solution of conditionally properly posed problems. In: *Numerical Solutions Of Partial Differential Equations II, SYNSPADE 1970*. New York–London: Academic Press, 1971.

Lavrentiev, M. M. and Vasil'ev, V. G.: On some ill-posed problems of mathematical physics. *Sib. Math. J.*, **VII**, 3 (1966) [Russian].

Lavrentiev, M. M., Romanov, V. G., and Vasil'ev, V. G.: *Multi-Dimensional Inverse Problems for Differential Equations*. Novosibirsk: Nauka, 1969 [Russian].

Landis, E. M.: On some properties of the solutions of elliptic equations. *Doklady AN SSSR*, **107**, 4 (1956) [Russian].

Levitan, B. M.: *Expansions in Eigenfunctions of Second Order Differential Equations*. Moscow: Gostekhizdat, 1950 [Russian].

Levitan, B. M. and Sargsyan, I. S.: *Introduction in the Spectral Theory*. Moscow: Nauka 1970 [Russian].

Lions, J. L. and Lattes, R.: *The Method of Quasi-Reversibility: Application to Partial Differential Equations*. New York: American Elsevier, 1969.

Marchuk, G. I.: Equations for the value of information from the meteorological satellites and formulations of inverse problems. *Kosmicheskie Issledovania*, **11**, 3 (1964) [Russian].

Marchuk, G. I. and Atanbayev, S. A.: Some problems in global regularization. *Doklady AN SSSR*, **199**, 3 (1970) [Russian].

Marchuk, G. I. and Vasil'ev, V. G.: On an approximate solution of operator equations of the first order. *Doklady AN SSSR*, **199**, 4 (1970) [Russian].

Marchuk, G. I. and Drobyshev, Yu. P.: Some problems in the linear theory of measurements. *Avtometriya*, 1967, No. 3 [Russian].

Marchuk, G. I. and Orlov, V. V.: On the conjugate functions theory. In: *Neitronnaya Fizika*. Moscow: Gosatomizdat, 1961 [Russian].

Mergel'yan, S. N.: The harmonic approximation and the approximate solution of the Cauchy problem for the Laplace equation. *Usp. Mat. Nauk*, **XI**, 5 (1956) [Russian].

Morozov, V. A.: Methods of solving unstable problems (Lecture notes). Rotaprint Mosc. Univ., 1967 [Russian].

Tikhonov A. N.: On stability of inverse problems. *Doklady AN SSSR*, **39**, 5 (1943) [Russian].

Tikhonov, A. N.: On the solution of ill-posed problems using the method of regularization. *Doklady AN SSSR*, **151**, 3 (1963) [Russian].

Tikhonov, A. N.: On regularization of ill-posed problems. *Doklady AN SSSR*, **153**, 1 (1963) [Russian].

Turchin, V. F. and Nozik, V. Z.: Statistical regularization of the solutions of ill-posed problems. *Izv. AN SSSR, Ser. FA i O*, **5**, 1 (1969) [Russian].

Usachev, L. N.: An equation of the cost of the kinetic reactor neutrons and the perturbation theory. In: *Reactor Design and the Theory of Reactors*. Moscow: Izd. AN SSSR, 1955 [Russian].

17. Numerical Methods in the Transport Theory

Bardos, P. G.: Equations du premier ordre à coefficients réels. *Ann. Scient. Ec. Norm. Sup.*, 4e serie, **3** (1970).
Bogolyubov, N. N.: *Problems of the Dynamical Theory in Statistical Physics*. Moscow: Gostekhtheoryzdat, 1946 [Russian].
Germogenova, T. A.: On the convergence of certain approximate methods for the transport equation. *Doklady AN SSSR*, **181**, 3 (1968) [Russian].
Germogenova, T. A.: Generalized solutions of boundary value problems for the transport equation. *Zh. Vych. Matem. Matem. Fiz.*, **9**, 3 (1969) [Russian].
Godunov, S. K.: The Energy Integral and Accuracy Estimates of Approximate Eigenvalues. *Zh. Vych. Matem. Matem. Fiz.*, **11**, 5 (1971) [Russian].
Godunov, S. K. and Sultangazin, U. M.: On the dissipativity of the boundary conditions of V. S. Vladimirov for a symmetric system of the method of spherical harmonics. *Zh. Vych. Matem. Matem. Fiz.*, **11**, 3 (1971) [Russian].
Gol'din, V. Ya.: The quasi-diffusion method for the kinetic equation. *Zh. Vych. Matem. Matem Fiz.*, **4**, 6 (1964) [Russian].
Jorgens, K.: An asymptotic expansion in the theory of neutron transport. *Comm. Pure Appl. Math.*, **11**, 2 (1958).
Karlson, B. and Bell, Dzh.: Solving the transport equation by the S_n method. In: *Physics of Nuclear Reactors* (Transl. from English), Moscow, 1963.
Kuznetsov, E. S. and Marchuk, G. I.: Numerical methods in the theory of radiative transfer. In: *Trudy IV Vsesyuz. Mat. S'ezda, Vol. II*. Moscow: Izd. AN SSSR, 1964 [Russian].
Lebedev, V. I.: On solving kinetic problems in the theory of transfer. Doct. Dissertation, Novosibrisk, 1967 [Russian].
Lebedev, V. I.: On the *KP*-method and the difference schemes for the kinetic equations. In: *Numerical Methods in the Transport Theory*. Moscow: Atomizdat 1969 [Russian].
Marchuk, G. I.: *Numerical Methods in the Design of Nuclear Reactors*. Moscow: Atomizdat, 1958 [Russian].
Marchuk, G. I. and Kochergin, V. P.: An effective method for two-dimensional diffusion equation using the cells of rectangular and hexagonal forms. *Atomic Energy*, **18**, 6 (1965) [Russian].
Marchuk, G. I. and Lebedev, V. I.: *Numerical Methods in Neutron Transport Theory*. Moscow: Atomizdat, 1971 [Russian].
Marchuk, G. I. and Sultangazin, U. M.: On convergence of the splitting-up method for equations of radiative transfer. *Doklady AN SSSR*, **161**, 1 (1965) [Russian].
Marchuk, G. I. and Sultangazin, U. M.: On solving the kinetic equation of radiative transfer by the splitting-up method. *Doklady AN SSSR*, **163**, 4 (1965) [Russian].
Marchuk, G. I. and Yanenko, N. N.: Solution of the multi-dimensional kinetic equation by the splitting-up method. *Doklady AN SSSR*, **157**, 6 (1964) [Russian].
Marek, I.: On a problem of mathematical physics. *Appl. Math.*, **11**, 89 (1966).
Nikolaishvili, SH. S.: An approximate solution of the transport equation by the method of moments. *Atomnaya energia*, **9**, 2 (1961) [Russian].
Nikolaishvili, Sh. S.: On the solution of the one-velocity transport equation using the Ivon–Martens approximations. *Atomnaya energia*, **20**, 4 (1966) [Russian].
Vladimirov, V. S.: Mathematical problems in the one-velocity theory of particle transport. *Trudy Mat. Inst. AN SSSR*, **61** (1961).
Vladimirov, V. S.: The numerical solution of the kinetic equation for a sphere. In: *Numerical Mathematics*, No. 3, Moscow: Izd. AN SSSR, 1958 [Russian].
Vladimirov, V. S.: On some variational methods of approximate solution of the transport equation. In: *Numerical Mathematics*, No. 7, Moscow: Izd. AN SSSR, 1961 [Russian].

Smelov, V. V.: *Lectures on the Theory of Neutron Transfer*. Novosibirsk, Izd. Novosib. Univ. i VC Sib. otd. AN SSSR, 1970 [Russian].

Sultangazin, U. M.: *Differential Properties of Solutions of the Mixed Cauchy Problem for the Non-Stationary Kinetic Equation*. Preprint VC Sib. otd. AN SSSR, Novosibirsk, 1971 [Russian].

Sultangazin, U. M.: Concerning the weak approximation method for the equations of spherical harmonics. Preprint VC Sib. Otd. AN SSSR, Novosibirsk, 1971 [Russian].

Sultangazin, U. M.: Weak convergence of the spherical harmonics method. Preprint VC Sib. Otd. AN SSSR, Novosibirsk, 1971 [Russian].

Shikhov, S. B.: Some problems of the mathematical theory of the critical state of a reactor. *Zh. Vych. Matem. Matem. Fiz.*, **7**, 1 (1967) [Russian].

18. The Monte Carlo Method

Buslenko, N. P., Golenko, D. I., Sobol, I. M., Sragovich, V. G., and Shreyder, Yu. A.: *The Monte Carlo Method*. Moscow: Fizmatgiz, 1962 [Russian].

Ermakov, S. M.: *The Monte Carlo Method and Related Problems*. Moscow: Nauka, 1971 [Russian].

Ermakov, S. M. and Zolotukhin, V. G.: The polynomial approximations and the Monte Carlo method. *Teoria Veroyatnostei Prim.*, **5**, 4 (1960) [Russian].

Fano, U., Spencer, L., and Berger, M.: *Gamma Radiative Transfer*. Moscow: Gosatomizdat, 1963 [Russian transl. from English].

Gel-fand, I. M., Frolov, A. S., and Chentsov, N. N.: Computation of the continual intervals by the Monte Carlo method. *Izv. Vuzov. Matematika*, **5** (1958) [Russian].

Kertis, D.: The Mone Carlo methods for the iterations of linear operators. *Usp. Matem. Nauk*, **XII**, 5 (1957) [Russian].

Metropolis, M. and Ulam, S.: The Monte Carlo method. *J. Amer. Stat. Assoc.*, **44**, 247 (1949).

Mikhailov, G. A.: The statistical modelling of processes of the radiative transfer in the atmosphere. Doct. Dissertation, Novosibirsk, 1971 [Russian].

Vladimirov, V. S., Sobol, I. M.: Computation of the smallest eigenvalue of the Peyerls equation by the Monte Carlo method. In: *Numerical Mathematics, No. 3*. Moscow: Izd. AN SSSR, 1958 [Russian].

19. The Method of Large Particles

Belotserkovskii, O. M. and Davydov, Yu. M.: The non-stationary method of "large particles" in the gas-dynamical design. *Zh. Vych. Matem. Matem. Fiz.*, **11**, 1 (1971) [Russian].

Dyachenko, V. F.: On a new numerical method for non-stationary problems of gas dynamics with two spatial variables. *Zh. Vych. Matem. Matem. Fiz.*, **5**, 4 (1965) [Russian].

Harlow, F.: The "Particle in the cell" method for the problems of hydrodynamics. In: *Numerical Methods in Hydrodynamics*, Moscow: Mir, 1967 [Russian].

Vedeshkina, K. A., Levina, Z. F., Lomnev, S. P., Prudkovskii, G. P., Rastopchina, T. V., Ruben, G. V., and Yurchenko, V. V.: *Solving Problems by the Method of "Large Particles."* Izd. VC AN SSSR, 1970 [Russian].

Yanenko, N. N., Anuchina, N. N., Petrenko, V. E., Shokin, Yu. I.: On the computation methods in gas dynamics with large deformations. *Numerical Methods in Continuum Mechanics*, **1**, 1 (1970) [Russian].

20. Optimization of Algorithms

Babuška, I. and Sobolev, S. L.: Optimization of numerical methods. *Appl. Mat.*, **10**, 2 (1965) [Russian].
Bakhvalov, N. S.: On optimal methods of problem solving. *Appl. Mat.*, **13**, 1 (1968) [Russian].
Chernousko, F. L., Banichuk, N. V., and Petrov, V. M.: The numerical solution of variational and boundary value problems by the local variations method. *Zh. Vych. Matem. Matem. Fiz.*, **6**, 6 (1966) [Russian].
Dahlquist, G.: Convergence and stability in the numerical integration of ordinary differential equations. *Math. Scand.*, **4**, 1 (1956).
Kolmogorov, A. N.: Discrete automata and finite algorithms. In: *Trudy IV Vsesoyuz. Mat. S'ezda, Vol. I.* Moscow: Izd. AN SSSR, 1963 [Russian].
Korobov, N. M.: Computation of multiple integrals by the method of optimal coefficients. *Vest. Mosc. Univ., Ser. Mat.*, **4**, 1959 [Russian].
Moiseev, N. N.: Numerical methods using variations in the state space and some control problems of large systems. In: *Papers of the International Congress of Mathematicians*, Moscow, 1966 [Russian].
Moiseev, N. N. and Krasovskii, N. N.: A theory of optimal controllable systems. *Izv. AN SSSR, Tekh. Kibernetika*, **5** (1967) [Russian].
Moor, R.: *Interval Analysis*. Prentice Hall, 1966.
Nickel, K.: Über die Notwendigkeit einer Fehlerschranken—Arithmetik für Rechnenautomaten. *Numer. Math.*, **9**, 1, 1966.
Nickel, R.: Bericht über neue Karlsruher Ergebnisse bei der Fehlererfassung von numerischen Prozessen. *Appl. Mat.*, **13**, 2 (1968).
Vinogradov, I. M.: On an estimate of trigonometric sums. *Izv. AN SSSR, Ser. Matem.*, **29**, 3 (1965) [Russian].

Index

Adaptive methods, 103
Alternating direction method, 113
Approximation, 15, 279
  global, 33
Asymptotic rate of convergence, 91

Biorthogonal bases, 3
Boundary conditions
  natural, 44
  principal, 44

Central differences scheme, 142
Chebyshev acceleration method, 92
Component-by-component method. *See*
  Splitting-up methods
Conditionally well-posed problems, 184, 287
Conditioning number of a matrix, 91
Conjugate function, 200, 202
Conjugate gradients method, 108
Convergence, 30, 91, 279
Crank–Nicholson scheme, 22, 142

Decomposition and decentralization method, 285
Displacement method, 92
Domain of an operator, 2

Energy method, 281
Equation of motion, 233
Equivalence theorem, 279
Evolution equation, 18
Explicit difference scheme, 18, 20

Factorization method, 139
Fast Fourier transform, 132
Finite elements method, 51, 283

Galerkin method, 45, 51
Gauss
  elimination method, 288
  transformation, 106
Gradient methods, 103
Grid size. *See* Mesh size

Hockney's method, 135

Ill-conditioned system, 99, 287
Ill-posed problem, 184, 287
Implicit difference scheme, 18, 21
Interpolation
  piece-wise bicubic, 84
  piece-wise cubic, 72
    smoothing, 76

Inverse problems, 184
  second approximation in, 208
Iterative process
  asymptotic rate of convergence of, 91
  nonstationary, 89
  optimization of, 88
  residual of, 88
  stationary, 89

Kellogg's lemma, 5

$L_2(D)$-space, 1
Lanczos transformation, 289
Least squares method, 46
Lyusternik method, 6

Matrix factorization method, 140
Mesh size, 11
Method of weights, 47
Monte Carlo method, 286

Net, 11
  function, 11
  point, 11
  region, 12
  space, 15
Nonstationary problem, 87, 142
Norm
  energy, 4
  $L_2(D)$, 1
  Sobolev space, 5

Operation, 133
Operator
  adjoint, 2
  closed, 2
  energy, 4
  Laplace, 9
    differential analog of, 12
  norm of, 3
  positive, 2
  positive definite, 2
  positive semidefinite, 2
  singular, 126
  source, 19
  spectral equivalent, 138
  transition, 19
Optimization, 88, 291
Over-relaxation method, 97

Particle-in-the-cell method, 285
Penalty method, 44
Perturbation theory, 203

Predictor-corrector method, 151
Projection on the net, 12

Quadrature formulae, 292
Quasisolution, 206

Radiative transfer, 259
Regularization family, 184, 287
Relaxation parameter, 104
Relaxation process, 98
Residual method
    one-step, 104
    n-step, 108
Ritz method, 43, 57
Round-off errors, 291
Running count scheme, 240

Scattering
    angle, 276
    function, 260
    nonisotropic, 276
Simple iterative method, 90
Sobolev space, 4
Spectral radius of a self-adjoint operator, 4
Spline, 74, 72
    interpolation method, 72

Splitting-up methods, 113, 146, 169, 180, 284
    component-by-component, 147, 154, 164, 170
    with variational optimization, 123
Stability
    absolute, 24
    conditional, 28
    countable, 22, 25
    L_2 criterion of, 280
    local, 280
    spectral criterion of, 24
Stabilization method, 147, 160, 174
Stationary problem, 87
Sturm–Liouville problem, 41

Transition factor of Fourier coefficients, 237
Transport equation, 261
Two-layer scheme, 22

Value, 200, 202
Variational difference schemes, 53
Variational methods, 42

Weak approximations, 285
Well-posed problem, 184, 287